CAMBRIDGE TROPICAL BIOLOGY SERIES

Vascular epiphytes

Vascular epiphytes

General biology and related biota

DAVID H. BENZING
Oberlin College

CAMBRIDGE UNIVERSITY PRESS
Cambridge
New York Port Chester Melbourne Sydney

CAMBRIDGE UNIVERSITY PRESS
Cambridge, New York, Melbourne, Madrid, Cape Town, Singapore, São Paulo

Cambridge University Press
The Edinburgh Building, Cambridge CB2 8RU, UK

Published in the United States of America by Cambridge University Press, New York

www.cambridge.org
Information on this title: www.cambridge.org/9780521266307

First published 1990
This digitally printed version 2008

A catalogue record for this publication is available from the British Library

Library of Congress Cataloguing in Publication data

Benzing, David H.

Vascular epiphytes.

(Cambridge tropical biology series)

Bibliography: p.

1. Epiphytes. I. Title.
QK922.B46 1990 582.5 88–35353

ISBN 978-0-521-26630-7 hardback
ISBN 978-0-521-04895-8 paperback

To A. S. Renfrow, for her skilled contribution and remarkable dedication to the production of this monograph and to most of my other publications and studies of epiphyte biology.

Contents

Preface

There are about 25,000 vascular species sharing the peculiar habit of rooting in tree crowns rather than on the ground, yet only an occasional epiphyte – the wide-ranging Spanish moss, for example – has attracted much scientific curiosity. Uncounted thousands of animal populations (mostly insects) regularly associate with these plants, sometimes because there are no alternatives for lodging, food, or other critical resources. Vascular epiphytes remain best known to horticulturists and systematists; the how and why of their growth in nature under such novel conditions have been mostly ignored. Other ecological groups such as carnivores, halophytes, mangroves, and parasites have been thoroughly covered in monographs despite their smaller numbers, more restricted distribution, and limited literature base. Also underrepresented is information on epiphyte-dependent fauna and effects on supporting trees. But times are changing. Improved climbing techniques allow extended observation and collection of representative fauna. Portable equipment for measuring such plant phenomena as gas exchange has opened the upper canopy to sophisticated analyses. Clearing land for roads, while destroying woodland, has fostered research in the field; so has establishment of permanent field stations, particularly in the neotropics. Results are heartening; what for many years was only a trickle of papers on nontaxonomic aspects of epiphyte biology and forest canopy fauna now approaches a flood. Three international symposia devoted to epiphytes have been held in just the past four years.

What is needed now for the student and nonspecialist alike is a comprehensive and, it is hoped, provocative volume that distills the primary literature as well as the considerable body of unpublished data on epiphytism and interactions between canopy flora and co-occurring biota. There are several ways to produce this product. I have chosen to focus on functional aspects of arboreal life, and examples were chosen without regard to systematics. However, the extensive indexes are designed to provide the specialist easy access to ants, Bromeliaceae, mistletoes, Orchidaceae, residents of phytotelmata, and pteridophytes. Three strictly botanical topics are emphasized: (1) aspects of the forest canopy that most constrain its resident flora; (2) corresponding adaptations in that flora; and (3) occurrence and diversity among the epiphytes. Following a preliminary overview, focus will center on photosynthesis, water balance, mineral nutrition, and reproduction. Next, the epiphytes' effects on hosting trees and the ways in which they utilize and serve other life forms are highlighted. Finally, epiphyte occurrence will be considered from ecological, geographic, and historical perspectives. Speculation is unavoidable where there are gaps in our understanding of many of the subjects covered here; it is nevertheless my hope that the reader will not be irritated by guesswork but rather will regard it as a stimulus to further research.

For more than 20 years, my interest in epiphytes has been variously assisted and promoted by numerous colleagues, students, and institutions. Travel to tropical habitats and laboratory research on bromeliads and orchids were made possible by grants and contracts from the National Science Foundation, the National Geographic Society, the National Park Service, and Oberlin College. Biologists who have been particularly helpful in providing information on specific plant groups or on other subjects included in this volume are Joe Arditti, Bradley Bennett, Cathy Burt, Diane Davidson, Calaway Dodson, Murray Evans, Susan Gardner, Alan Gentry, Amy Jean Gilmartin, Camilla Huxley, John Longino, Harry Luther, Michael Madison, Craig Martin, Nalini Nadkarni, Mauritzio Paoletti, Alec Pridgeon, Jack Putz, Suzanne Renner, Russ Sinclair, Ben and Debbie Stinner, Warren Ullum, John Utley, and Norris Williams. Oberlin College students have helped conduct many of my investigations on epiphyte structure and function. Especially noteworthy were the contributions of Andrew Bent, Ray Broggini, Chris Dahle, Rick Davidson, Karen Henderson, Page Owen, Will Pockman, Joanne Sulak, and John Titus. David Bermudes, Janice Derr, Ned Friedman, and Jeff Seemann not only provided valuable assistance while undergraduates at Oberlin but continue as collaborators from other institutions. Special thanks are owed to Calaway Dodson, David Miller, and Michael Zimmerman who read chapters and provided valuable input; also to Michael Madison and Barry Tomlinson for their critical reviews of the entire manuscript.

Abbreviations

A_{max}	maximum photosynthetic capacity
ATP	adenosine triphosphate
C_3	see RPP
C_4	the Hatch and Slack photosynthetic pathway
CAM	Crassulacean acid metabolism
Chl	chlorophyll
C_a	partial pressure of carbon dioxide outside the leaf
C_i	concentration of carbon dioxide inside the leaf
CS	continuously supplied
ΔH^+	diurnal change in titratable acidity
$\delta^{13}C$	carbon isotope ratio; ^{13}C enrichment in parts per thousand
MPa	megapascals
MUE	mineral use efficiency
NADP	nicotinamide adenine dinucleotide phosphate
PAR	photosynthetically active radiation
PEPc	phosphoenolpyruvate carboxylase
PPFD	photosynthetic photon flux density
PPNUE	potential photosynthetic nitrogen use efficiency
PS	pulse-supplied
RH	relative humidity
r_m	Malthusian coefficient
RPP	reductive pentose phosphate or C_3 pathway
RuBPc/o	ribulose phosphate carboxylase/oxygenase
RWC	relative water content as percentage of saturation
ST	spirally thickened (idioblasts)
TR	transpiration ratio
Ψ_{crit}	leaf water potential at stomatal closure
Ψ_L	bulk leaf water potential
Ψ_P	turgor
Ψ_S	solute potential
VAM	vesicular–arbuscular mycorrhizas
WUE	water use efficiency
WVPD	water vapor pressure deficit

Glossary

Accidental epiphyte: A typically terrestrial species with occasional members that grow to maturity anchored in a tree crown.

Allelopathy: The process whereby one plant inhibits or kills another through the production of toxic compounds.

Allogamy: Seed production resulting from pollen flow between different plants.

Anemochory: The dispersal of seeds by wind.

Ant-fed ant-house epiphyte: A species that produces hollow organs (domatia) specifically for housing ant colonies.

Ant nest-garden epiphyte: A species that regularly and often exclusively roots in arboreal ant nests.

Atmospheric epiphyte: A species directly dependent on the atmosphere for moisture and nutrient ions (e.g., xerophytic *Tillandsia*).

Autogamy: Seed production resulting from self-pollination.

Axenic: Applied to tropical trees that by nature do not support epiphytes.

Bromelioid: An adjective applied to taxa assigned to subfamily Bromelioideae of Bromeliaceae.

CAM-cycling: A photosynthetic syndrome characterized by diurnal CO_2 fixation and nocturnal recapture of respired CO_2.

CAM-idling: A photosynthetic syndrome characterized by continuous stomatal closure and energy maintenance through internalized CO_2 recycling; a common stress response of CAM plants.

Carton: The material constructed of plant fiber, seeds, feces, and diverse other materials by ants to produce their nests and enclosed runways.

Chasmogamy: The condition allowing pollen transport to and from an opened flower.

Clade: A group of species that share a single ancestral lineage.

Cladogenesis: The process of creating two species from one; the splitting of a single lineage into two.

Cleistogamy: The condition of a flower that sets seed without presenting anthers or stigma to pollinators.

CS epiphyte: An epiphyte with relatively continuous access to external supplies of moisture and nutritive ions.

Decarboxylase: An enzyme that catalyzes the release of CO_2 from an organic acid.
Deciduous: An adjective describing leaves that abscise in less than one year.
Domatium: A plant cavity regularly occupied by ants.
Epiparasite: A parasite that taps its host via a fungal intermediate. The term is also used to describe vascular parasites that attack other parasitic plants.
Epiphyll: A nonvascular plant that inhabits the surfaces of foliage.
Epivelamen: The delicate layer of cells forming the outer boundary of some velamentous roots.
Eutroph: A plant native to fertile substrata.
Evergreen: An adjective describing green organs – usually foliage – that remain functional for at least a full year.
Everwet forest: A forest that receives enough rainfall through the year to support predominantly drought-sensitive vegetation (includes rain forest).
Exodermis: The specialized outer cell layer of the root cortex that is especially well developed in epiphytic Orchidaceae.
Facultative CAM: Carbon fixation via CAM or C_3 photosynthesis depending on environmental conditions.
Facultative epiphyte: A species that regularly grows epiphytically and rooted in earth-soil, often emphasizing one or the other habit in a particular habitat.
Glycophyte: A plant that does not accumulate high concentrations of sodium chloride.
Guild: A group of co-occurring but not necessarily related species that utilize one or more common resources.
Halophyte: A plant native to saline habitats.
Haustorium: The invasive appendage of a parasitic plant.
Hemiepiphyte: A plant that maintains vascular connection with earth-soil over part of its life. The epiphytic stage may occur early (primary hemiepiphyte) or later (secondary hemiepiphyte).
Hemiparasite: An haustorial parasite with significant photosynthetic capacity.
Histosol: A soil composed primarily of organic matter (e.g., peat).
Homoeosis: Assumption by one member of a series of a form or structure characteristic of another member of the series.
Homoiohydry: The condition of maintaining tissue hydrature relatively independent of ambient humidity.
Homoptera: The taxonomic order of arthropods including plant-sucking forms exemplified by aphids and scale insects.

Hypodermis: A subepidermal zone of usually achlorophyllous, thin-walled, collapsible water storage cells in leaves.

Iteroparity: The type of reproductive timing illustrated by plants that fruit repeatedly rather than once.

Lineage: The unbroken succession of generations that constitutes the history of a taxon through geologic time.

Mist epiphyte: A species equipped to live where moisture is often available as mist.

Mistletoe: A woody hemiparasite of tree branches.

Monocarpy: The type of life cycle characterized by a single reproductive effort just before death.

Mycotroph: A plant with fungus-assisted nutrition (e.g., orchid seedlings) but excluding species with typical mycorrhizas.

Myrmecochory: The dispersal of seeds by ants.

Myrmecophyte: A plant that is regularly associated with ant colonies and utilizes that symbiosis to some degree.

Nitrogenase: The enzyme complex responsible for reducing N_2 to organic nitrogen.

Nutritional piracy: The process whereby epiphytes intercept nutrients moving between the supporting tree crown and the forest floor.

Oligotroph: A plant native to nutrient-deficient substrata.

Paedomorphy: Exhibition of adult traits that reflect juvenile conditions in an ancestor.

Peloton: A dense mass of collapsed chitinous walls of an invading fungus that characterizes infected plant cells of aging orchid mycorrhizas.

Phorophyte: A support for vascular epiphytes.

Phylloclade: A branch that resembles, and functions like, a leaf.

Phytoalexin: Low-molecular-weight, toxic organic compounds synthesized in certain plant tissues in response to infection by pathogenic microbes.

Phytotelmata: Natural plant cavities that hold water and support resident aquatic organisms.

Phytotelm plant: A plant that produces phytotelmata.

Pluvial forest: Continuously wet forest that receives rainfall exceeding 5000 mm/yr (includes rain forest).

Pneumathode: A localized zone in the rhizodermis of a velamentous root that repels moisture and apparently promotes ventilation of the root interior.

Poikilohydry: The condition of maintaining tissue water content at levels strongly influenced by ambient humidity.

PS epiphyte: A pulse-supplied epiphyte with only sporadic access to external supplies of moisture and required nutritive ions.

Reproductive index: The proportion of the mature plant body committed to seeds and associated reproductive tissue.

Ruderal: A fast-maturing plant native to temporary but resource-rich habitats.

Sciophyte: A plant tolerant of deep shade.

Sclerophylly: The condition describing evergreen foliage containing much sclerified tissue.

Seasonal forest: Forest occurring in areas characterized by enough seasonal variation in rainfall to require plant tolerance for moderate to severe drought.

Shootless orchid: A species with much reduced vegetative shoots and green roots that have replaced foliage as photosynthetic organs (members of subtribe Sarcanthinae).

Spathe: A sheathing bract or pair of bracts enclosing an inflorescence.

Strangler: A primary hemiepiphyte that produces a robust root system that eventually provides self-support and may kill the host tree (e.g., *Ficus*).

Suspended humus: The mantle of degrading vegetation and living plants that covers bark and serves as a rooting medium for epiphytes native to humid forests.

Synusia: A group of co-occurring plants of the same life form.

Tillandsioid: An adjective applied to taxa assigned to subfamily Tillansioideae of Bromeliaceae.

Tilosome: A fibrous outgrowth produced by velamen cells positioned above transfer cells in the aerial roots of some orchids and other monocots.

Transfer cell: A cell specialized to perform solute transport and absorption (e.g., passage cells of the root exodermis and endodermis).

Transpiration ratio: A coefficient produced by dividing the mass of water lost in transpiration by the simultaneous gain in weight attributable to photosynthesis.

Trash-basket epiphyte: A species that creates relatively dry collections of intercepted litter in catchments fashioned of roots or leaves (e.g., *Catasetum, Platycerium*).

Trophic myrmecophyte: See ant-fed ant-house epiphyte.

Velamen: A spongy, multilayered rhizodermis composed of dead cells that surround the cortex of aerial and some terrestrial roots, particularly those of orchids.

Vivipary: The germination of an embryo before the enclosing seeds are shed; the term is sometimes also used to describe the proliferation of asexual progeny on an inflorescence.

1 Epiphytism: a preliminary overview

Historical notes

Columbus is credited with the first recorded comment on canopy-adapted vegetation; he wrote that tropical trees "have a great variety of branches and leaves, all of them growing from a single root" (Gessner 1956). The earliest known picture of an epiphyte – or, for that matter, reference to American botany – appears in *The Badianus Manuscript,* a Mexican herbal of 1552, written and probably illustrated by the Aztec Indian physician Martinus de la Cruz and translated into Latin by his Indian colleague Juannes Badianus at the College of Santa Cruz (Emmart 1940). The subject was *Vanilla fragrans,* a vining hemiepiphytic orchid. The fruit of this species (*tlilxochitl* in Aztec, meaning "black flower") was an ingredient in the doctor's prescription for "The Traveler's Safeguard," a mixture of pulverized herbs wrapped in a magnolia leaf and hung around the neck so that the voyager could "catch and inhale the very redolent odor . . ."

By the eighteenth century, ships' captains and explorers the world over were carrying ornamental plants, epiphytes included, back to Europe. Within decades, a brisk trade had developed; many additional aroids, bromeliads, cacti, orchids, and ferns were imported. Showiness, small size, and easy culture encouraged fads that drove prices to exorbitant levels and prompted more than one collector to lie about where he found his specimens. But scientific interest in these plants did not keep pace; other groups such as carnivores, halophytes, ruderals, and succulents are far better known today. A. F. W. Shimper's monograph *Die epiphytische Vegetation Amerikas* was a noteworthy exception. This treatise (1888) and related works (e.g., 1884, 1903) published over a long career were major contributions to the study of epiphytes, and in fact remain arguably the most comprehensive treatment to date. Schimper's skillful accounts of fascinating and unusual adaptations in several epiphytic taxa contain remarkable insight for that time.

Over the next 75 years, occasional papers covered the functional and ecological aspects of epiphytism: absorption by certain bromeliad leaf trichomes (Mez 1904); fauna and nutrition of tank bromeliads (Picado 1913);

water balance in orchid roots, particularly the role of the velamen–exodermis complex (Haberlandt 1914); osmotica in epiphyte versus host foliage (Harris and Lawrence 1916; Harris 1918); autecology of the epiphytic fern *Polypodium polypodioides* (Fig. 1.6; Pessin 1925); and an overview of New Zealand's canopy-adapted flora (Oliver 1930). Richards (1952) covered epiphytes in *The Tropical Rain Forest,* and Curtis (1952) cited 170 references on epiphytism in his reviews of common life systems in plants. Other classic writings include Went's (1940) attempt to explain the factors responsible for host selection, and the identification by Sanford (1969, 1974) and Johansson (1974) of the influence of climate on orchid distribution in Nigeria. Broader coverages are those of Gessner (1956) on water economy and Madison's (1977) overview. Benzing and Seemann (1978) and Benzing (1983, 1984) considered interactions between epiphytes and their supports and dependent fauna – subjects that provide the second focus for this monograph. A recent surge of reports on epiphyte biology reflects increased interest in tropical forests and an improved climate for biological research in parts of Latin America, Malaysia, Singapore, and several other strategically located areas. Journals now contain papers on physiology, functional anatomy, life history, associated biota, and the effects of epiphytes on nutrient cycling in hosting ecosystems.

Character and importance of epiphytic vegetation

Epiphytes are responsible for much of the biotic diversity that makes humid tropical forests the most complex of all the world's terrestrial ecosystems (Gentry and Dodson 1987a). Canopy-based species constitute fully one third – perhaps up to 50% – of the total vascular flora in some pluvial neotropical forests (Fig. 8.4). In addition, the capacity of these sites to accommodate fauna, including the majority of a suspected 30,000,000 insect species worldwide (Erwin 1983), can be attributed in significant measure to the shelter and sustenance provided by epiphytic vegetation. Effects on community dynamics are no less impressive. Green biomass (and presumably photosynthetic capacity as well) of nonvascular and higher plants anchored in tree crowns can rival – probably even exceed – that of phorophytes. Epiphytes display various mechanisms, some novel, for countering drought and acquiring essential ions; such specialization is unexcelled in the plant world. Some animals are preyed upon, others become symbiotic mutualists; out-of-the-ordinary substrata are utilized. Of all wide-ranging plant groups, the tropical forest canopy supports one of the least studied and most important.

A few days' exploration can still yield undescribed epiphytes, primarily in Orchidaceae but also in such sizable families as Araceae and Gesneriaceae.

The geologic record of epiphytism

Paleontologists have found little evidence of pre-Cenozoic or even early- to mid-Tertiary epiphytism. Araceae and several other families that today contain many epiphytes were well differentiated by the end of the Eocene, but no fossil remains can be unquestionably assigned to their canopy-based lineages. Except for a few doubtful Eocene fossils from southern Europe, Orchidaceae have not been reported in pre-Quaternary sediments (Schmid and Schmid 1977). *Tillandsia*-like pollen has been reported for the Panamanian Eocene (A. Graham pers. comm.), but there is no way to determine the habits of the bromeliads that produced it.

Animal dispersal in so many epiphyte-containing angiosperm families also indicates that tree crowns were colonized only recently by flowering plants. Fleshy fruit, and the bats, birds, and mammals that eat it, first became diverse and abundant in the Paleocene/Eocene (Tiffney 1984). Involvement of early pteridophytes and primitive seed plants in epiphytism is even more obscure. Petrified stems of arborescent lycopods, calamites, and other potential pre-Cretaceous hosts show penetration by alien roots, but these intrusions probably occurred after death; the invading axes appear to belong to other terrestrials. The absence of verified epiphytic angiosperms in ancient geologic deposits and their concentration today in a few large advanced families point to a recent massive expansion (Table 1.1). The present active state of evolution of many tropical orchid clades and other species-rich, canopy-based genera (e.g., *Anthurium, Rhododendron, Drymonia, Peperomia, Tillandsia;* Table 1.2), as well as their concentration in geologically young montane habitats, suggests that much epiphyte diversity dates from the Pliocene/Pleistocene.

Canopy colonization was also limited before the rise of Magnoliophyta because this division had to supply most of the substrata. The broadleaf forests at low humid latitudes, where so many epiphytes anchor today, began to develop near the Cretaceous/Tertiary boundary. That was only about 65 million years ago after a shift toward wetter climate fostered extensive multistratal vegetation. Early angiosperms were small and probably restricted to savanna, riverine, and other disturbed habitats. Extant relatives provide a little insight on the nature of early conifers as arboreal habitats: A few modern taxa (*Taxodium* and some *Pinus* spp.) support heavy epiphyte

Table 1.1. **The systematic distribution of vascular epiphytes**

Taxa	Genera[a]	Species[a]
Division Pteridophyta	92/239	2,593/9,000
Class Filicopsida	90/235	2,388/7,749
Subclass Polypodiidae	88/233	2,380/7,740
Order Ophioglossales	1/3	8/56
Family Ophioglossaceae	1/3	8/56
Genus *Ophioglossum* L.		8/30
Order Polypodiales	87/224	2,372/7,565
Family Schizaeaceae	1/4	2/170
Genus *Schizaea* J. Sm.		2/30
Family Hymenophyllaceae	2/2	400/600
Genus *Hymenophyllum* J. Sm.		250/300
Trichomanes L.		150/300
Family Vittariaceae	9/9	112/112
Genus *Ananthacorus* Underw.& Maxon		1/1
Anetium (Kunze) Splitg.		1/1
Antrophyum Kaulf.		40/40
Hecistopteris J. Sm.		1/1
Monogramma Schkurh.		2/2
Polytaenium Desv.		10/10
Scoliosorus Moore		1/1
Vaginularia Fee		6/6
Vittaria J. Sm.		50/50
Family Dennstaedtiaceae	2/18	3/370
Genus *Lindsaea* Dryander ex J. Sm.		2/150
Oenotrichia Copel.		1/4
Family Dryopteridaceae	10/55	292/1,920
Genus *Arthropteris J. Sm.*		15/15
Dryopteris Adans.		1/150
Elaphoglossum Schott		250/500
Lastreopsis Ching		1/35
Lomariopsis Fee		1/45
Oleandra Cav.		20/40
Polystichum Roth		1/160
Psammiosorus C. Christ.		1/1
Rumohra Raddi		1/2
Teratophyllum Holtt.		1/9
Family Aspleniaceae	1/7	400/675
Genus *Asplenium* L.		400/650
Family Davalliaceae	8/9	139/150
Genus *Araiostegia* Copel.		12/12
Davallia J. Sm.		40/40
Davallodes (Copel.) Copel.		11/11
Humata Cav.		50/50
Nephrolepis Schott		15/20
Parasorus Alderwerelt		1/1
Scyphularia Fee		8/8
Trogostolon Copel.		2/2
Family Blechnaceae	1/9	1/175
Genus *Stenochlaena* J. Sm.		1/5

Table 1.1. **(cont.)**

Taxa	Genera[a]	Species[a]
Family Polypodiaceae	53/65	1,023/1,100
Genus *Acrosorus* Copel.		5/5
Aglaomorpha Schott		4/4
Amphoradenium Desv.		6/6
Anarthropteris Copel.		1/1
Arthromeris (Moore) J. Sm.		6/9
Belvisia Mirbel		15/15
Calymmodon Presl		25/25
Campyloneurum Presl		25/25
Christiopteris Copel.		4/4
Colysis Presl		2/30
Crypsinus Presl		40/40
Dendroconche Copel.		2/2
Diblemma J. Sm.		1/1
Dictymia J. Sm.		3/3
Drymotaenium Makino		2/2
Drynaria (Bory) J. Sm.		20/20
Drynariopsis (Copel.) Ching		1/1
Eschatogramme Trev.		4/4
Goniophlebium (Bl.) Presl		20/20
Grammatopteridium Alderwerelt		2/2
Grammatis Sw.		400/400
Holcosorus Moore		3/3
Holostachyum (Copel.) Ching		1/1
Lecanopteris Reinw.		15/15
Lemmaphyllum Presl		4/4
Leptochilus Kaulf.		1/1
Loxogramme (Bl.) Presl		25/25
Marginariopsis C. Christ.		1/1
Merinthosorus Copel.		2/2
Microgramma Presl		13/13
Microsorium Link		40/40
Nematopteris Alderwerelt		1/1
Neocheiropteris C. Christ.		3/3
Neurodium Fee		1/1
Niphidium J. Sm.		4/4
Oleandropsis Copel.		1/1
Oreogrammatis Copel.		1/1
Paragramma (Bl.) Moore		2/2
Paraleptochilus Copel.		2/2
Photinopteris J. Sm.		1/1
Platycerium Desv.		15/15
Pleopeltis/Kunth in HBK		10/10
Polypodiopteris Reed		3/3
Polypodium L.		140/150
Prosaptia Presl		20/20
Pteropsis Desv.		6/6
Pycnoloma C. Christ.		3/3
Pyrrosia Mirbel		100/100

Table 1.1. **(cont.)**

Taxa	Genera[a]	Species[a]
Scleroglossum Alderwerelt		6/6
Selliguea Bory		4/5
Solanopteris Copel.		4/4
Thayeria Copel.		1/1
Thylacopteris Kunze ex Mett.		2/2
Subclass Psilotidae	2/2	8/9
Order Psilotales	2/2	8/9
Family Psilotaceae	2/2	8/9
Genus *Psilotum* Sw.		2/2
Tmesipteris Bernh.		6/7
Class Lycopodiopsida	2/4	205/1,251
Order Lycopodiales	1/2	200/401
Family Lycopodiaceae	1/2	200/401
Genus *Lycopodium* L.		200/400
Order Selaginellales	1/1	5/700
Family Selaginellaceae	1/1	5/700
Genus *Selaginella* Beauv.		5/700
Division Cycadophyta	2/10	2/155
Class Cycadopsida	2/10	2/155
Order Cycadales	2/10	2/155
Family Zamiaceae	2/7	2/125
Genus *Zamia* L.		2/35
Division Gnetophyta	1/3	3/66
Class Gnetopsida	1/3	3/66
Order Gnetales	1/1	3/30
Family Gnetaceae	1/1	3/30
Genus *Gnetum* L.		3/30
Division Magnoliophyta	782/11,836	20,859/221,868
Class Magnoliopsida	262/9,409	4,251/167,893
Subclass Magnoliidae	4/496	717/11,761
Order Magnoliales	1/177	6/2,948
Family Winteraceae	1/7	6/100
Genus *Drimys* J. R. & G. Forst. s.l.		6/70
Order Piperales	2/20	710/1,782
Family Piperaceae	2/10	710/3,100
Genus *Peperomia* Ruiz & Pav.		700/1,000
Piper L.		10/2,000
Order Ranunculales	1/148	1/3,148
Family Ranunculaceae	1/50	1/2,000
Genus *Thalictrum* L.		1/150
Subclass Hamamelidae	8/174	563/3,373
Order Urticales	7/112	562/2,130
Family Moraceae	4/40	522/1,000
Genus *Antiaropsis* K. Schum.		1/1
Coussapoa Aubl.		20/45.
Ficus L.		500/800
Pourouma Aubl.		1/50
Family Urticaceae	3/45	40/700
Genus *Elatostema* Gaudich.		10/200

Table 1.1. **(cont.)**

Taxa	Genera[a]	Species[a]
Pilea Lindl.		20/400
Procris Comm. ex Juss.		10/20
Order Myricales	1/3	1/50
Family Myricaceae	1/3	1/50
Genus *Myrica* L.		1/35
Subclass Caryophyllidae	20/500	152/10,864
Order Caryophyllales	20/458	152/9,464
Family Cactaceae	18/115	150/1,500
Genus *Aporocactus* Lem.		6/6
Cryptocereus Alex.		1/2
Disocactus Lindl.		7/7
Eccremocactus Britton & Rose		3/3
Epiphyllum Haworth		21/21
Heliocereus (Berg.) Britton & Rose		5/5
Hylocereus (Berg.) Britton & Rose		18/20
Lymanbensonia Kimm.		1/1
Mediocactus Britton & Rose		2/2
Nopalxochia Britton & Rose		1/1
Pfeiffera Salm-Dyck		1/1
Rhipsalis Gaertn.		58/65
Schlumbergera Lem.		6/6
Selenicereus (Berg.) Britton & Rose		13/17
Strophocactus Britton & Rose		1/1
Weberocereus Britton & Rose		3/3
Werckleocereus Britton & Rose		2/2
Wilmattea Britton & Rose		1/1
Family Caryophyllaceae	2/75	2/2,000
Genus *Arenaria* L.		1/250
Stellaria L.		1/150
Subclass Dilleniidae	60/1,460	925/24,643
Order Theales	13/176	181/3,385
Family Marcgraviaceae	7/7	89/122
Genus *Marcgravia* L.		50/55
Marcgraviastrum Bedell, ined.		10/15
Norantea Aubl.		1/2
Ruyschia Jacq.		7/7
Sarcopera Bedell, ined.		4/10
Schwartzia Vell.		8/14
Souroubea Aubl.		9/19
Family Clusiaceae	6/50	92/1,200
Genus *Clusia* L.		85/145
Clusiella Planch. & Triana		3/7
Havetiopsis Planch. & Triana		1/5
Odematopus Planch. & Triana		1/10
Quapoya Aubl.		1/3
Renggeria Meisn.		1/3
Order Malvales	3/225	6/3,300
Family Elaeocarpaceae	1/10	1/400
Genus *Sericolea* Schlecht.		1/20

Table 1.1. **(cont.)**

Taxa	Genera[a]	Species[a]
Family Bombacaceae	2/25	5/200
Genus *Ceiba* Mill.		1/10
Spirotheca Ulbrich		4/4
Order Nepenthales	1/8	6/193
Family Nepenthaceae	1/1	6/75
Genus *Nepenthes* L.		6/75
Order Violales	1/276	30/4,818
Family Begoniaceae	1/4	30/1,000
Genus *Begonia* L.		30/900
Order Ericales	37/174	673/4,044
Family Epacridaceae	1/30	1/400
Genus *Prionotes* R. Br.		1/1
Family Ericaceae	36/122	672/3,500
Genus *Agapetes* D. Don ex G. Don		60/80
Anthopterus W. J. Hook.		3/6
Anthopteropsis A. C. Sm.		1/1
Calopteryx A. C. Sm.		1/2
Cavendishia Lindl.		75/100
Ceratostema Juss.		16/23
Costera J. J. Sm.		8/8
Demosthenesia A. C. Sm.		6/11
Didonica Luteyn & Wilbur		2/2
Dimorphanthera F. Muell.		25/71
Diogenesia Sleum.		5/13
Diplycosia Bl.		61/98
Disterigma Niedenzu ex Drude		15/30
Gaultheria Kalm ex L.		8/200
Gonocalyx Planch. & Lind. ex A. C. Sm.		6/8
Killipiella A. C. Sm.		2/2
Lateropora A. C. Sm.		2/3
Lyonia Nutt.		1/35
Macleania W. J. Hook.		25/45
Mycerinus A. C. Sm.		1/3
Oreanthes Benth.		4/4
Orthaea Kl.		20/31
Pellegrinnia Sleum.		1/4
Pernettyopsis King & Gamble		1/1
Plutarchia A. C. Sm.		6/12
Psammisia Kl.		25/55
Rhododendron L.		112/850
Rusbya Britton		1/1
Satyria Kl.		20/23
Semiramisa Kl.		2/4
Siphonandra Kl.		1/1
Sphyrospermum Poepp. & Endl.		18/22
Themistoclesia Kl.		22/31
Thibaudia Ruiz & Pav.		20/60
Utleya Wilbur & Luteyn		1/1
Vaccinium L.		95/450

Table 1.1. **(cont.)**

Taxa	Genera[a]	Species[a]
Order Ebenales	1/87	1/1,752
Family Sapotaceae	1/70	1/800
Genus *Bumelia* Sw.		1/60
Order Primulales	4/64	28/2,100
Family Myrsinaceae	4/30	28/1,000
Genus *Cybianthus* Mart.		5/40
Embelia Burm.		5/130
Grammadenia Benth.		6/15
Myrsine L.		12/200
Subclass Rosidae	68/3,194	791/57,047
Order Rosales	10/317	21/6,696
Family Cunoniaceae	2/25	3/350
Genus *Ackama* A. Cunn.		1/3
Weinmannia L.		2/170
Family Pittosporaceae	1/9	5/200
Genus *Pittosporum* Banks ex Soland.		5/150
Family Grossulariaceae	1/25	1/350
Genus *Phyllonoma* Willd. ex Schult.		1/8
Family Crassulaceae	3/25	5/900
Genus *Echeveria* DC.		2/150
Kalanchoe Adans.		1/200
Sedum L.		2/600
Family Saxifragaceae	2/40	4/700
Genus *Hydrangea* L.		2/80
Quintinia A. DC.		2/20
Family Rosaceae	1/100	3/3,000
Genus *Pyrus* L.		3/30
Order Myrtales	37/445	671/7,205
Family Alzateaceae	1/1	1/1
Genus *Alzatea*		1/1
Family Myrtaceae	2/140	7/3,000
Genus *Mearnsia* Merr.		4/7
Metrosideros Banks ex Gaertn.		3/60
Family Onagraceae	1/17	15/675
Genus *Fuchsia*		15/100
Family Melastomataceae	33/180	648/4,770
Genus *Adelobotrys* DC.		21/25
Anerincleistus Korth.		1/1
Backeria Bakh. f.		2/2
Blakea P. Br.		98/100
Calvoa J. D. Hook		4/18
Catanthera F. Muell.		16/16
Clidemia D. Don		11/145
Creochiton Bl.		4/6
Dalenia Korth.		2/2
Dicellandra J. D. Hook.		1/3
Diplectria Reichb.		4/4
Dissochaeta Bl.		20/20
Graffenrieda DC.		2/40

Table 1.1. **(cont.)**

Taxa	Genera[a]	Species[a]
Gravesia Naud.		13/110
Hypenanthe Bl.		4/4
Kendrickia J. D. Hook.		1/1
Leandra Raddi		4/200
Macrolenes Naud. ex Miq.		20/20
Medinilla Gaudich.		300/400
Miconia Ruiz & Pav.		11/1,000
Monolena Triana		6/15
Myrianthemum Gilg		1/1
Neodissochaeta Bakh. f.		10/10
Omphalopus Naud.		1/1
Ossaea DC.		2/100
Pachycentria Bl.		8/8
Phainantha Gleason		4/4
Plethiandra J. D. Hook.		6/6
Pleiochiton Naud.		7/7
Pogonanthera Bl.		1/1
Preussiella Gilg		2/2
Topobea Aubl.		59/60
Triolena Naud.		2/18
Order Cornales	1/16	3/140
Family Cornaceae	1/11	3/100
Genus *Griselinia* G. Forst.		3/6
Order Celastrales	3/119	4/2,149
Family Celastraceae	2/50	3/800
Genus *Euonymus* L.		2/175
Microtropis Wall.		1/70
Family Aquifoliaceae	1/4	1/400
Genus *Ilex* L.		1/400
Order Rhamnales	2/67	4/1,670
Family Vitaceae	2/11	4/700
Genus *Pterisanthes* Bl.		2/20
Tetrastigma Planch.		2/90
Order Sapindales	3/500	3/5,346
Family Aceraceae	1/2	1/112
Genus *Acer* L.		1/110
Family Burseraceae	1/18	1/600
Genus *Dacryodes* Vahl		1/50
Family Anacardiaceae	1/70	1/600
Genus *Spondias* L.		1/12
Order Geraniales	1/25	5/2,154
Family Balsaminaceae	1/2	5/451
Genus *Impatiens* L.		5/450
Order Apiales	11/370	80/3,700
Family Araliaceae	9/70	78/700
Genus *Didymopanax* Decne. & Planch.		1/40
Motherwellia F. Muell.		1/1
Oreopanax Decne. & Planch.		1/120
Pentapanax Seem.		2/15
Polyscias J. R. & G. Forst		5/80
Pseudopanax C. Koch		2/6

Table 1.1. **(cont.)**

Taxa	Genera[a]	Species[a]
Schefflera J. R. & G. Forst.		60/200
Sciadophyllum P. Br.		5/30
Tupidanthus J. D. Hook & Thoms.		1/1
Family Apiaceae	2/300	2/3,000
Genus *Hydrocotyle* L.		1/100
Myrrhidendron Coulter & Rose		1/5
Subclass Asteridae	102/3,585	1,103/60,205
Order Gentianales	14/547	163/5,502
Family Loganiaceae	2/20	21/500
Genus *Desfontainia* Ruiz & Pav.		1/5
Fragraea Thunb.		20/35
Family Gentianaceae	3/75	4/1,000
Genus *Leiphaimos* Cham. & Schlecht.		1/40
Macrocarpaea Gilg		2/35
Voyria Aubl.		1/8
Family Apocynaceae	1/200	1/2,000
Genus *Mandevilla* Lindl.		1/114
Family Asclepiadaceae	8/250	137/2,000
Genus *Ceropegia* L.		3/160
Conchophyllum Bl.		1/10
Cynanchum L.		2/150
Dischidia R. Br.		60/90
Dischidiopsis Schlecht.		9/9
Heynella Backer		1/1
Hoya R. Br.		60/200
Marsdenia R. Br.		1/10
Order Solanales	12/182	56/5,099
Family Solanaceae	12/85	56/2,800
Genus *Dyssochroma* Miers		2/2
Ectozoma Miers		1/1
Hawkesiophyton A. T. Hunz.		3/3
Juanulloa Ruiz & Pav.		10/10
Lycianthes Hassl.		2/200
Markea L. C. Rich.		8/8
Merinthopodium Donn. Sm.		5/5
Rahowardiana D'Arcy		1/1
Schultesianthus A. T. Hunz.		5/5
Solandra Sw.		1/10
Solanum L.		15/1,700
Trianaea Planch. & Linden		3/3
Order Lamiales	1/403	2/7,805
Family Verbenaceae	1/100	2/2,600
Genus *Clerodendrum* L.		2/400
Order Scrophulariales	37/758	615/11,465
Family Scrophulariaceae	1/190	3/4,000
Genus *Wightea* Wall.		3/3
Family Gesneriaceae	30/120	560/2,500
Genus *Aeschynanthus* Jack		80/80
Agalmyla Bl.		15/15
Alloplectus Mart.		25/65

Table 1.1. (cont.)

Taxa	Genera[a]	Species[a]
Alsobia Hanst.		2/2
Asteranthera Hanst.		1/1
Boea Comm. ex Lam.		2/25
Capanea Planch.		8/11
Codonanthe (Mart.) Hanst.		17/17
Codonanthopsis Mansf.		3/3
Columnea L.		70/70
Cyrtandra J. R. & G. Forst		10/600
Dalbergaria Tussac		65/65
Dichrotrichum Reinw.		4/4
Didymocarpus Wall.		1/120
Drymonia Mart.		100/110
Fieldia A. Cunn.		1/1
Heppiella Regel		1/23
Loxostigma C. B. Cl.		3/4
Lysionotus G. Don		2/2
Mitraria Cav.		1/1
Monopyle Benth.		1/23
Nematanthus Schrader		26/26
Neomortonia Wiehler		3/3
Paradrymonia Hanst.		8/28
Pentadenia (Planch.) Hanst.		23/24
Rufodorsia Wiehler		4/4
Sarmienta Ruiz & Pav.		1/1
Sinningia Nees		3/60
Streptocarpus Lindl		10/132
Trichantha W. J. Hook.		70/70
Family Acanthaceae	2/250	8/2,500
Genus *Glockeria* Nees		1/10
Louteridium S. Watson		7/7
Family Bignoniaceae	2/100	29/800
Genus *Gibsoniothamnus* L. O. Wms.		11/11
Schlegelia Miq.		18/18
Family Lentibulariaceae	2/5	15/200
Genus *Pinguicula* L.		2/60
Utricularia L.		12/150
Order Campanulales	5/93	24/2,490
Family Campanulaceae	5/70	24/2,000
Genus *Burmeistera* Karst. & Triana		5/82
Canarina L.		1/3
Centropogon Presl		7/300
Clermontia Gaudich.		10/27
Cyanea Gaudich.		1/60
Order Rubiales	25/451	223/6,503
Family Rubiaceae	25/450	223/6,500
Genus *Amaracarpus* Bl.		3/60
Balmea Martinez		1/1
Coprosma J. R. & G. Forst.		6/90
Cosmibuena Ruiz & Pav.		6/15
Didymochlamys J. D. Hook		2/2

Table 1.1. **(cont.)**

Taxa	Genera[a]	Species[a]
Hillia Jacq.		20/20
Hydnophytum Jack		75/80
Lecananthus Jack		1/2
Lucinaea DC.		15/25
Malanea Aubl.		2/27
Manettia Mutis ex L.		5/130
Myrmecodia Jack		40/45
Myrmedoma Becc.		2/2
Myrmephytum Becc.		2/2
Nertera Banks & Soland.		6/12
Ophiorrhiza L.		5/150
Posoqueria Aubl.		1/15
Proscephaleium Korth.		1/1
Psychotria L.		7/700
Randia L.		2/250
Ravnia Oerst.		4/4
Relbunium Benth. & J. D. Hook.		2/30
Schradera Vahl		12/15
Squamellaria Becc.		2/2
Timonius DC.		1/150
Order Asterales	8/1,100	20/20,000
Family Asteraceae	8/1,100	20/20,000
Genus *Anaphylis* DC.		1/35
Dahlia Cav.		1/20
Eupatorium L. (s.l.)		7/1,200
Liabum Adans.		2/90
Pseudogynoxys (Greenm.) Cabrera		1/21
Rensonia S. F. Blake		1/1
Senecio L.		5/2,000
Tuberostylis Steetz		2/2
Class Liliopsida	520/2,427	16,608/53,975
Subclass Arecidae	21/329	1,439/6,461
Order Cyclanthales	7/10	86/200
Family Cyclanthaceae	7/10	86/200
Genus *Asplundia* Harling		60/90
Dicranopygium Harling		5/50
Evodianthus Oerst.		1/1
Ludovia Brongn.		3/3
Sphaeradenia Harling		15/40
Stelestylis Drude		1/3
Thoracocarpus Harling		1/1
Order Pandanales	1/3	4/732
Family Pandanaceae	1/3	4/732
Genus *Pandanus* L.		4/550
Order Arales	13/116	1,349/2,529
Family Araceae	13/110	1,349/2,500
Genus *Amydrium* Schott		4/4
Anthurium Schott		750/1,000
Epipremnum Schott		15/15
Monstera Adans.		29/30

Table 1.1. **(cont.)**

Taxa	Genera[a]	Species[a]
Pedicellarum Hotta		1/1
Philodendron Schott		300/350
Pothos L.		50/75
Remusatia Schott		1/4
Rhaphidophora Hassk.		100/100
Rhodospatha Poepp.		14/20
Scindapsus Schott		20/30
Stenospermation Schott		30/30
Syngonium Schott		35/35
Subclass Commelinidae	10/703	15/14,977
Order Commelinales	5/71	10/1,004
Family Rapateaceae	2/16	6/100
Genus *Epidryos* Maguire		3/3
Stegolepis Kl. ex Koern.		3/23
Family Commelinaceae	3/50	4/700
Genus *Campelia* L. C. Rich.		1/3
Cochliostema Lem.		2/2
Cyanotis D. Don.		1/50
Order Cyperales	5/570	5/12,000
Family Cyperaceae	3/70	3/4,000
Genus *Cephalocarpus* Nees		1/7
Cyperus L.		1/550
Pseudoeverardia Gilly		1/1
Family Poaceae	2/500	2/8,000
Genus *Microlaena* R. Br.		1/10
Tripogon Roem. & Schult.		1/20
Subclass Zingiberidae	33/134	1,170/4,520
Order Bromeliales	26/45	1,144/2,500
Family Bromeliaceae	26/45	1,144/2,500
Genus *Acanthostachys* Link. Kl.& Otto		2/2
Aechmea Ruiz & Pav.		120/150
Araeococcus Brongn.		4/4
Billbergia Thunb.		45/50
Brocchinia Schult.f.		3/18
Bromelia L.		3/40
Canistrum E. Morr.		3/7
Catopsis Griseb.		20/20
Glomeropitcairnia Mez		2/2
Guzmania Ruiz & Pav.		120/140
Hohenbergia Bak.		20/40
Hohenbergiopsis L. B. Sm. & R. Read		1/1
Lymania R. Read		4/4
Mezobromelia L. B. Sm. & R. Read		4/4
Navia J. H. Schult.		2/60
Neoregelia L. B. Sm.		65/75
Nidularium Lem.		15/22
Pitcairnia L'Herit.		75/280
Protea Brongn. & C. Koch		5/7
Pseudaechmea L. B. Sm. & R. Read		1/1
Quesnelia Gaudich		6/12

Table 1.1. **(cont.)**

Taxa	Genera[a]	Species[a]
Ronnbergia E. Morr. & Andre		6/7
Streptocalyx Beer		14/15
Tillandsia L.		400/450
Vriesea Lindl.		200/260
Wittrockia Lindl.		4/6
Order Zingiberales	7/89	26/2,020
Family Zingiberaceae	5/47	20/1,000
Genus *Alpinia* Roxb.		1/100
Brachychilum (R. Br. ex Wall.) Petersen		1/1
Burbidgea J. D. Hook.		5/5
Hedychium Koen.		12/50
Riedelia Oliv.		1/50
Family Costaceae	1/4	4/175
Genus *Costus* L.		4/150
Family Marantaceae	1/30	2/400
Genus *Maranta* L.		2/23
Subclass Liliadae	456/1,199	13,984/27,516
Order Liliales	15/451	31/8,248
Family Liliaceae	10/280	24/4,000
Genus *Astelia* Banks & Soland.		6/25
Clivia Lindl.		1/3
Collospermum Skotts.		5/5
Curculigo Gaertn.		1/10
Cyrtanthus Ait.		1/47
Dianella Lam.		2/30
Hippeastrum Herb.		2/75
Pamianthe Stapf		1/3
Rhodocodon Baker		1/8
Smilacina Desf.		4/25
Family Agavaceae	2/18	3/600
Genus *Agave* L.		1/300
Yucca L.		2/40
Family Smilacaceae	2/12	3/300
Genus *Lapageria* Ruiz & Pav.		1/1
Luzuriaga Ruiz & Pav.		2/3
Family Dioscoreaceae	1/6	1/630
Genus *Dioscorea* L.		1/600
Order Orchidales	441/748	13,953/19,268
Family Burmanniaceae	1/20	2/130
Genus *Burmannia* L.		2/57
Family Orchidaceae	440/725	13,951/19,128
Genus *Abdominea* J. J. Sm.		2/2
Acampe Lindl.		6/6
Acineta Lindl.		10/10
Acostaea Schltr.		8/8
Acriopsis Reinw. ex Bl.		12/12
Ada Lindl.		9/9
Adenoncos Bl.		17/17
Adrorhizon J. D. Hook.		1/1
Aerangis Rchb. f.		60/60

Table 1.1. **(cont.)**

Taxa	Genera[a]	Species[a]
Aeranthes Lindl.		30/30
Aerides Lour.		19/19
Aganisia Kaempf. ex Spreng.		1/1
Aglossorhyncha Schltr.		6/6
Agrostophyllum Bl.		60/60
Alamania La Ll. & Lex.		1/1
Ambrella H. Perrier		1/1
Amesiella Schltr. ex Garay		1/1
Amparoa Schltr.		2/2
Ancistrochilus Rolfe		2/2
Ancistrorhynchus Finet		13/13
Andreettaea Luer		1/1
Angraecopsis Krzl.		14/14
Angraecum Bory		206/206
Anguloa Ruiz & Pav.		10/10
Ansellia Lindl.		2/2
Anthosiphon Schltr.		1/1
Antillanorchis Garay		1/1
Appendicula Bl.		100/100
Arachnis Bl.		2/2
Armodorum Breda		2/2
Arpophyllum La Ll. & Lex.		5/5
Artorima Dressl. & Poll.		1/1
Ascocentrum Schltr.		8/8
Ascochilopsis Carr		1/1
Ascochilus Ridl.		6/6
Ascoglossum Schltr.		1/1
Aspasia Lindl.		6/6
Barbosella Schltr.		27/27
Barkeria Knowles & Westc.		14/14
Barombia Schltr.		11
Basiphyllaea Schltr.		3/3
Batemannia Lindl.		4/4
Beadlea Small		1/54
Beclardia A. Rich		1/1
Beloglottis Schltr.		1/7
Benthamia A. Rich.		6/26
Biermannia King & Pantl.		8/8
Bifrenaria Lindl.		27/27
Bogoria J. J. Sm.		4/4
Bollea Rchb. f.		7/7
Bolusiella Schltr.		10/10
Bonniera Cordemoy		2/2
Brachypeza Garay		7/7
Brachtia Rchb. f.		6/6
Brassavola R. Br.		23/23
Brassia R. Br.		38/38
Bromheadia Lindl.		11/11
Broughtonia R. Br.		6/6
Bulbophyllum Thouars		1,000/1,000

Table 1.1. **(cont.)**

Taxa	Genera[a]	Species[a]
Bulleyia Schltr.		1/1
Cadetia Gaud.		67/67
Calymmanthera Schltr.		5/5
Calyptrochilum Krzl.		2/2
Campylocentrum Benth.		45/45
Capanemia Barb. Rodr.		16/16
Cardiochilus Cribb		2/2
Catasetum L. C. Rich. ex Kunth		76/76
Cattleya Lindl.		45/45
Caucaea Schltr.		1/1
Caularthron Raf.		3/3
Centroglossa Barb. Rodr.		6/6
Ceratochilus Bl.		2/2
Ceratostylis Bl.		70/70
Chamaeangis Schltr.		15/15
Chamaeanthus Schultr. ex J. J. Sm.		10/10
Chamelophyton Garay		1/1
Chaseella Summerh.		1/1
Chaubardia Rchb. f.		3/3
Chaubardiella Garay		6/6
Chauliodon Summerh.		1/1
Cheiradenia Lindl.		2/2
Chilopogon Schltr.		3/3
Chiloschista Lindl.		15/15
Chitonanthera Schltr.		7/7
Chitonochilus Schltr.		1/1
Chondrorhyncha Lindl.		16/16
Chroniochilus J. J. Sm.		5/5
Chrysocycnis Lind. & Rchb. f.		5/5
Chysis Lindl.		6/6
Chytroglossa Rchb. f.		4/4
Cirrhaea Lindl.		3/3
Cischweinfia Dressl. & N. Wms.		6/6
Claderia Hook. f.		2/2
Cleisomeria Lindl. ex G. Don		2/2
Cleisocentron Bruhl		3/3
Cleisostoma Bl.		95/95
Clowesia Lindl.		5/5
Cochleanthes Raf.		20/20
Cochlioda Lindl.		7/7
Coelia Lindl.		5/5
Coeliopsis Rchb. f.		2/2
Coelogyne Lindl.		100/100
Comparettia Poepp. & Endl.		11/11
Constantia Barb. Rodr.		4/4
Cordiglottis J. J. Sm.		7/7
Coryanthes W. J. Hook.		20/20
Cottonia Wight		1/1
Cryptarrhena Lindl.		4/4
Cryptocentrum Benth.		14/14

Table 1.1. **(cont.)**

Taxa	Genera[a]	Species[a]
Cryptochilus Wall.		6/6
Cryptophoranthus Barb. Rodr.		36/36
Cryptopus Lindl.		3/3
Cryptopylos Garay		1/1
Cycnoches Lindl.		17/17
Cymbidiella Rolfe		3/3
Cymbidium Sw.		50/50
Cypholoron Dodson & Dressl.		2/2
Cyrtidium Schltr.		4/4
Cyrtopodium R. Br.		12/12
Cyrtorchis Schltr.		18/18
Dendrobium Sw.		900/900
Dendrochilum Bl.		120/120
Dendrophylax Rchb. f.		5/5
Diadenium Poepp. & Endl.		2/2
Diaphananthe Schltr.		45/45
Dichaea Lindl.		45/45
Dickasonia L. O. Wms.		1/1
Dilochia Lindl.		3/5
Dilomilis Raf.		4/4
Dimerandra Schltr.		2/2
Dimorphorchis D. Don		2/2
Dinklageella Mansf.		1/1
Diothonaea Lindl.		7/7
Diplocaulobium Krzl.		94/94
Diplocentrum Lindl.		2/2
Diploprora J. D. Hook		1/1
Dipodium R. Br.		12/12
Dipteranthus Barb. Rodr.		2/2
Dipterostele Schltr.		2/2
Distylodon Summerh.		1/1
Dodsonia Ackerman		2/2
Domingoa Schltr.		2/2
Dracula Luer		93/93
Dresslerella Luer		8/8
Dressleria Dodson		4/4
Dryadella Luer		31/31
Dryadorchis Schltr.		2/2
Drymoanthus Nicholls		2/2
Drymoda Lindl.		2/2
Dunstervillea Garay		1/1
Dyakia E. A. Christ., ined.		1/1
Earina Lindl.		7/7
Eggelingia Summerh.		2/2
Elleanthus Presl		70/70
Eloyella P. Ortiz		3/3
Encheiridion Summerh.		1/1
Encyclia W. J. Hook		130/130
Eparmatostigma Garay		1/1

Table 1.1. **(cont.)**

Taxa	Genera[a]	Species[a]
Epiblastus Schltr.		20/20
Epidanthus L. O. Wms.		3/3
Epidendrum L.		500/500
Epigeneium Gagnep.		12/12
Eria Lindl.		500/500
Eriopsis Lindl.		2/3
Erycina Lindl.		2/2
Esmeralda Rchb. f.		2/2
Eulophiella Rolfe		2/2
Eurychone Schltr.		2/2
Fernandezia Ruiz & Pav.		9/9
Flickingeria Hawkes		70/70
Galeandra Lindl.		20/20
Gastrochilus D. Don		38/38
Genyorchis Schltr.		6/6
Glomera Bl.		50/50
Glassorhyncha Ridl.		70/70
Gomesa R. Br.		9/9
Gongora Ruiz & Pav.		40/40
Grammangis Rchb. f.		2/2
Grammatophyllum Bl.		12/12
Graphorkis Thouars		5/5
Grobya Lindl.		3/3
Grosourdya Rchb. f.		8/8
Gynoglottis J. J. Sm.		1/1
Hagsatera G. Tomayo		2/2
Haraella Kudo		1/1
Harrisella Fawc. & Rendle		4/4
Hederorkis Thouars		2/2
Helcia Lindl.		1/1
Helleriella Hawkes		3/3
Hexisea Lindl.		5/5
Hintonella Ames		1/1
Hippeophyllum Schltr.		6/6
Hoehneella Ruschi		2/2
Hofmeisterella Rchb. f.		1/1
Holcoglossum Schltr.		8/8
Homalopetalum Rolfe		4/4
Houlletia Brongn.		8/8
Huntleya Batem. ex Lindl.		10/10
Hybochilus Schltr.		2/2
Hygrochilus Pfitz.		1/1
Hymenorchis Schltr.		9/9
Imerinaea Schltr.		1/1
Ionopsis Kunth		3/3
Isabelia Barb. Rodr.		2/2
Ischnocentrum Schltr.		1/1
Ischnogyne Schltr.		1/1
Isochilus R. Br.		13/13

Table 1.1. **(cont.)**

Taxa	Genera[a]	Species[a]
Jacquiniella Schltr.		11/11
Jumellea Schltr.		60/60
Kefersteinia Rchb. f.		25/25
Kegeliella Mansfeld		4/4
Koellensteinia Rchb. f.		1/11
Lacaena Lindl.		3/3
Laelia Lindl.		69/69
Lemurella Schltr.		3/3
Lemurorchis Krzl.		1/1
Leochilus Knowles & Westc.		16/16
Lepanthes Sw.		500/500
Lepanthopsis Ames		25/25
Leptotes Lindl.		5/5
Liparis L. C. Rich.		300/350
Listrostachys Rchb. f.		3/3
Lockhartia W. J. Hook.		29/29
Loefgrenianthus Hoehne		1/1
Lopharis Raf.		25/25
Lueddemannia Lind. & Rchb. f.		1/1
Luisia Gaud.		47/47
Lycaste Lindl.		43/43
Lycomormium Rchb. f.		5/5
Macradenia R. Br.		11/11
Macroclinium Dodson		25/25
Macropodanthus L. O. Wms.		1/1
Malleola J. J. Sm.		34/34
Masdevallia Ruiz & Pav.		400/400
Maxillaria Ruiz & Pav.		600/600
Mediocalcar J. J. Sm.		20/20
Megalotus Garay		1/1
Meiracyllium Rchb. f.		2/2
Mendoncella Hawkes		11/11
Mesospinidium Rchb. f.		7/7
Mexicoa Garay		1/1
Microcoelia Lindl.		26/26
Micropera Lindl.		19/19
Microsaccus Bl.		14/14
Microtatorchis Schltr.		49/49
Miltonia Lindl.		12/12
Miltoniopsis Godefr.-Lebeuf		6/6
Mobilabium Rupp		1/1
Monomeria Lindl.		4/4
Mormodes Lindl.		64/64
Mormolyca Fenzl		6/6
Myoxanthus Poepp. & Endl.		42/42
Mystacidium Lindl.		5/5
Nabaluia Ames		1/1
Nageliella L. O. Wms.		2/2
Neobathiea Schltr.		7/7

Table 1.1. **(cont.)**

Taxa	Genera[a]	Species[a]
Neocogniauxia Schltr.		2/2
Neodryas Rchb. f.		4/4
Neofinetia Hu		1/1
Neogardneria Schltr.		1/1
Neogyna Rchb. f.		1/1
Neokoehleria Schltr.		7/7
Neomoorea Rolfe		1/1
Neowilliamsia Garay		5/5
Nephrangis Summerh.		1/1
Nidema Britt. & Millsp.		2/2
Notylia Lindl.		46/46
Oberonia Lindl.		300/300
Octarrhena Thwaites		35/35
Octomeria R. Br.		134/134
Odontoglossum Kunth (s.l.)		140/140
Oeonia Lindl.		6/6
Oeoniella Schltr.		3/3
Oerstedella Rchb. f.		28/28
Oliveriana Rchb. f.		4/4
Omoea Bl.		2/2
Oncidium Sw.		430/432
Orleanesia Barb. Rodr.		7/7
Ornithocephalus W. J. Hook.		28/28
Ornithochilus Wall. ex Lindl.		1/1
Ornithophora Barb. Rodr.		2/2
Otochilus Lindl.		4/4
Otoglossum (Schltr.) Garay & Dunsterv.		8/8
Oxyanthera Brongn.		6/6
Pabstia Garay		5/5
Pachyphyllum Kunth		25/25
Palumbina Rchb. f.		1/1
Paphinia Lindl.		8/8
Paphiopedilum Pfitz.		33/70
Papilionanthe Schltr.		10/10
Papillalabium Dockr.		1/1
Papperitzia Rchb. f.		1/1
Paraphalaenopsis Hawkes		4/4
Pedilochilus Schltr.		15/15
Pelatantheria Ridl.		3/3
Pennilabium J. J. Sm.		10/10
Peristeranthus T. E. Hunt		1/1
Peristeria W. J. Hook		8/8
Perrierella Schltr.		1/1
Pescatorea Rchb. f.		14/14
Phalaenopsis Bl.		46/46
Phloeophila Hoehne & Schltr.		7/7
Pholidota Lindl. ex W. J. Hook.		40/40
Phragmipedium Rolfe		5/15
Phragmorchis L. O. Wms.		1/1

Table 1.1. **(cont.)**

Taxa	Genera[a]	Species[a]
Phreatia Lindl.		190/190
Phymatidium Lindl.		7/7
Physosiphon Lindl.		6/6
Physothallis Garay		2/2
Pinelia Lindl.		3/3
Pityphyllum Schltr.		4/4
Platyglottis L. O. Wms.		1/1
Platyrhiza Barb. Rodr.		1/1
Platystele Schltr.		58/58
Plectorrhiza Dockr.		3/3
Plectrelminthus Raf.		1/1
Plectrophora Focke		6/6
Pleurothallis R. Br.		1,500/1,500
Poaephyllum Ridl.		3/3
Podangis Schltr.		1/1
Podochilus Bl.		75/75
Polycycnis Rchb. f.		20/20
Polyotidium Garay		1/1
Polyradicion Garay		4/4
Polystachya W. J. Hook.		150/150
Pomatocalpa Breda		46/46
Ponera Lindl.		9/9
Porpax Lindl.		8/8
Porphyrodesme Schltr.		3/3
Porphyroglottis Ridl.		1/1
Porroglossum Schltr.		21/21
Porrorhachis Garay		2/2
Promenaea Lindl		15/15
Pseudacoridium Ames		1/1
Pseuderia Schltr.		4/4
Pseudolaelia Campos-Porto & Brade		6/6
Psychopsis Raf.		4/4
Psygmorchis Dodson & Dressl.		6/6
Pteroceras Hasselt ex Hassk.		41/41
Pterostemma Krzl.		1/1
Quekettia Lindl.		5/5
Quisqueya D. Dod		4/4
Rangaeris Summerh.		6/6
Rauhiella Pabst & Braga		1/1
Reichenbachanthus Barb. Rodr.		5/5
Renanthera Lour.		14/14
Renantherella Ridl.		2/2
Restrepia Kunth		32/32
Restrepiella Garay & Dunsterv.		1/1
Restrepiopsis Luer		17/17
Rhaesteria Summerh.		1/1
Rhinerrhiza Rupp		2/2
Rhipidoglossum Schltr.		4/4
Rhynchogyna Seidenf. & Garay		2/2

Table 1.1. **(cont.)**

Taxa	Genera[a]	Species[a]
Rhyncholaelia Schltr.		2/2
Rhynchophreatia Schltr.		5/5
Rhynchostylis Bl.		3/3
Ridleyella Schltr.		1/1
Robiquetia Gaud.		39/39
Rodriguezia Ruiz & Pav.		34/34
Rodrigueziopsis Schltr.		2/2
Rossioglossum (Schltr.) Garay & Kennedy		5/5
Rudolfiella Hoehne		2/2
Rusbyella Rolfe		2/2
Saccoglossum Schltr.		2/2
Saccolabiopsis J. J. Sm.		13/13
Saccolabium Bl.		4/4
Salpistele Dressl.		6/6
Sanderella O. Ktze.		2/2
Sarcochilus R. Br.		14/14
Sarcostoma Bl.		2/2
Saundersia Rchb. f.		1/1
Scaphosepalum Pfitz.		26/26
Scaphyglottis Poepp. & Endl.		52/52
Scelochilus Kl.		34/34
Schlimmia Planch. & Lind. ex Lindl. & Paxt.		5/5
Schoenorchis Bl.		22/22
Schomburgkia Lindl.		17/17
Scuticaria Lindl.		6/6
Sedirea Garay & Sweet		2/2
Seidenfadenia Garay		1/1
Sepalosiphon Schltr.		1/1
Sievekingia Rchb. f.		15/15
Sigmatogyne Pfitz.		2/2
Sigmatostalix Rchb. f.		35/35
Sirhookera O. Ktze.		2/2
Smithsonia Saldanha		3/3
Smitinandia Holtt.		3/3
Sobennikoffia Schltr.		3/3
Sobralia Ruiz & Pav.		96/96
Solenangis Schltr.		2/2
Solenidium Lindl.		3/3
Sophronitis Lindl.		7/7
Sphyrarhynchus Mansfeld		1/1
Sphyrastylis Schltr.		6/6
Stanhopea Frost ex W. J. Hook.		55/55
Stelis Sw.		300/300
Stellilabium Schltr.		16/16
Stenia Lindl.		1/1
Stenorrhynchus L. C. Rich		1/9
Stereochilus Lindl.		5/5
Stolzia Schltr.		4/4
Summerhayesia Cribb		2/2

Table 1.1. **(cont.)**

Taxa	Genera[a]	Species[a]
Sunipia Buc.-Ham. ex J. E. Sm.		25/25
Symphyglossum Schltr.		4/4
Systeloglossum Schltr.		5/5
Taeniophyllum Bl.		187/187
Taeniorrhiza Summerh.		1/1
Telipogon Kunth		82/82
Tetramicra Lindl.		11/11
Teuscheria Garay		6/6
Thecostele Rchb. f.		5/5
Thelasis Bl.		10/10
Thrixspermum Lour.		165/165
Thysanoglossa Porto & Brade		1/1
Trachoma Garay		6/6
Tervoria Lehmann		4/4
Trias Lindl.		2/2
Triceratorhynchus Summerh.		1/1
Trichocentrum Poepp. & Endl.		23/23
Trichoceros Kunth		8/8
Trichoglottis Bl.		80/80
Trichopilia Lindl.		21/21
Trichosalpinx Luer		84/84
Tridactyle Schltr.		35/35
Trigonidium Lindl.		12/12
Trisetella Luer		15/15
Trizeuxis Lindl.		1/1
Tuberolabium Yamamoto		5/5
Uncifera Lindl.		7/7
Vanda Jones		45/45
Vandopsis Pfitz.		18/18
Ventricularia Garay		1/1
Warmingia Rchb. f.		2/2
Xenikophyton Garay		1/1
Xylobium Lindl.		22/22
Ypsilopus Summerh.		2/2
Zygopetalum W. J. Hook.		40/40
Zygosepalum Rchb. f.		7/7
Zygostates Lindl.		12/12

[a]For genera and species the number of epiphytic taxa is followed by the total number of taxa.
Sources: Kress 1986; Gentry and Dodson 1987b; and Casper 1987, for *Pinguicula*.

loads in contrast to the sparsity on most tropical gymnosperms, cycads included. But so far one can only guess at what sorts of plant life grew attached to the bennettitaleans, gnetaleans, seed ferns, and members of other mostly extinct pre-Tertiary spermatophyte groups.

Table 1.2. **Genera containing at least 100 epiphytic species and their geographic zone**

Division/family/ genus	Total number of species	Number of epiphytic species	Geographic distribution
A. Pteridophyta			
Aspleniaceae			
Asplenium	650	400	pantropical
Hymenophyllaceae			
Hymenophyllum	300	250	pantropical
Trichomanes	300	150	pantropical
Lycopodiaceae			
Lycopodium	400	200	cosmopolitan
Polypodiaceae			
Elaphoglossum	500	250	pantropical
Grammatis	400	400	pantropical
Polypodium	150	140	pantropical
Pyrrosia	100	100	paleotropical
B. Magnoliophyta: monocots			
Araceae			
Anthurium	850	600	neotropical
Philodendron	475	350	neotropical
Rhaphidophora	100	100	paleotropical
Bromeliaceae			
Aechmea	150	120	neotropical
Guzmania	140	120	neotropical
Tillandsia	450	400	neotropical
Vriesea	260	200	neotropical
Orchidaceae			
Angraecum	206	206	African
Appendicula	100	100	Australasian
Bulbophyllum	1000	1000	pantropical
Coelogyne	100	100	Asian
Dendrobium	900	900	Australasian
Dendrochilum	120	120	Australasian
Encyclia	130	130	neotropical
Epidendrum	800+	720	neotropical
Eria	550	500	Australasian
Lepanthes	600	600	neotropical
Liparis	350	300	cosmopolitan
Masdevallia	400	400	neotropical
Maxillaria	600	570	neotropical
Oberonia	300	300	paleotropical
Octomeria	130	130	neotropical
Odontoglossum	300	285	neotropical
Oncidium	500	475	neotropical
Phreatia	190	190	Australasian

Table 1.2. **(cont.)**

Division/family/ genus	Total number of species	Number of epiphytic species	Geographic distribution
Pleurothallis	1600	1500	neotropical
Polystachya	210	200	pantropical
Stelis	600	540	neotropical
Taeniophyllum	120	120	paleotropical
C. Magnoliophyta: dicots			
Ericaceae			
Rhododendron	850	112	pantropical
Gesneriaceae			
Columnea	265	262	neotropical
Drymonia	100	100	neotropical
Melastomataceae			
Medinilla	400+	300+	paleotropical
Moraceae			
Ficus	800	500	pantropical
Piperaceae			
Peperomia	1000	700	pantropical

Source: After Gentry and Dodson 1987b.

Global occurrence

Epiphytes inhabit an unusually extensive biotope; virtually every humid tropical forest contains them, sometimes in such great variety that single trees host dozens of species. Occurrence on drier sites is less common and usually involves fewer taxa, but not necessarily lower abundance; certain Mexican and Peruvian cactus/scrub forests, for instance, support dense populations of a few stress-tolerant bromeliads and orchids. Epiphytes are best represented on stable, relatively temperate sites where pre- and lower-montane, wet, and pluvial forest is well developed (Fig. 8.4; Table 8.6). Reasons for this as well as for their disproportionate presence in New as opposed to Old World tropical forests are discussed in Chapter 8.

Frost, a powerful impediment to vascular epiphytism, sets different limits in the two hemispheres. Only a few bromeliads, ferns, and orchids penetrate higher American latitudes along marine coasts. *Tillandsia usneoides* (Fig. 7.2) and *Polypodium polypodioides* (Fig. 1.6) are regularly subjected to short bouts of subfreezing weather in the southeastern United States. Subtropical Japan has a modest epiphyte flora. Warmth from oceans allows penetration of still higher latitudes in South America and Australasia. Lichens and bryophytes are much hardier and occur in almost all temperate forests; however, unlike many of their tropical counterparts, few grow as epiphylls (Fig. 7.6). A number of mistletoes (Fig. 6.1J) are uncharacteristically tolerant to cold, but they are the ones whose bodies reside largely within the host.

Systematic occurrence

About 10% of all vascular plant species anchor on bark too often to be accidental epiphytes (Tables 1.1, 1.3), but distribution among higher taxa is uneven. Epiphytism is pronounced among ferns – about 29% regularly occur in tree crowns – but of the microphyllous pteridophytes, only *Lycopodium* (Fig. 1.18) is extensively epiphytic. *Psilotum* and *Tmesipteris* frequently anchor in pockets of suspended humus. Few gymnosperms are arboreal, in part no doubt because of their relatively immobile seeds, but also because of wind pollination – a mechanism that mandates abundant pollen production and dense populations. Surprisingly, rather massive *Zamia pseudoparasitica* and a similar unnamed congeneric in Ecuador somehow manage to thrive in the canopies of a few humid Panamanian and northern Andean forests. Even small branches occasionally provide support, apparently after brightly colored seeds are partially eaten and discarded by frugivorous birds, including toucans. Orchidaceae have been more successful than any other lineage in colonizing tree crowns. About two out of three epiphytes are orchids; at least 70% of the family are canopy-adapted. Two other monocot groups with pronounced epiphytic bias are Araceae (especially *Anthurium, Philodendron,* and *Rhaphidophora*) and Bromeliaceae, about half of whose species are epiphytic. Canopy-dwelling dicots are overrepresented by Cactaceae, Ericaceae, Gesneriaceae, Melastomataceae, and Piperaceae. In all, 84 vascular families (Table 1.3) contain at least one epiphytic member. But there are puzzling omissions: Some very large ecologically diverse taxa (e.g., Fabaceae, Lamiaceae, Poaceae, and Scrophulariaceae) are absent, or nearly so, from canopy floras. Generally, epiphytic angiosperms are concentrated in families considered advanced in terms of reproductive morphology. The same is true of ferns, but not without exceptions. Several members of the relic order Ophioglossales (Table 1.1) root in such humic impoundments as those of persistent palm leaf bases.

Epiphytism as a coherent ecological category

A few writers of popular literature attribute conscious intent to the epiphytes' historical move up the forest profile. This obviously holds little

Table 1.3. **Taxonomic distribution of vascular epiphytes**

Major groups	Taxonomic categories	Number of taxa containing epiphytes in each category	Percentage of taxa containing epiphytes in each category
	classes	6	75
	orders	44	45
All vascular plants	families	84	19
	genera	876	7
	species	23,456	10
	classes	2	67
	orders	5	50
Ferns and allies	families	13	34
	genera	92	39
	species	2,593	29
	classes	2	67
	orders	2	33
Gymnosperms	families	2	13
	genera	2	3
	species	5	<1
	subclasses	6	100
	orders	28	44
Angiosperms (dicots)	families	52	16
	genera	262	3
	species	4,251	3
	subclasses	4	80
	orders	9	47
Angiosperms (monocots)	families	17	26
	genera	520	21
	species	16,608	31

Source: Kress 1986.

appeal for the biologist; invasion of the forest canopy was just one facet of the massive radiation, commencing over 400 million years ago, of increasingly versatile tracheophytes. But this vertical journey obliged some major trade-offs. Benefit gained from sunnier sites often incurred heightened constraints on water and nutrient balance, and by extension, on reproduction. Ground-free status has been no simple achievement and has provoked fascination for many years. Insight into its mechanisms has been slow in coming, however. And there is another mystery. Why so many plants root exclusively on bark, yet never invade host vasculature as the mistletoes do, remains one of the most intriguing questions of tropical botany.

Plants with common ways of living usually share key qualities that set them apart from other vegetation. Occurrence on similar substrata under similar climate fosters evolutionary convergence. Botanical carnivores occupy impoverished sunny sites and possess traps in which captured fauna are processed for food; lianas feature slender habit and novel vascular anatomy and photomorphogenesis; vernal ephemerals in temperate deciduous forests deploy simplified shoots bearing heliophilic, short-lived foliage. Yet approximately 25,000 epiphytic species (mistletoes included) exhibit little obvious unifying basis; no growth form, seed type, kind of pollen vector, water–carbon balance regimen, nutrient source, or resource procurement mechanism is shared by all. Nor does a common type of ancestral habitat explain their anchorage in tree crowns – one might even say their total intolerance of terrestrial soil. Furthermore, characteristics of epiphytic and ground flora overlap broadly, as do important aspects of their habitats. Cultured seeds of many obligate epiphytes produce healthy adults on appropriate artificial soil mixtures; there is nothing mandatory about anchorage in canopies when reduced to this level of simplicity.

The epiphytic habitat

Growing conditions in canopies – rooting media and microclimate – are diverse and often similar to those on the ground. Aerial substrata in arid woodlands, like dry sterile soils, probably impose comparable stresses on resident vegetation. Data presented later suggest that the sodden rotting trunks, ant nests, and continuously moist, debris-filled knotholes of an aseasonal forest offer to certain epiphytes resources equal to those available in equable terrestrial habitats. In effect, the similarity between the two habitats complicates definition of the epiphytic biotope and leaves in doubt the reason why so many species root exclusively in tree crowns.

Within their broad milieu, canopy-adapted species seem to be no more versatile than soil-rooted plants; most will survive only under narrowly prescribed circumstances (e.g., rigid confinement to twigs vs. larger axes, humus as opposed to "unconditioned" bark, dim instead of brighter loci). Microsites are most diverse and abundant in multistratal forests. There, insolation is strong, and temperature and air saturation fluctuate in the upper canopy. Nearer the ground, photosynthetic photon flux density (PPFD; Fig. 1.1) is the major variable because sun flecks provide most of the irradiance (Chazdon and Fetcher 1984b). Survival at canopy margins

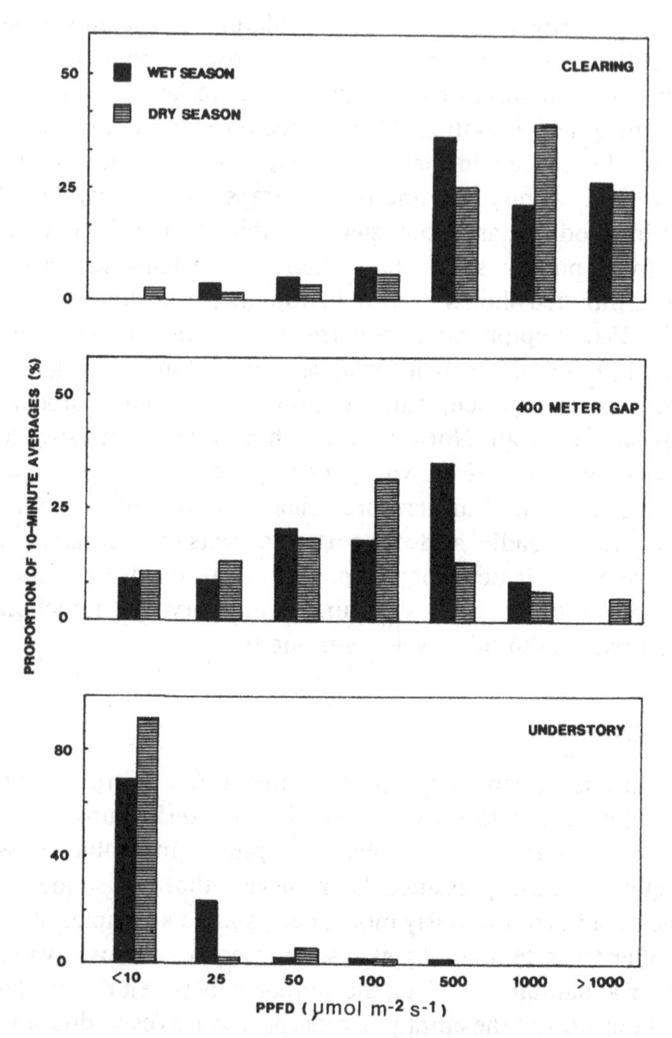

Figure 1.1. The percentage of PPFD (determined from averaged 10-minute integrated readings) falling into seven classes on a wet-season (solid bars) and a dry-season (hatched bars) median day in a Costa Rican rain forest. PPFD (μmol $m^{-2}s^{-1}$) ranges of the classes were (1) <10; (2) 10–24; (3) 25–49; (4) 50–99; (5) 100–499; (6) 500–1000; and (7) >1000. (From Chazdon and Fetcher 1984a.)

requires considerable succulence, a capacity to impound precipitation (root balls, leaf cavities), or access to absorptive substrata (mats, ant nests). Dry, open woodlands are even harsher habitats for the epiphyte, but patchiness, disturbance, and gravity, as well as potentially adverse chemistry, must be overcome everywhere.

Fragmentation of substratum within and among tree crowns ensures that seed will probably land on the inhospitable ground rather than on a suitable canopy surface; mounting the necessary fecundity to accommodate these losses tests populations already hampered by aridity and nutrient scarcity. Moreover, successfully established progeny often suffer lethal disturbance: Bark fragments exfoliate, colonized twigs and branches fall, and infested phorophytes eventually collapse. Terrestrial microbes and adsorbents are not available to mitigate allelopathy between epiphyte and tree. On the positive side, epiphytes escape production of much of the mechanical and vascular tissue required by trees to secure light: It seems fair to label them mechanical parasites. They are also spared ground fires and the attention of flightless, weak-climbing predators.

Whatever the benefits and disadvantages, forest canopies offer diverse opportunities for varied life-styles. Just how fully tracheophytes have responded to these opportunities is reflected in the myriad habits and associated water-balance mechanisms, nutritional modes, and reproductive systems exhibited by arboreal species.

Classification of epiphytes

Segregation of epiphytes has been based on many parameters: the nature of adaptation and fidelity to supporting vegetation; growth habit; climatic tolerance; type of substratum; and mechanisms for securing basic resources. Because detailed comparisons of all these systems could alone fill several chapters, a synthesis is presented here that borrows heavily from worthy predecessors (Schimper 1888; Hosokawa 1943; Richards 1952, Wallace 1981). Wherever possible, terminology used in earlier accounts has been preserved. These reworked classifications emphasize function as well as form, using additional refinements in older categories that are made possible by recent developments in plant physiology. As in all earlier taxonomies, species assigned the same identity according to one set of criteria may fall into different categories when compared on other grounds. Scheme I groups the epiphytes according to fundamental types of phorophyte use and deals with peculiarities of that phenomenon in subcategories.

Scheme I. Categories based on relationship to the host

A. Autotrophs: plants supported by woody vegetation; no nutrients extracted from host vasculature
   1. Accidental
   2. Facultative

3. Hemiepiphytic
   a. Primary
      (1) Strangling
      (2) Nonstrangling
   b. Secondary
4. "Truly" epiphytic (the "holoepiphytes" of Schimper)

B. Heterotrophs: plants subsisting on xylem contents and sometimes receiving a substantial part of their carbon supply from a host
   1. Parasitic (mistletoes)

Accidental epiphytes possess no special modifications for canopy life, yet they occasionally grow to maturity in forests without ever rooting in the ground. Birds and wind promote colonization wherever moist cavities exist, be they in tree crowns, stone fences, derelict buildings, or rock cervices. Diversity is broadest and individuals are most abundant in humid forests. Autotrophs anchored in the crowns of north-temperate and Boreal trees always belong to this group.

Facultative epiphytes inhabit forest canopies and the ground interchangeably (Fig. 1.13). Depending on local conditions, a single species may anchor in earth or on bark or on both media in the same community. Group 2A is best represented on humid sites where tree branches and soil alike support thick, moisture-retaining mantles of bryophytes, lichens, vascular plants, and associated litter (Fig. 1.20). This group also occupies dry sites where, again, canopy and terrestrial media provide similar – in this instance, demanding – growing conditions. Locations featuring the greatest physical disparity between bark and soil substrata are least conducive to facultative epiphytism.

Primary hemiepiphytes, some of which are stranglers, have no access to the ground early on, but later, after elongate feeder roots grow down to the trunk's base, growth becomes more vigorous (Fig. 1.8). In time, the phorophyte can become enmeshed in anastomosing roots and may eventually die as a result of girdling and competition. Should a support decay, a strangling hemiepiphyte with its vigorous vascular cambium (e.g., large-leaf species of *Ficus*) becomes free-standing. Nonstranglers include small-leaf *Ficus* and most canopy-based members of *Clusia*. Secondary hemiepiphytes begin life rooted in earth near a phorophyte and become arboreal when attachment to the tree has been achieved and the vine's older stems and roots decay (Fig. 1.9). The common monocot pattern of steady basal dieback is conducive to secondary hemiepiphytism and explains the preponderance of Liliopsida in

Figures 1.2–1.7. Selected epiphyte types: (1.2) *Tillandsia fasciculata* with impounded debris in a South Florida swamp forest (×½); (1.3) *Ionopsis utricularioides* growing on guava in Ecuador (×⅕); (1.4) *Hydnophytum formicarium* illustrating interior of tuberous hypocotyl (×⅕); (1.5) a trash-basket *Anthurium* in Amazonian rain forest (×⅕); (1.6) the resurrection fern *Polypodium polypodioides* growing on *Quercus virginiana* in South Florida (×0.4); (1.7) *Campylocentrum fasciola* on *Theobroma* in Ecuador (×⅓).

this group. A capacity for vascular renewal via stem thickening favors the liana habit among vining dicot species.

True epiphytes routinely spend their entire lives without contacting either forest floor or host vasculature (Figs. 1.3, 1.5, 1.10, 1.11). This group contains the most specialized canopy dwellers, those whose supplies of water and mineral ions are often obtained through unusual plant form and physiology.

The heterotrophs are distinguished from all previous groups by parasitism via haustoria (Fig. 1.12). Mistletoes retain their photosynthetic capacity, although specialized Viscaceae (Figs. 6.1J, 6.2E) draw substantial photosynthate from hosts; a few (e.g., *Tristerix aphyllus*) are entirely endophytic except for reproductive shoots.

A second scheme categorizes epiphytes by growth habit, a criterion that parallels such other plant characteristics as type of nutrient and water economy. The main distinction in this instance is secondary thickening which, in turn, often correlates with size and, to a lesser extent, longevity. Much more elaborate classifications based on gross form have been erected by others (e.g., Hosokawa 1943; Wallace 1981).

Scheme II
A. Trees
B. Shrubs
C. Suffrutescent to herbaceous forms
  1. Tuberous
    a. Storage: woody and herbaceous
    b. Myrmecophytic: mostly herbaceous
  2. Broadly creeping: woody or herbaceous
  3. Narrowly creeping: mostly herbaceous
  4. Rosulate: herbaceous
  5. Root/leaf tangle: herbaceous
  6. Trash-basket: herbaceous

Most forest dominants are trees which grow up from the ground without passing through an epiphytic phase, but some start out as primary strangling hemiepiphytes. A number of true and facultative epiphytes are shrubs, a few growing several meters tall (e.g., *Blakea, Rhododendron*). All but the most specialized mistletoes are shrubby. Occasional woody forms (e.g., *Ficus* as juveniles, *Markea,* some Ericaceae) and pseudobulbous orchids (Figs. 1.10, 1.15) produce storage tubers. Swollen hypocotyls of suffrutescent *Myrmecodia* and *Hydnophytum* (Fig. 1.4) also house ants that provide mineral ions (Huxley 1978). Thickened rhizomes of some ferns (Fig. 4.24E–G) perform

Figures 1.8–1.13. Selected epiphyte types: (1.8) *Ficus aurea,* a strangler fig growing on *Taxodium distichum* in South Florida; (1.9) an araceous secondary hemiepiphyte after it has lost contact with the soil in an Ecuadoran wet forest; (1.10) *Encyclia tampensis* (×⅓); (1.11) *Tillandsia paucifolia* shedding wind-dispersed seeds (×⅓); (1.12) *Phoradendron flavescens* parasitizing *Prunus avium* in central Kentucky (×⅓); (1.13) *Tillandsia fasciculata* growing on and beneath *Quercus virginiana* in South Florida.

Figures 1.14–1.19. Selected epiphyte types: (1.14) trap-bearing stems of *Utricularia humboldtii* in leaf axils of *Brocchinia tatei* on Cerro Neblina, Venezuela (×½); (1.15) young leafy *Catasetum* growing on a rotten limb in a Mexican wet forest (×¼); (1.16) *Psygmorchis glossomystax* anchored to guava twigs (×⅛); (1.17) an older ant nest-garden supporting luxuriant bromeliads, *Peperomia*, and gesneriads in Ecuadoran wet forests (×1/10); (1.18) a *Lycopodium* sp. (×⅓); (1.19) *Campyloneurum angustifolium* with a well-developed root ball growing on *Acer rubrum* in a South Florida swamp forest (×⅛).

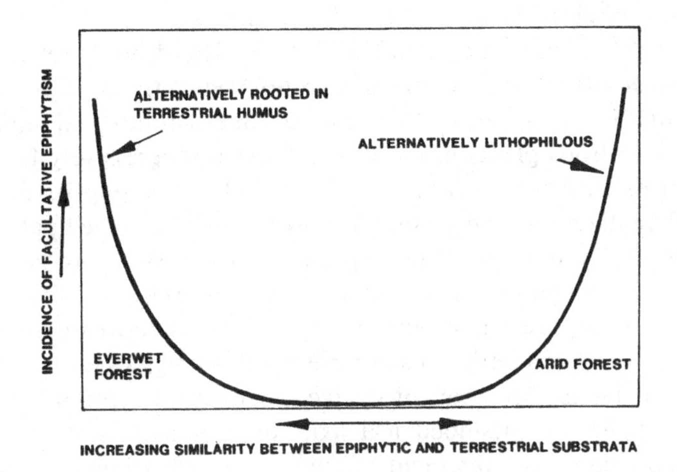

Figure 1.20. A graphic model depicting conditions most conducive to facultative epiphytism.

much the same way. Broadly creeping taxa with herbaceous or woody stems are well represented in Asclepiadaceae (Fig. 4.24C,D), Ericaceae, Gesneriaceae (Fig. 1.17), and several fern families (Fig. 1.6). Rosulate shoots are best developed in Bromeliaceae (Fig. 1.2); in *Anthurium* (Fig. 1.5) and (fewer) *Philodendron;* in Liliaceae *(Astelia* and *Collospermum);* in Commelinaceae *(Cochliostema);* in Gesneriaceae *(Paradrymonia),* and in the fern genera *Platycerium* (Fig. 4.19), *Drynaria,* and *Asplenium.* Root/leaf tangle epiphytes are illustrated by orchids with dangling roots (Fig. 1.7) and atmospheric bromeliads with numerous filiform leaves (Figs. 3.11, 7.2). Masses of upward-growing roots qualify various members of *Anthurium* (Fig. 1.5), *Catasetum, Cyrtopodium, Dendrobium,* and several other orchid genera for classic trash-basket status.

Humidity and light are the two most decisive factors governing epiphyte location; scheme III addresses the former variable.

Scheme III

A. Poikilohydrous: many bryophytes and lower plants; an unknown number of ferns; and very few, if any, angiosperms
B. Homoiohydrous
   1. Hygrophytes
   2. Mesophytes
   3. Xerophytes
      a. Drought endurers
      b. Drought avoiders
   4. Impounders

In this scheme, desiccation tolerance distinguishes two major groups: the "poikilohydrous" (desiccation-tolerant) and "homoiohydrous" (relatively desiccation-intolerant) species. Members of group A, sometimes called "resurrection" plants (Fig. 1.6), closely track changes in environmental humidity. The homoiohydrous plants of group B possess a progressively better-insulated water balance system from B1 through B3a. Subgroup B3 plants are further differentiated by leaf texture, longevity, and associated patterns of carbon gain and water use. These species either possess productive, ephemeral, desiccation-prone foliage suitable only for wet season activity (B3b; Fig. 1.15), or they are active year round by virtue of desiccation-resistant leaves or green stems with considerable water storage capacity (B3a; Figs. 1.10, 1.11, 3.1A–E). Members of sizable subgroup B4 maintain external moisture supplies in expanded leaf axils or root masses (Figs. 1.2, 1.5, 1.19). A more complete treatment of these water balance mechanisms and related characteristics of the epiphytes that possess them is presented later.

The second decisive habitat requirement – light – was studied by, among others, Colin Pittendrigh (1948) who segregated the bromeliads of Trinidad into three categories (scheme IV) based on apparent affinity for fully exposed, intermediate, and deeply shaded microsites.

Scheme IV

A. Exposure types: largely restricted to sites in full or nearly full sun
B. Sun types: tolerant of medium shade
C. Shade-tolerant types: tolerant of deep shade

More recent surveys of epiphytic floras elsewhere indicate that Pittendrigh's three-tiered scheme has general applicability. Less clear is its stability and functional basis. Pittendrigh himself suspected that certain sciophytic species are restricted to shaded habitats more by moisture needs than by the damage strong irradiance might inflict. Acclimatization must occur to some degree among canopy-adapted plants as it does in terrestrial vegetation. But *Tillandsia usneoides* (Fig. 7.2) shows little physiological/anatomical response to growth in deep shade, a subject pursued further in the next chapter. There is no proof that any epiphyte is as shade-demanding as certain understory herbs in humid forests.

Epiphytes can be divided into main categories on the basis of how adaptable they are to media provided by, or associated with, their phorophyte (scheme V). In this sense, schemes V and I show a bit of overlap.

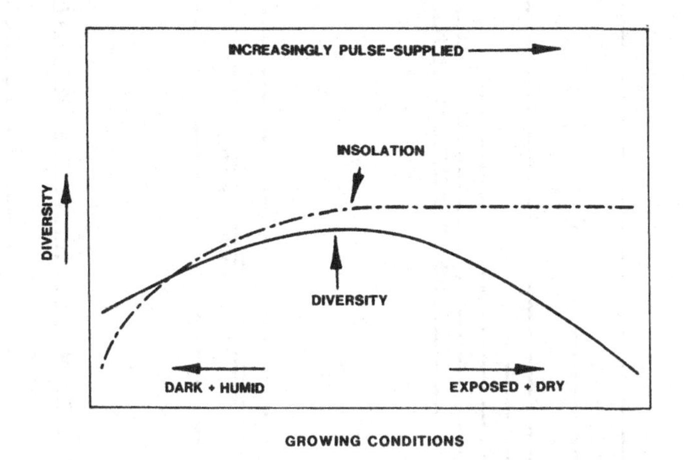

Figure 1.21. A graphic model depicting physical constraints in canopy habitats that influence epiphyte diversity.

Scheme V

A. Relatively independent of rooting medium (obtain moisture and nutritive ions primarily from other sources)
   1. Mist and atmospheric forms with minimal attachment to bark
   2. Twig/bark inhabitants
   3. Species that create substitute soils (impounders) or attract ant colonies (ant-house epiphytes)

B. Tending to utilize a specific type of rooting medium for moisture and nutritive ions
   1. Humus-adapted
      a. General types that root on shallow humus mats
      b. Deep humus types that penetrate knotholes or rotting wood
      c. Ant nest-garden and plant-catchment inhabitants (e.g., *Platycerium* nest endemics, *Utricularia humboldtii* in tanks of *Brocchinia tatei,* numerous species comprising ant nest-gardens)
   2. Mistletoes

Group A species utilize phorophytes mainly for anchorage; there is little opportunity for influence from the supporting surface. Roots of certain aroids and orchids (Fig. 1.7) hang free (group A1) and intercept mist droplets or throughfall. Twig/bark forms (A2) draw most nutritive ions and moisture from flowing precipitation and leachates. Only naked twigs and unadorned bark are colonized, as if moisture-retaining media must be avoided (Figs. 1.3, 1.11). Indeed, prolonged contact with wet materials kills these species in culture; heavily trichomed tillandsioid bromeliads (Figs.

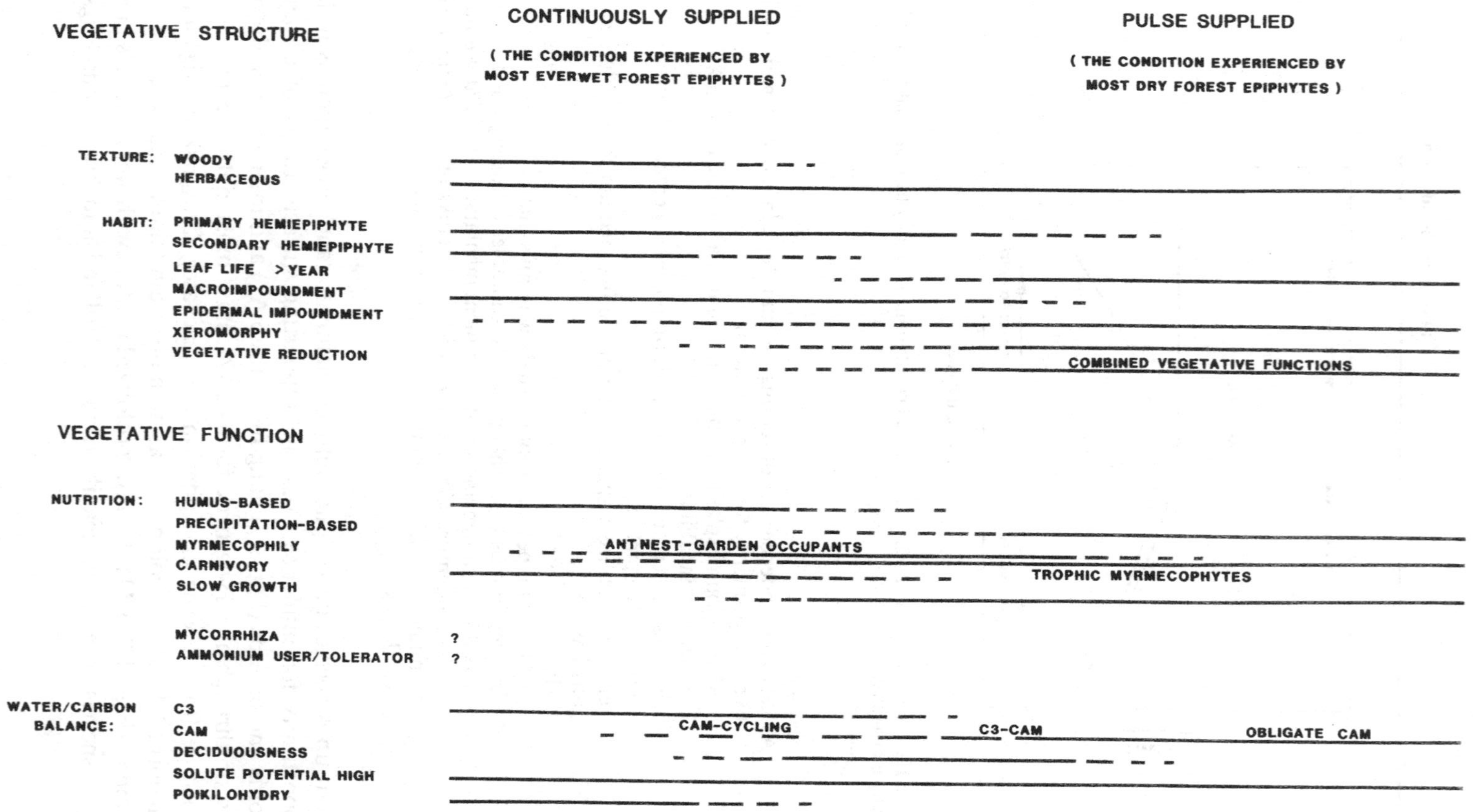

CONTINUOUSLY SUPPLIED

(THE CONDITION EXPERIENCED BY MOST EVERWET FOREST EPIPHYTES)

PULSE SUPPLIED

(THE CONDITION EXPERIENCED BY MOST DRY FOREST EPIPHYTES)

VEGETATIVE STRUCTURE

TEXTURE: WOODY
HERBACEOUS

HABIT: PRIMARY HEMIEPIPHYTE
SECONDARY HEMIEPIPHYTE
LEAF LIFE >YEAR
MACROIMPOUNDMENT
EPIDERMAL IMPOUNDMENT
XEROMORPHY
VEGETATIVE REDUCTION

COMBINED VEGETATIVE FUNCTIONS

VEGETATIVE FUNCTION

NUTRITION: HUMUS-BASED
PRECIPITATION-BASED
MYRMECOPHILY
CARNIVORY
SLOW GROWTH

ANT NEST-GARDEN OCCUPANTS

TROPHIC MYRMECOPHYTES

MYCORRHIZA ?
AMMONIUM USER/TOLERATOR ?

WATER/CARBON BALANCE: C3
CAM
DECIDUOUSNESS
SOLUTE POTENTIAL HIGH
POIKILOHYDRY

CAM-CYCLING

C3-CAM

OBLIGATE CAM

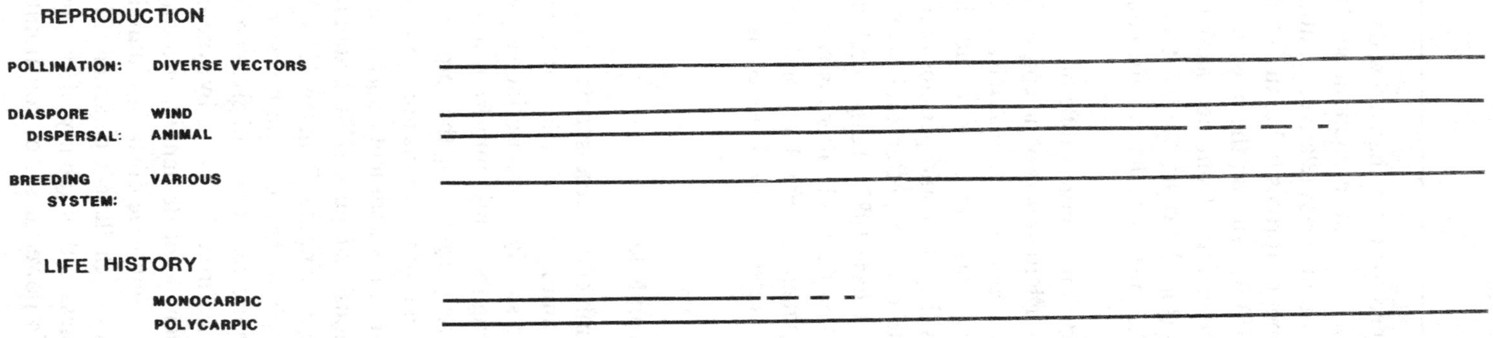

Figure 1.22. Character states of vascular epiphytes dictated by moisture supply in their microhabitats. (From Benzing 1987a.)

1.11, 3.11) succumb if shoots remain moistened for more than a few days. Water and nutrient ions are drawn from root or shoot impoundments by members of group A3 (Figs. 1.2, 1.5, 1.19). Group B1 taps relatively continuous resource pools; possessing no impoundments, it roots in humus. Only knotholes or absorbent, rotten limbs (Fig. 1.15) will sustain the most transpiration-prone, drought-deciduous taxa. Inhabitants of ant nest-gardens (Figs. 1.17, 5.10, 5.11) and plant catchments (Fig. 1.14) are the most substratum-specific of this second group, as are the mistletoes that require connections to host vasculature.

Epiphytes may also be distinguished by additional, more subtle but nevertheless important, growth phenomena. Although members of no epiphytic species complete life cycles as quickly as do the ephemeral terrestrials, their maturation periods vary widely. Most precocious are some oncidioid orchids; *Psygmorchis* (Fig. 1.16) and *Ionopsis* (Fig. 1.3), and perhaps other miniatures as well, develop quickly enough to colonize, often exclusively, short-lived twigs and even individual leaves (Chase 1987). Larger orchids and other epiphytes in the same forests ripen first fruit only after many seasons. The atmospheric bromeliad *Tillandsia paucifolia* (Fig. 1.11) probably represents the typical pulse-supplied (PS) forms described below, requiring more than five years to mature (pers. obser.).

Continuously supplied versus pulse-supplied epiphytes

Resource availability and plant mechanisms influencing access to water and nutrients underlie another useful distinction among canopy-adapted flora. Two related agencies – climate and resource supply – impose increasing plant stress and reduce epiphyte diversity along the continuum from more to less equable forest canopies (Fig. 1.21), thereby segregating the epiphytes into two functional groups. Where moisture and nutrient ions are more or less steadily available in rooting media or plant impoundments, species on the left of the continuum are labeled continuously supplied (CS). Data are scarce, but debris trapped by tank bromeliads (Benzing and Renfrow 1974a), as well as that used by early stages of hemiepiphytic figs on palm hosts and most epiphytes in pluvial forests, may be quite nutritive and frequently moist (Putz and Holbrook 1986). Distinguished from CS forms by type of habit and tolerance – if not actual requirement – for demanding growing conditions are the PS forms. For this second assemblage of species, unreliability of resource supply (particularly moisture) challenges persistence in the face of mortality imposed by the patchiness and instability of canopy substrata. Few lineages have adapted to this powerful set of constraints

(Benzing 1978a). Those that have include xeromorphic Bromeliaceae, particularly members of *Tillandsia,* and various orchids that anchor on exposed bark surfaces. The epiphytic contingent of most families falls left of center in Figure 1.21

Several characters in epiphytic flora (Fig. 1.22) track humidity and fertility along the ecological continuum just described. At one extreme, equable-forest species differ little in form and probably physiology from adjacent ground-rooted vegetation; not surprisingly, epiphytism in such communities is often facultative. At the other extreme, canopy-adapted taxa specialized to counter the strongest ecoclimatic constraints possess elaborate, sometimes unique, mechanisms for tapping unusual resource pools, prolonging contact with passing canopy fluids, promoting resource use efficiency, and maximizing fecundity. These species will receive the lion's share of attention in the chapters devoted to vegetative form and performance.

2 Photosynthesis

All epiphytes photosynthesize, but certain life stages of several taxa are heterotrophic – for example, gametophytes of arboreal *Lycopodium, Ophioglossum, Psilotum,* and *Tmesipteris.* Similarly, orchid seedlings remain achlorophyllous for weeks to months, subsisting on substrates provided by symbiotic fungi (Figs. 4.10, 4.12). Flow of fungal metabolites into adult orchids may also occur, but claims for epiparasitism in canopy-based Orchidaceae (e.g., Ruinen 1953; Johansson 1977) need confirmation using labeled nutrients. As for dwarf mistletoes, utilization of host substrates has been great enough to allow considerable leaf reduction; most of the vegetative body is endophytic.

The question now is how autotrophy operates in canopy-adapted vegetation. This chapter will examine (1) photosynthetic pathways among epiphytes, two of which are certain, the third equivocal; (2) accommodation of carbon balance to mineral scarcity and shade; (3) segregation of co-occuring populations along light gradients; (4) photosynthetic phenomena peculiar to certain specialized taxa; (5) ancestral habitats; and (6) the economics of epiphyte foliage versus that of phorophytes. The interrelationship between photosynthesis and water balance is covered more thoroughly in Chapter 3.

Photosynthetic pathways

The reductive pentose phosphate (C_3) pathway

Machinery for trapping radiant energy, perfected in plants long before land was colonized, required tailoring as the terrestrial flora developed. Light intensity and quality, supplies of moisture, nitrogen (N), and presumably other key nutritive elements, influenced selection during the subsequent radiation, up to and including colonization of tree crowns. Whereas the fundamental mechanisms of photosynthesis reflect ancient origins, today the forms and performances of green organs in epiphytes are diverse and occasionally unique. Some species still behave like the first, and most of the extant, terrestrials. They take up carbon dioxide (CO_2) by day and immediately process it through the ubiquitous C_3 (or RPP for reductive pentose phosphate) pathway. Here, the key enzyme is ribulose biphosphate carboxylase/oxygenase (RuBPc/o). Although this protein has a rather low affinity for CO_2, vigorous growth is possible for C_3 species under a variety of exposures and temperatures as long as moisture supplies are reasonably abundant. In fact, strictly C_3 forms are found in more kinds of habitats than are members of either of the other two photosynthetic categories.

The Hatch and Slack (C_4) pathway

A relatively small group – mostly short-lived dicots and monocots in the grass and sedge families – photosynthesize via the C_4 pathway. This mechanism fixes CO_2 via a three-step sequence: (1) Incoming atmospheric CO_2 is captured in mesophyll cells via phosphoenolpyruvate carboxylase (PEPc) to produce four-carbon products (mostly malate or aspartate) which move through the symplast to tightly packed, green bundle sheaths; (2) here CO_2 is liberated by one of three different decarboxylating systems, depending on the species; (3) CO_2 is then refixed to hexose via the RPP pathway. The proximity of sheaths to vascular tissue allows rapid export of carbohydrate. Other products return to the mesophyll for regeneration of CO_2 receptors.

Compared to C_3 species under similar warm, exposed conditions, typical C_4 plants conserve water better. Moreover, record photosynthetic rates are possible in optimum environments. This superior performance is due to the greater affinity PEPc has for CO_2 compared to RuBPc/o and the isolation of RuBPc/o in a special gastight leaf compartment. By concentrating CO_2 in bundle sheaths where hexose is produced, C_4 plants suppress the oxygenase activity of RuBPc/o, thereby diminishing photorespiration. This capacity of PEPc to maximize the C_a/C_i ratio (C = partial pressure of CO_2; i = intercellular space; a = atmosphere) is reflected in a very depressed CO_2 compensation point; diffusive conductance in leaves, hence transpiration, is normally lower in a C_4 versus a C_3 species for the same amount of processed carbon.

Crassulacean acid metabolism (CAM)

Biochemical near-equivalents of C_4 taxa are the CAM species, but they succeed with far less moisture. Both carboxylases are maintained in large, hypervacuolate, green mesophyll cells; the two enzymes operate under different conditions, however, so they are not active simultaneously. A useful tool for determining biochemical pathway (and environmental condi-

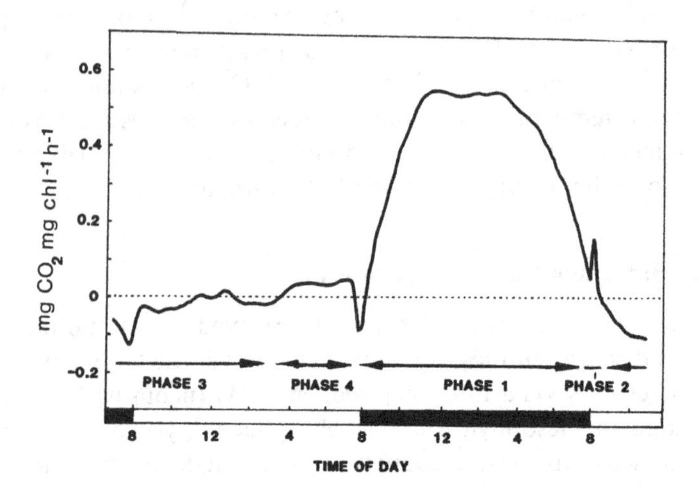

Figure 2.1. Carbon dioxide exchange by *Tillandsia usneoides* illustrating the four phases of CAM. (Data from Martin and Siedow 1981.)

tions during photosynthesis) is $\delta^{13}C$ (isotope ratio of carbon products), expressed in parts per thousand (‰) and calculated according to the following formula:

$$\delta^{13}C = \left[\frac{^{13}C/^{12}C \text{ sample biomass}}{^{13}C/^{12}C \text{ standard}}\right]^{-1} \times 1000‰$$

Both C_4 and CAM plants discriminate less against the heavier carbon isotope ^{13}C than do C_3 forms: the value of $\delta^{13}C$ for biomass produced by strictly CAM and C_4 species is only about $-12‰$ compared to approximately $-30‰$ in C_3 forms. This ratio has also proved useful for studying mistletoe biology (see Chapter 6).

Unlike C_3 and C_4 plants which assimilate CO_2 by day, CAM plants do so predominantly at night (Figs. 2.1, 2.2). Carbohydrate reserves provide energy to open stomata and capture CO_2; fixation via PEPc to malic acid (phase I) follows. A brief burst of CO_2 uptake around sunrise (phase II) remains little understood. Now in the light, malic acid is decarboxylated, and liberated CO_2 is photofixed via RuBPc/o to hexose (phase III). Well after midday, when acid stores become depleted, stomata may reopen to permit direct C_3 photosynthesis (phase IV). The reason CAM works is that PEPc activity is minimal while malic acid concentration is elevated. Without such enzyme regulation, daytime competition between the two carbox-

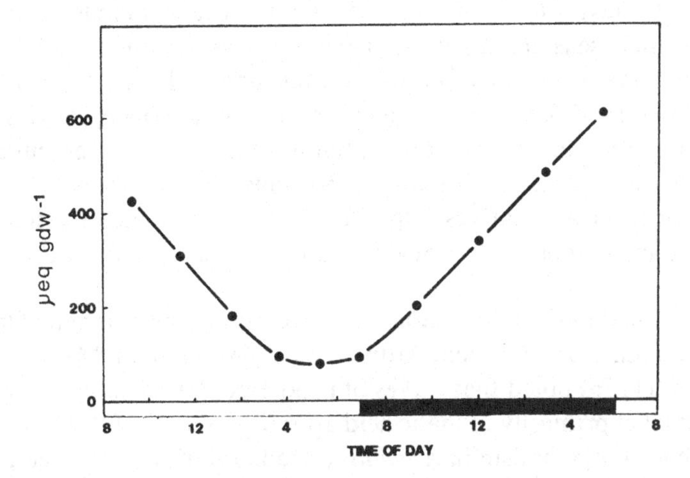

Figure 2.2. The diurnal course of titratable acidity in *Tillandsia usneoides.* (gdw, grams dry weight.) (From Martin et al. 1981.)

ylases would create a wasteful, futile carbon cycle, a problem C_4 plants avoid by isolating the two key proteins in separate tissues.

Phase III operates internally throughout the hottest, driest part of the day; diurnal transpiration in CAM plants is mostly cuticular, thus held well below the norms for C_3 and C_4 plants under comparable environmental conditions. A trade-off is inevitable, however: CAM-plant growth is relatively slow and usually demands considerable light. The photosynthetic organs of typical CAM species are evergreen (long-lived) and succulent or coriaceous. More than two dozen angiosperm families, as well as some ferns, lycopods, and gymnosperms, include one or more members that are wholly or partially CAM types. The suggestion that epiphytes capable of nocturnal CO_2 fixation are benefited because night air is CO_2-enriched by respiring C_3 canopy tissue (Knauft and Arditti 1969) has not been confirmed.

Pathways in canopy habitats

Evidence for C_4 photosynthesis in canopy-adapted flora is scanty. Approximately a dozen families contain C_4 members, but few of these have managed even minor incursions into canopy habitats. Amaranthaceae, Azoiaceae, Chenopodiaceae, Euphorbiaceae, Portulacaceae, and Zygophyllaceae, among others, contain no epiphytes at all. Asteraceae, Cyperaceae, and Poaceae include a few canopy-adapted species, none so far identified as C_4, plants. Most C_4 taxa and C_3–C_4 intermediates are ephemerals native to ruderal or strongly seasonal habitats, or they exist as dominant herbaceous perennials in some of the most productive, resource-rich sites known (e.g., *Spartina* in tidal mud flats). Short life cycles are rare among epiphytes as a whole, but probably not for lack of C_4 photosynthesis with its potential to support high vigor. Rather, climatic conditions that typically favor C_4 metabolism over the alternatives – specifically, high irradiance and temperature, and at most moderate drought – do not usually coincide in tree crowns.

There are hints that C_4 photosynthesis may exist among epiphytic Orchidaceae (Avadhani, Goh, Rao, and Arditti 1982). Avadhani and Goh (1974) and Avadhani (1976) noted that leaves of irradiated *Arundina graminifolia* incorporated label primarily in malic acid after brief pulses of $^{14}CO_2$. Aspartic and malic acid metabolism in *Arachnis* (Avadhani, Khan, and Lee 1978) and a *Schombocattleya* cultigen (Rubenstein, Hunter, McGowan, and Withner 1976) suggests that other orchids perform combined C_4–CAM photosynthesis. More definitive criteria are needed if the predominant carbon fixation pathway is to be identified in these equivocal cases.

Continuously humid microsites in tropical forest canopies are often cool and either shaded by foliage or shrouded by recurrent clouds or fog, thus most appropriate for C_3 activity. Arid locations offer conditions most congenial to CAM; this pathway may be, in fact, better represented in the epiphytes than in any other ecological type, even desert inhabitants. The CAM pathway is by far the most capable of accommodating the fickle moisture supplies of most tree crowns. Moreover, capacity for high quantum yields in shade, and tolerance for full insolation, further equips the CAM plant for canopy existence (Osmond 1987). Photosynthesis proceeds in facultative species in the relatively productive C_3 mode until mounting moisture deficits trigger fallback to nocturnal CO_2 fixation. The bromeliad *Guzmania monostachia* is known to be a C_3–CAM switcher (Medina, Delgado, Troughton, and Medina 1977) at least in parts of its extensive range. Intermediate values for $\delta^{13}C$ suggest that other family members as well (Table 2.1) are responsive to moisture status (including filled or empty tanks) or that they can engage in considerable late-day C_3 (phase IV) activity. In comparison, values for the most obligate CAM types varied remarkably little among conspecifics collected in Trinidad at different sites (e.g., just $-0.8‰$ and $-1.2‰$ for the bromeliads *Aechmea nudicaulis* and *A. aquilega,* respectively; Smith, Griffiths, and Lüttge 1986a).

An emergency measure available to many CAM plants under severe drought is CAM-idling. Here, energy reserves (thus carbon balance) can be maintained for months in spite of water deficits great enough to impede stomatal movement. During this period, CO_2 generated in the leaf interior is simply recycled internally day and night until wet weather returns. Some semixeric, primarily C_3, species utilize PEPc at night in a process called CAM-cycling, presumably to enhance water economy. Well-watered *Peperomia camptotricha* recycled an amount of CO_2 equivalent to about 17% of net carbon uptake (Patel and Ting 1987). Ting, Bates, Sternberg and DeNiro (1985) have proposed an intriguing possibility – that CAM-cycling simply poises a stressed plant for CAM-idling. Epiphytes should certainly be tested because so many of them seem to engage in the process.

The discovery that some CAM epiphytes refix much more respired than atmospheric CO_2 at night – between 47% and 95% more in one survey (Table 3.1) – raises questions about energy economy. All of these plants expend considerable reserves in order to generate PEP, ATP, and reducing power for synthesis and mobilization of malic acid, but why is there such a wide range? Three factors are influential, none of which is fully understood. The ratio of atmospheric to internal sources of CO_2 depends on the species, its moisture status, and the environment, and costs are even more variable than data from a large sample of Trinidadian bromeliads suggest (Table 3.1). Whereas less than 50% of the early morning acid content in well-watered *Tillandsia schiedeana*, for instance, had come from recycled CO_2, that figure rose to about 90% after 23 days of droughting (Martin and Adams 1987). Peak acidity, however, diminished considerably during that time, evidence of increasing stress and lapse into CAM-idling. In another case, water stress had less influence on the origin of acid stores (Lüttge 1987). Whereas total nocturnal acidification in *Tillandsia flexuosa* in a north coast Venezuelan habitat diminished 35% from midwet to middry seasons, the amount of malic acid produced from recycled CO_2 changed very little (76% vs. 73%). The acid production of nearby *Schomburgkia humboldtiana* was also lessened by the time of the dry season reading, but meanwhile nightly CO_2 contributions from the atmosphere had actually increased from 65% to 70% (H. Griffiths et al. 1989). The work of Griffiths et al. (1986) showing that well-watered bromeliads sometimes fix considerably more internal than external CO_2 is perplexing at first, but may simply reflect the tendency for such a factor as superoptimal nighttime temperature to suppress nocturnal uptake of CO_2. Perhaps even more fundamental biochemical mechanisms contribute, however. Lüttge and Ball (1987) felt that the generally high Q_{10} of dark respiration in CAM plants and peculiarities of malate synthesis in Bromeliaceae might account for the high recycling values recorded in some members of this family. Generation of PEP using free hexose to supplement glu-

Table 2.1. **Carbon isotope ratio ($\delta^{13}C$), deuterium/oxygen ratio (δD), organ assayed, habit, and habitat for various canopy-adapted species**

	$\delta^{13}C$ (‰)	δD	Organ	Habit[a]	Habitat
I. Pteridophytes					
Aspleniaceae					
Asplenium australasicum[f]	−28.0		leaf	E	rain forest
Lomariopsidaceae					
Elaphoglossum queenslandicum[f]	−32.0		leaf	E	submontane rain forest
Lycopodiaceae					
Lycopodium phlegmaria[f]	−27.7; −30.9		leaf	E	lowland rain forest
Lycopodium proliferum[f]	−24.3		leaf	E	cultivation
Ophioglossaceae					
Ophioglossum pendulum[f]	−31.8		leaf	E	lowland rain forest
Polypodiaceae					
Drynaria rigidula[f]	−27.6		leaf	E	rain forest
Dictymia brownii[f]	−29.9		leaf	E	temperate rain forest
Platycerium hillii[f]	−24.6		leaf	E	rain forest
Pyrrosia confluens[f]	−19.2		leaf	E	temperate rain forest
Pyrrosia longifolia[f]	−14.2		leaf	E	lowland rain forest
Psilotaceae					
Psilotum complanatum[f]	−30.7		green stem	E	cultivation
Vittariaceae					
Vittaria elongata[f]	−30.1		leaf	E	lowland rain forest
II. Flowering plants					
Apocynaceae					
Parsonsia straminea[f]	−29.5		leaf	CL	dry subtropical rain forest
Araceae					
Pothos longipes[f]	−33.6		leaf	CL	lowland rain forest
Raphidophora pachyphylla[f]	−30.1		leaf	CL	lowland rain forest
Anthurium crenatum[d]	−28.9		leaf	E	moist forest
Araliaceae					
Schefflera actinophylla[f]	−33.4		leaf	E	lowland rain forest
Asclepiadaceae					
Dischidia nummularia[f]	−15.7		leaf	E	open eucalypt forest
Hoya nicholsoniae[f]	−18.3		leaf	E	lowland rain forest
Bromeliaceae					
Aechmea aquilega[b]	−14.5		leaf	E	see Table 2.3
Aechmea fendleri[b]	−13.3		leaf	E	see Table 2.3
Aechmea lingulata[b]	−12.2		leaf	E	see Table 2.3
Aechmea nudicaulis[b]	−14.7		leaf	E	see Table 2.3
Guzmania monostachia[b]	−31.5		leaf	E	see Table 2.3
Tillandsia elongata[b]	−26.4		leaf	E	see Table 2.3
Tillandsia fendleri[b]	−31.8		leaf	E	humid forest
Tillandsia utriculata[b]	−11.2		leaf	E	see Table 2.3

Table 2.1. **(cont.)**

	$\delta^{13}C$ (‰)	δD	Organ	Habit[a]	Habitat
Vriesea amazonica[b]	−28.0		leaf	E	see Table 2.3
Vriesea splitgerberi[b]	−32.4		leaf	E	see Table 2.3
Cactaceae					
Zygocactus truncatus[e]	−16.4	+50	green stem	E	moist forest
Hylocereus trigonus[d]	−13.3		green stem	H	moist forest
Clusiaceae					
Clusia rosea[d]	−17.9		leaf	H	moist forest
Ericaceae					
Dimorphanthera sp.[f]	−25.9		leaf	E	cultivation
Liliaceae					
Geitonoplesium cymosum[f]	−29.2		leaf	CL	dry subtropical rain forest
Moraceae					
Ficus crassipes[f]	−28.4		leaf	H	rain forest
Orchidaceae					
Bulbophyllum aurantiacum[f]	−12.4		leaf	E	rain forest
Cadetia taylori[f]	−27.9		leaf	E	cultivation
Chiloschista phyllorhiza[f]	−14.5		root	E	cultivation
Dendrobium canaliculatum[f]	−13.1		leaf	E	open eucalypt forest
Eria fitzalani[f]	−28.4		leaf	E	lowland rain forest
Liparis persimilis[f]	−24.1		leaf	E	cultivation
Phalaenopsis amabilis[f]	−14.1		leaf	E	cultivation
Taeniophyllum malianum[f]	−15.8		root	E	lowland rain forest
Vanda whiteana[f]	−14.8		leaf	E	cultivation
Piperaceae					
Peperomia johnsonii[f]	−30.0		leaf	E	submontane rain forest
Peperomia tetraphylla[f]	−29.0		leaf	E	cultivation
Peperomia camptotricha[e]	−27.7	+23	leaf	E	cultivation
Potaliaceae					
Fragraea berteriana[f]	−26.3		leaf	E	cultivation
Rubiaceae					
Hydnophytum formicarium[f]	−21.8		leaf	E	lowland rain forest
Myrmecodia muelleri[f]	−22.4		leaf	E	lowland rain forest
Trimonius singularis[f]	−28.7		leaf	E	lowland rain forest
Urticaceae					
Procris cepalida[f]	−33.8		leaf	E	lowland rain forest
Vacciniaceae					
Agapetes meiniana[f]	−27.8		leaf	E	cultivation

[a]E, epiphytic; H, secondary hemiepiphytic; CL, climbing.
Sources: [b]Griffiths and Smith 1983; [c]Sipes and Ting 1985; [d]Ting et al. 1985a,b; [e]Sternberg et al. 1984; [f]Winter et al. 1983.

can reserves required more respiration by *Aechmea fasciata* and two terrestrial bromeliads than was necessary for several Crassulaceae. Gas exchange and acid metabolism must be monitored in the field more systematically to develop an understanding of carbon budgets in CAM epiphytes, especially in warm habitats. Griffiths (1988) initiated this type of investigation using *Aechmea nudicaulis* and *A. fendleri*. Specific combinations of night temperature, leaf-air WVPD, and daily photon flux affected the proportions of recycled CO_2 and water use efficiency (WUE) differently in these two ecologically distinct bromeliads. Results were difficult to interpret, but they left no doubt that factors other than moisture stress influence the magnitude of CAM and its sources of CO_2.

Carbon can be processed in different ways by the foliage of a single epiphyte. *Peperomia camptotricha* from southern Mexico's everwet forests displayed a wide array of photosynthetic patterns depending on leaf age, moisture status, and environment (Ting et al. 1985a). Young foliage CAM-cycled, older organs took up about equal amounts of CO_2 by day and night. When moisture stress was applied in the laboratory, first CAM and then CAM-idling developed in young and old organs. Young leaves were insensitive to photoperiod, but mature unstressed foliage tended toward CAM during short days. (In native habitats, decreasing day length may serve as a sign that moisture supply is about to change). Leaves of *Peperomia scandens* exhibited a constitutive progression from C_3 to CAM: Shifts occurred in fully hydrated specimens under controlled day lengths and temperature cycles (Holthe, Sternberg, and Ting 1987). Also discovered was variation in CAM activity along parts of a single bromeliad leaf and among leaves of one rosette (Lüttge, Stimmel, Smith, and Griffiths 1986), probably reflecting intercalary expansion and maturation from leaf tip to base. Less heterogeneity of this type would be expected within a dicot leaf.

Schäfer and Lüttge (1986) demonstrated in the laboratory both ontogenetic variation and stress response in *Kalanchoe uniflora,* a humid montane forest epiphyte from Madagascar. Young well-watered foliage showed C_3 photosynthesis; more mature leaves also took up some CO_2 at night. Only about 10% of the entire plant's net CO_2 gain could, however, be attributed to CAM, and water loss was substantial. After four days of droughting, young foliage had switched almost exclusively to CAM whereas older, generally less productive, leaves were less affected. Better water economy was partly due to lower dark respiration rates in stressed young organs; two days following rewatering, plants had returned to earlier gas-exchange patterns.

Much of what is known about CAM dynamics in canopy habitats is based on studies of the Australian orchids *Dendrobium speciosum* and *Plectorrhiza tridentata* (Wallace 1981), the Malaysian orchids *Dendrobium tortile,*

D. crumenatum, and *Eria velutina* (Sinclair 1983a,b; 1984), a selection of Trinidad bromeliads (Smith, Griffiths, Bassett, and Griffiths 1985; Griffiths et al. 1986, Lüttge et al. 1986b; Smith et al. 1986a,b; Griffiths 1988), several ferns (e.g., Hew 1984), and *Tillandsia* of Bromeliaceae, especially *T. usneoides* (Martin, Christensen, and Strain 1981; Martin and Siedow 1981; Martin, Eades, and Pitner 1986). The contributions of Martin et al. are among the most enlightening to the ecophysiologist because they report coordinated laboratory and field results.

Tillandsia usneoides acted like a conventional CAM plant in most respects. Acid rhythms and gas exchange patterns showed the characteristic four phases (Figs. 2.1, 2.2): Titratable acid was maximal in early morning, lowest late in the afternoon; CO_2 consumption peaked near midnight and then subsided steadily until about dawn. There was the predictable brief surge in CO_2 uptake at first light each morning. Stomata reopened in midafternoon, and CO_2 influx resumed at a low rate. Specimens in a North Carolina forest gained dry weight and engaged most heavily in CAM from midspring through midfall. Acid fluctuation and CO_2 uptake were virtually nil in midwinter. As in other such plants, CAM activity was very temperature-sensitive, operating best in late spring at 25°–30°C. If night temperature was held at 20°C in the laboratory, diurnal maxima up to 35°C had no damping effect on CO_2 consumption; however, as nights grew warmer, CO_2 uptake fell, beginning with the disappearance of phase IV. If the daytime high reached 20°C, CAM could be sustained to a nocturnal low of 5°C. Regardless of other environmental conditions, fixation ceased below 5°C or if day/night temperature fluctuations were less than 5°C.

In some respects other than gas exchange and acid metabolism, *Tillandsia usneoides* is unconventional. Heavily shaded Spanish moss continued to match the carbon gain of fully exposed controls in contrast to the usual CAM plant behavior (Martin et al. 1986). It seems reasonable to lay the responsibility for this anomaly on plant architecture. Whereas the stout bodies of true succulents contain a large proportion of nonphotosynthetic tissue – in roots, thick-leaf interiors, and stems – *T. usneoides* has an unusually finely divided shoot (Fig. 4.25D). Therefore, its shade tolerance may simply derive from a relatively high surface/volume ratio in a totally photosynthetic (rootless) biomass. As expected, the PAR required to achieve ΔH^+_{max} is generally low compared to other CAM plants (Griffiths 1988).

The response of *Tillandsia usneoides* to moisture is also atypical. Wetting suppressed photosynthesis in Spanish moss through a novel mechanism (Fig. 2.3). If strands were moistened, CO_2 absorption fell off; water held against stomata by a confluent trichome layer impeded gas diffusion (Figs. 3.6, 3.7, 3.11, 4.25B–D). Similar findings have been reported for several

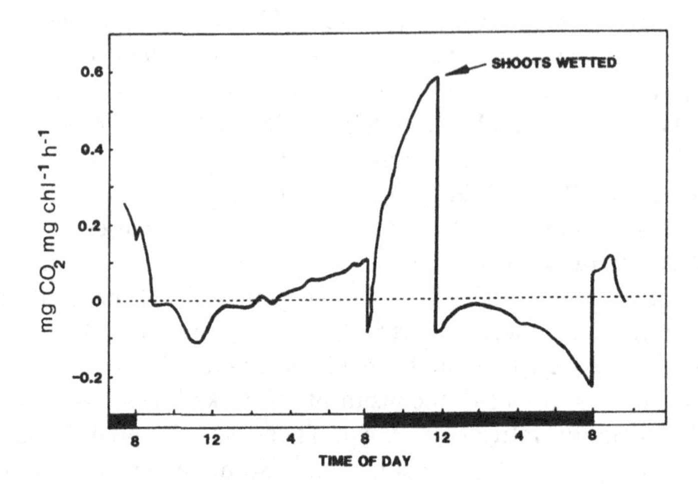

Figure 2.3. Effects of surface wetting on the CO_2 exchange of *Tillandsia usneoides.* (From Martin and Siedow 1981.)

other *Tillandsia* species. Lightly trichomed relatives (Fig. 3.10) with greater water repellence were unaffected by irrigation (Benzing and Renfrow 1971a; Benzing, Seemann, and Renfrow 1978). Reported enhancement of CO_2 uptake by water vapor probably involves the guard-cell turgor/foliar conductance mechanism. Stomata, including those of *T. recurvata,* a close relative of *T. usneoides,* are delicate humidity sensors (Lange and Medina 1979).

In striking contrast to the above, partially dehydrated *Tillandsia usneoides* has exhibited stronger CAM than have fully hydrated controls (Medina et al. 1977; Martin et al. 1981, but see Martin and Schmitt, 1989). This phenomenon is not so easily explained, but is it possible that mild dehydration is optimal for carbon gain in hygromorphic *T. usneoides* and similar congeners, and that this status is promoted by their curiously delicate epidermal barriers (Fig. 3.1D)? To be sure, some other atmospheric tillandsias produce stouter epidermal boundaries and desiccate more slowly, but they occur in warm, arid habitats – for example, *T. concolor* (Fig. 3.1C) and *T. circinnatoides* in microphyllous forests and the cactus/scrub deserts of central Mexico. Atmospheric *T. schiedeana,* another tillandsia native to Mexico, with hard brittle foliage containing a well-differentiated hypodermis, definitely did not exhibit more vigorous CAM after modest drying (Martin and Adams 1987).

Two Australian orchids studied by Wallace (1981) are ostensibly conventional CAM epiphytes with only the normal amount of phase IV C_3 activity.

Dendrobium speciosum is a large trash-basket species that draws water from a spongy root system impoundment. In sharp contrast, *Plectorrhiza tridentata* is a smaller, wiry-stemmed, twig epiphyte anchored by a few strands of its root-tangle system; it acquires moisture from mist and rain. Both species grow in the same relatively arid subtropical forest (1000–1100 mm/yr rainfall). Foliar conductance and acid content were measured throughout 24-h cycles every three months for one year. Dark CO_2 fixation was evident in every run, but with differences between and within species; attempts were made to interpret results.

Plectorrhiza tridentata, which has less water storage capacity than the *Dendrobium,* proved to be vulnerable to summertime desiccation that lowered leaf conductance and reduced nocturnal acidification. *Dendrobium speciosum* never incurred equal stress, although one specimen growing on a rock did exhibit some foliar necrosis which Wallace attributed to excess insolation. All but the lithophytic *Dendrobium* and all examined *Plectorrhiza* were free of similar damage, a healthy condition ascribed to less violent extremes (particularly of temperature) in canopy versus rocky sites.

Evidence for the pervasiveness of CAM in epiphytes is based mostly on $\delta^{13}C$ discrimination (Table 2.1). Photosynthetic pathways have been extensively examined in Bromeliaceae and Orchidaceae in relation to this index. Winter, Wallace, Stocker and Roksandic (1983) have cast the widest net by determining these values for about a third of Australia's vascular epiphyte flora (127 of 380 species representing 17 families). Of those tested $\delta^{13}C$ fell between $-10.5‰$ and $-19.1‰$ ($\bar{x} = -15.2 \pm 1.95$) in 61 species; another 60 exceeded $-22/7‰$. Of 19 dicots included in the survey, eight showed signs of significant CAM. Unfortunately, carbon isotope data does not record CAM-idling, CAM-cycling, or $C_3 \rightarrow$CAM switching in facultative taxa. A peculiar pattern of deuterium enrichment does accompany CAM-cycling (Sternberg, DeNiro, and Ting 1984), but so far this tool has been used to evaluate only a few epiphytes (Table 2.1). Deuterium enrichment combined with a C_3-like $\delta^{13}C$ value identifies the CAM-cycler.

Distribution of CAM within certain families offers insight on the origin of this pathway and its association with canopy life. All three bromeliad subfamilies probably evolved CAM independently at least once (Fig. 2.4). Arguments about an epiphytic origin for CAM in Tillandsioideae continue (Benzing, Givnish, and Bermudes 1985). High light demand and CAM purportedly preceded canopy colonization by subfamily Bromelioideae (Pittendrigh 1948). Extant bromelioids are exclusively CAM plants with the exception of a few shade-tolerant mesophytic species (e.g., terrestrial *Cryptanthus* and epiphytic *Wittrockia*) which show evidence of near reversion to the

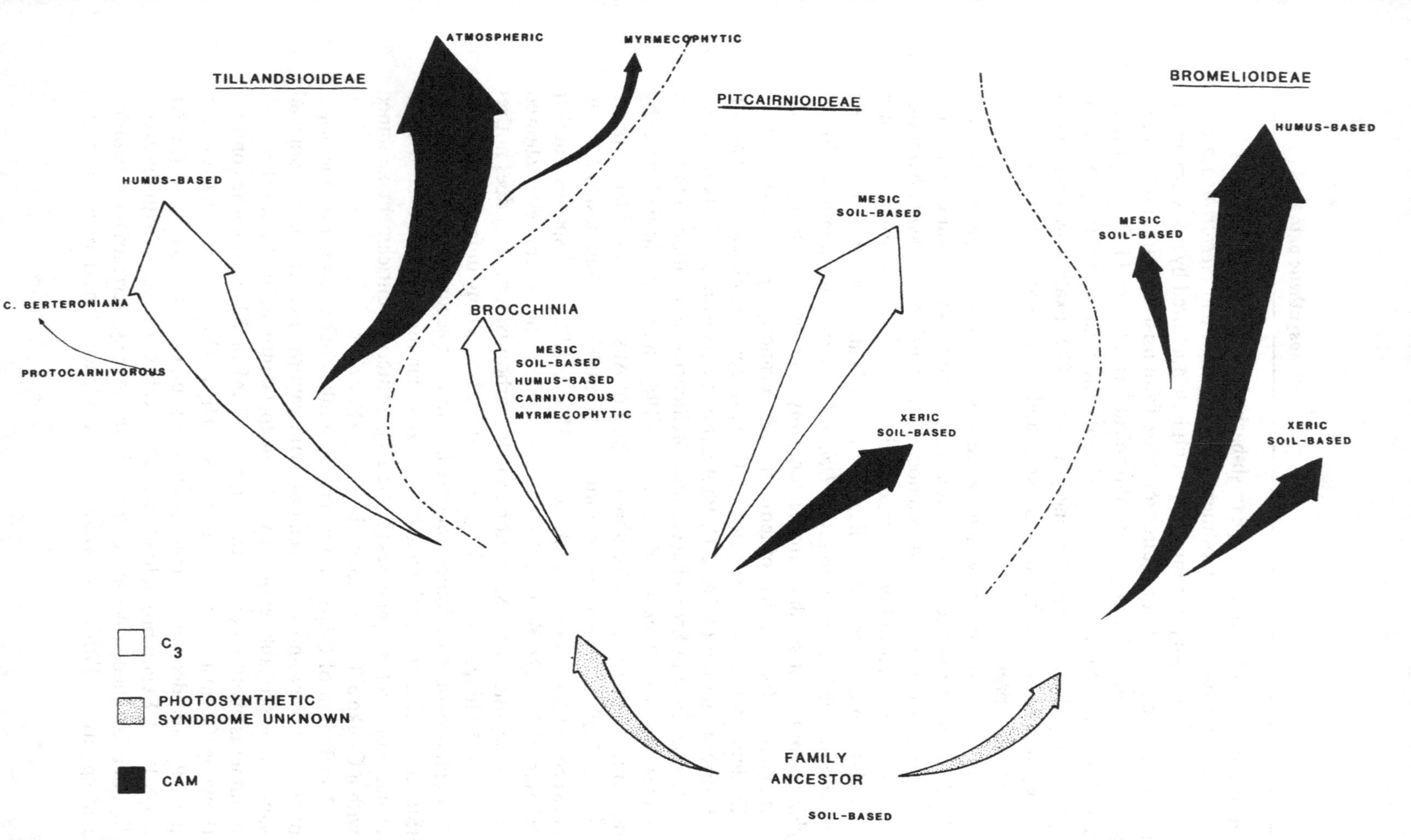

Figure 2.4. Phylogeny of Bromeliaceae depicting ecological divergence, nutritional mechanisms, and photosynthetic pathways among constituent lineages. (From Benzing et al. 1985.)

basic C_3 condition. (All CAM pitcairnioids are terrestrial.) A great many orchid lineages must also have acquired CAM separately. Epiphytic cacti most likely retain their CAM capacity from terrestrial relatives. *Peperomia,* a large predominantly epiphytic genus, includes both strict C_3 species and a variety of photosynthetic intermediates (Ting et al. 1985a).

The incidence of CAM in epiphytes and terrestrials alike is associated with particular habits. For instance, CAM is seldom associated with extensive woodiness. Mostly-woody to all-woody families such as Melastomataceae and Ericaceae contain C_3 but not CAM epiphytes. They, like terrestrial relatives with similar habits, probably tolerate moisture stress via the same mechanisms serving sclerophyllous shrubs everywhere. Winter et al. (1983) judged Australian *Agapetes, Ficus, Fragraea, Procris,* and *Schefflera,* all of which have active vascular cambia, to be incapable of CAM (Table 2.1). Most of these taxa are restricted, however, to pluvial or wet forest where severe drought may never occur. Only a couple of the massive stranglers of *Clusia* (but not *Ficus*) have so far proved capable of net nocturnal CO_2 fixation ($\delta^{13}C = -17.9‰$ for *Clusia rosea;* Ting et al. 1985b). *Clusia rosea* is especially notable for two additional reasons: Citric acid accounted for much of the titratable acidity, and ΔH^+_{max} reached 768 mmol m^{-3} – the highest of any value recorded for a CAM plant (Popp et al. 1987). Certain herbaceous forms can be equally deceptive in another way. Gesneriaceae and *Peperomia* have pronounced epiphytic tendencies, but *Boea* of Gesneriaceae and three Australian peperomias proved to be C_3 plants despite considerable succulence, modest intercellular volume in leaf mesophyll, and habitats that usually signal CAM capacity (Winter et al. 1983). The same was true of *Aeschynanthus pulcher* and *Columnea linearis* (Guralnick, Ting, and Lord 1986).

A novel type of CAM is practiced by other gesneriads *(Codonanthe)* and several peperomias. For example, like *Codonanthe paula* (Fig. 3.1E), *Peperomia camptotricha* exhibits a distinctive three-tiered leaf anatomy, and its associated spatial and temporal separation of C_4 and C_3 activities led Nishio and Ting (1987) to label this epiphyte a CAM–C_4 intermediate. Sequestered in the densely packed, vacuolate mesophyll was 51% of the PEPc, and in the thin, deeper green, palisade layer, 72% of the RuBPc/o. Phosphoenolpyruvate carboxykinase, the major decarboxylase present, was also largely localized in the spongy layer, indicating that CO_2 rather than malate or some other C_4-cycle product moves to the palisade for production of hexose. Incorporation of $^{14}CO_2$ during light and dark runs was greatest where the appropriate enzymes were most active, further documenting the division of labor in these leaves. Also indicative of unusual physiology was net CO_2 uptake day and night. More study of carbon flow among leaf compartments in this species is needed, including evaluation of its WUE. At present, *P. camptotricha* appears to reap the benefits of both CAM and C_4 photosynthesis. Yet gesneriads and *Peperomia* are not noted for rapid growth, and any unusual photosynthetic pathways possessed by these species are likely to have evolved primarily to enhance stress tolerance.

Photosynthesis in nonfoliar organs

Stems

Stems of epiphytes are frequent sites of photosynthesis, but appearances can be deceiving. With few exceptions (e.g., sympodial *Bulbophyllum ultissimum,* whose tiny stout shoots support virtually no leaf development), orchid pseudobulbs (Fig. 1.10) probably add little to carbon balance. But stems of epiphytic Cactaceae and some monopodial orchids are clearly the sole producers of carbohydrate. For example, *Schlumbergera* and *Zygocactus* bear cladophylls (Fig. 3.1A) as flat and leaflike as the most xeromorphic foliage. Other cacti display finely divided terete branchlets (e.g., some *Rhipsalis, Hatiora*); stems of more xeromorphic relatives are substantially stouter, spinose, and angular to terete. Quite a few of these better-armed species are hemiepiphytic or facultative in tree crowns (e.g., *Hylocereus, Selenicereus*). Aphyllous *Vanilla barbellata* (Fig. 4.27F) has a moderately thick round stem, yet grows profusely in heavy shade cast by mangrove and evergreen swamp forests in southern Florida. Fruiting there is infrequent, however; a specimen that flowers has usually scrambled over its host. This may be yet another case of too little radiation retarding sexual reproduction, as it does for so many hemiepiphytic aroids in deep forest. Other potentially interesting taxa (e.g., deciduous *Philodendron*) await study.

Reproductive structures

Ripening capsules can constitute a large fraction of the reproducing orchid shoot. Thin wings with no obvious function beyond increasing photosynthetic capacity adorn some fruits (e.g., *Encyclia cochleata;* Fig. 5.2H). Maturation is slow; although final size is reached during the first few months, nearly a year is often required for dehiscence. In *Encyclia tampensis,* a green fruit wall more sparsely equipped with stomata (3–7 mm^{-2}) than are the leaves (46 mm^{-2} on the abaxial surface) recaptures important amounts of CO_2 respired by developing seeds (Benzing and Pockman 1989). Several crop plants, despite less impetus to conserve water, also recycle CO_2 trapped in fruit locules, and may exhibit positive carbon balance. Reasons why green tissue in *E. tampensis* capsules, pseudobulbs, and roots mediate regenerative rather than net photosynthesis are discussed elsewhere (Benzing and Pockman 1989). Long-lived green flowers of certain moth-pollinated genera (e.g., *Epidendrum*) last for weeks if unpollinated. Dueker and Arditti (1968) reported that green perianths of a *Cymbidium* photosynthesize, but too feebly to do much good. Other species may effectively use longer-lived, more chlorophyllous flowers, however. More permanent parts of an inflorescence are sometimes sites of significant carbon gain. *Chiloschista lunifera,* a canopy-adapted orchid with almost no leaf and little stem tissue, retains its flattened green inflorescence axis for several months whether or not capsules are set. Massive vining inflorescences of *Oncidium* sect. *serpentaria* species must provide much of a reproducing adult's photosynthate and may be particularly important for reproduction. Persistent green spathes of Araceae and Cyclanthaceae seem well suited to match the flag leaves of certain cereals in providing much of the carbon needed to mature nearby fruit.

Roots

Photosynthetic roots deserve special mention in this discussion because they are featured by so many epiphytic orchids and aroids. Plastids responsible for the green color in an orchid root reside in cortical (Fig. 2.16) and less often stelar (Fig. 2.14) parenchyma, including the endodermis. Chloroplasts are scattered and relatively few per cell, much as they are in a CAM plant's succulent leaf and stem. If exposed on all sides, roots are uniformly pigmented. Those growing against solid objects remain achlorophyllous or pale green where contact blocks light. Embedded roots, of course, are incapable of autotrophy. Most orchid roots are oval to round, but those of shootless *Campylocentrum pachyrrhizum* and several *Phalaenopsis* have become almost planar. *Taeniophyllum rhizophyllum* roots are, as the name implies, even more leaflike. Structural features that might logically reflect trophic capacity do in fact seem to be related to photosynthetic vigor. One of these is root mantle development. Shootless species and relatives with noticeably reduced leaf surfaces never possess more than two to three layers of velamen. *Campylocentrum pachyrrhizum* sloughs its delicate rhizodermis on the exposed side, a process that promotes light penetration (Figs. 2.10, 2.16).

A second notable feature is the pneumathode (Fig. 4.27A). When roots are dry, the velamen offers little resistance to ventilation, but the exodermis is restrictive (Fig. 3.19G). When the velamen is wet, however, gases can still reach the exodermis via air-filled pneumathodes; here, certain U cells that

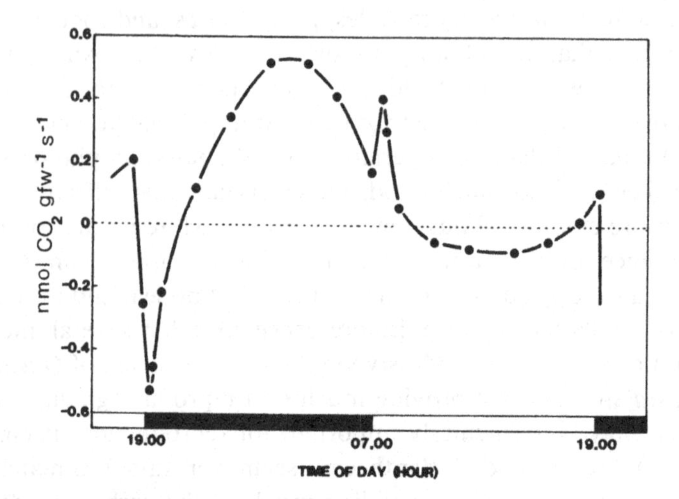

Figure 2.5. The diurnal course of CO_2 exchange by the shootless orchid *Campylocentrum tyrridion.* (gfw, grams fresh weight.) (From Winter et al. 1985.)

promote aeration (Fig. 3.19A,B,D,E) have thinner boundaries, and inner tangential walls may be missing entirely. A third factor affecting trophic capacity is the volume of cortical airspace, which ranges from the high of a C_3 plant's leaf to the low of a CAM organ (Benzing, Friedman, Peterson, and Renfrow 1983). Roots of shootless and transitional species exhibit the largest intercellular volume, doubtless a requirement for substantial CO_2 supply to deep-seated green cells. Cortical parenchyma cells in *Kingidium taeniale* are surprisingly similar to the irregularly shaped mesophyll cells in many C_3 leaves (Fig. 3.19A).

One of the most important unknown aspects of green root biology is its impact on plant economy. Is gas exchange across the velamen–exodermis interface regulated or passive? Does the pneumathode–aeration cell combination function as a stoma or as a lenticel? Shootless *Chiloschista usneoides* apparently has no stomatal analog in its roots but avoids excessive carbon loss by regenerating CO_2 from malic acid just rapidly enough to refix it with available photosynthetically active radiation (PAR; Cockburn, Goh, and Avadhani 1985). *Campylocentrum tyrridion* was less effective in retaining CO_2, although the loss was less than 25% of the total fixed during the preceding dark period (Fig. 2.5; Winter 1985; Winter et al. 1985).

Separate analyses of individual species must be carried out if one is to establish how green aerial roots affect whole-plant carbon balance. Studies of gas exchange, $^{14}CO_2$ incorporation, enzyme activity, and acid fluctuation show that roots of shootless orchids and such related semileafless species as *Kingidium taeniale* are indeed primary trophic organs; none, however, fixed CO_2 as vigorously as leaves from species with fully developed shoots (Benzing and Ott 1981). The shootless orchid's unusually high N and chlorophyll levels and the presence of chloroplasts in just about every root parenchyma cell (Fig. 2.14, 2.16) challenge the claim that shoot reduction has paralleled increased tapping of host substrates via root-inhabiting fungi (Johansson 1977). Although roots of leafy taxa (e.g., *Epidendrum radicans*) showed weaker activity than did those of shootless specimens (CO_2 exchange was continuously negative with efflux slowest during the day), this does not mean that they have no trophic significance to a leafy plant. On the contrary, roots sometimes comprise half or more of the vegetative body (Fig. 2.18) and are often exposed to light, especially during juvenile stages. At the very least, green roots reduce maintenance costs by recovering considerable dark- and light-respired carbon.

Adaptations to specific exposures

Light harvest

Photon capture and processing (transduction) by green plants is a complicated and only partially understood phenomenon. A quantum of energy (i.e., a photon) is first trapped by antennae containing various pigments. The photon's energy is then funneled via resonance transfer from one molecule to another and ultimately to a reaction center containing a special chlorophyll (Chl) *a* molecule. These pigment–protein assemblies for trapping and processing photons are called "quantasomes"; they occur in vast numbers in the lamellae of chloroplasts (thylakoids).

A different reaction center serves each of two distinct photosystems which together drive the overall photochemical reaction. Photosystem I reduces nicotinamide adenine dinucleotide phosphate (NADP) to NADPH, using hydrogen from water. Photosystem II synthesizes adenosine triphosphate (ATP), releasing molecular oxygen in the process. The two photosystems are connected by a series of molecules that act as electron carriers. The familiar "Z" scheme illustrated in plant physiology textbooks graphically depicts the photolysis of water, electron flow, and ultimately energy storage through the coupled syntheses of ATP and NADPH. Some of the captured energy may be used to process nitrate (NO_3^-) to ammonium (NH_4^+) ions, but most is utilized to transform CO_2 into photosynthate.

Variation in quantasome size depends on the number of molecules present. In exposed habitats, quantasomes will be smaller. In the shade, antenna size (thus quantasome size) increases, the better to harvest scarce photons; the Chl *a*/*b* ratio decreases. Investments in carboxylases, electron carriers, and other components of the energy-transducing apparatus are also cut; dark respiration lessens as metabolic rate is generally reduced. A lower specific leaf weight (dry weight per surface area) is an especially effective economy measure in low-energy sites. Other distinctions between sun and shade photosynthesis are based on more complex, poorly understood mechanisms (Chazdon and Pearcy 1986). Shade leaves quickly respond to sunflecks and can store enough acquired energy to continue fixation briefly after shade returns. Sun leaves are better able to dissipate excess energy and avoid photoinhibition of photosystem II.

Temperate-forest understory herbs sometimes alter the light responses of their foliage throughout the growing season; successive, or even the same, leaves switch from high- to low-light-adapted photosynthesis as overhanging branches leaf out in late spring. As yet no similar mechanisms have been identified in epiphytes, but they certainly seem probable. After all, many vining hemiepiphytes show dramatic shifts in leaf size, shape, and, more than likely, photosynthetic properties as well. Hemiepiphytic *Monstera dubia* produces only thin, small, sessile foliage appressed against the substratum during early stages of its climb. Once exposure reaches some threshold level, much larger petiolate organs, held well above the trunk surface, develop. Some Marcgraviaceae exhibit similar juvenile morphology on the main axis but, in the upper canopy, generate freely suspended, determinate, axillary branches equipped with robust leaves. Photomorphogenesis that allows vines such as those of *Monstera gigantea* to locate and ascend tree trunks (Strong and Ray 1975) is probably part of many more secondary hemiepiphytes' makeup.

A recent study (Winter, Osmond, and Hubick 1986) of the efficiency of photon harvest in Australian *Pyrrosia* and several other shade-adapted, nonfern, CAM, or CAM-cycling epiphytes produced unexpected results. Quantum yields (in this instance expressed as moles of quanta absorbed per mole of oxygen evolved) indicated that these species were about as efficient in dim light as were typical C_3 plants. Thick foliage and CAM, usually considered to be a poor combination for life in shade, had no diminishing effect. Performance varied a great deal among species, however. In another investigation of C_3 and CAM epiphytes (Lüttge, Ball, Kluge, and Ong 1986), quantum yields (here expressed as moles of CO_2 absorbed per mole of quanta) ranged from 0.002 to 0.070 (Table 2.4). Although energy is captured with comparable efficiency in dim light, under stronger illumination C_3 species usually achieve higher photosynthetic yields than do CAM types (Osmond 1978; Osmond 1987).

Shade acclimatization

The epiphyte most thoroughly examined for its ability to adjust to changing light regimens is *Tillandsia usneoides* (Martin, McLeod, Eades, and Pitzer 1985; Martin et al. 1986). Plant material collected in the field was first tested, then maintained in culture under various light exposures. At 7–15% of full sunlight (100–200 μmol $m^{-2}s^{-1}$ PAR at midday), subjects differed less than expected from fully exposed (1500–1600 μmol $m^{-2}s^{-1}$) controls in those features that constitute shade adaptation. The former group did, however, contain more Chl, although Chl *a*/*b* ratios were unaffected. High irradiance promoted greater starch deposition, but this was the only difference in chloroplast ultrastructure. In addition, internode length, leaf size, stomatal density, and morphology of trichomes and guard cells changed little, if at all, with PAR level. Nocturnal acidity, a reasonable index of photosynthetic vigor for CAM species, measured about 60% of that in fully irradiated greenhouse subjects.

In a second study, *Tillandsia usneoides* collected from what was clearly the darkest of three exposures in a South Carolina forest showed higher ΔH^{+}_{max} as compared to greenhouse specimens. It did not, however, deviate substantially from subjects exposed to stronger PAR at the other two sites. Stomata were somewhat more numerous and trichome shields only slightly smaller, compared to plants that had grown under fuller exposures. Chlorophyll data, however, were similar to those noted in greenhouse runs. Martin et al. concluded that *T. usneoides* acclimatizes across a broad PAR range without substantial morphological or metabolic alteration, and that a full day of about 10 mol m^{-2} integrated PAR is optimal for carbon gain. Although the general applicability of their findings remains questionable for reasons noted above, evidence that numerous additional epiphytes tend to be shade-tolerant, and perhaps sometimes even shade-requiring, is mounting. Adams (1988) reported similar versatility involving more plant parameters that he concluded indicated near-complete ability to accommodate to the unusually broad diurnal and seasonal PPFD fluctuations within tree crowns. Shade- versus sun-grown *Dendrobium speciosum* and *Pyrrosia confluens* exhibited thin leaves, more concentrated Chl, and reduced respiration. Light response curves, quantum yields, and photosystem II fluorescence at 77K demonstrated acclimatization to low, but not to unfiltered, irradiance. Photoinhibition of fully exposed foliage was reversed in shade, apparently due to decreased, nonradiative energy dissipation rather than to repair of damaged photoreaction centers.

Schäfer and Lüttge (1988) also recorded a more multidimensional response to exposure history in *Kalanchoe uniflora* cuttings propagated under low and high light regimens. Dark respiration was reduced, as were light compensation and saturation points; apparent photon yields were greater in shade- versus sun-grown foliage. Weaker illumination elevated chlorophyll contents, but decreased leaf thickness. Light-saturated photosynthesis was diminished by growth in fuller sun, apparently because mesophyll conductance – usually an indicator of carboxylation capacity – was less developed. No evidence indicated that this epiphyte can function like a true shade species, nor are highest exposures advantageous, even with a liberal supply of N. In fact, cultivation under strong PPFD produced a condition that the authors described as "chronic photoinhibition." Acclimatization was possible but only within an intermediate range of illumination, a finding consistent with occurrence in native habitats. Sun-induced water stress may help exclude *K. uniflora* from the most exposed microsites because its capacity for CAM is so limited. But many additional C_3 and CAM forms must be studied in equal detail if the critical question is to be answered, namely, Is light or humidity the more important factor in epiphyte distribution?

Optical and morphological enhancement of photosynthesis

For a small subset of epiphytes in deep evergreen forests, an optical phenomenon purportedly promotes shade tolerance. Some aroids and gesneriads, as well as the bromeliads *Aechmea fulgens, Nidularium bruchellii* (Fig. 4.25G), and several species of *Vriesea* and *Tillandsia,* possess abaxial mirrors which are thought to reflect unabsorbed photons back into overlying chlorenchyma (Lee, Lowry, and Stone 1979). Similar cyanic epidermal layers are fairly common among low-growing herbs in the same forests. Maroon to red backing is most appropriate, given the long-wavelength enrichment as well as overall depletion of light passing through dense overhead foliage. Not surprisingly, the bicolored leaves of such plants usually grow horizontally with minimal overlap (Fig. 2.17) – more about this later. Essentially monolayered *Nidularium bruchellii* exhibits a faint bluish-green irridescence reminiscent of the thin-film optical interference phenomenon reputedly responsible for greater relative absorption of red light in deep-shade-adapted *Selaginella willdenovii* (Lee and Lowry 1975). Cultivated in better exposure, both species exhibit much less irridescence, and electron microscopy shows the adaxial cuticle of *S. willdenovii* to lack the lamination characteristic of shade-grown material.

Shade tolerance could be enhanced in some aroids by velvety leaf surfaces (e.g., *Anthurium pallidiflorum*) which feature domed, adaxial, epidermal cells thought to gather shade-light. On a larger scale, prominent, lenslike, adaxial, hypodermal tissues in certain *Peperomia* and gesneriad leaves (Fig. 3.1E) could do likewise and store water as well. Leaves of various *Peperomia* species range from typical thin bifacial types to the succulent subunifacial forms (Kaul 1977). Light piping through a clear, internal, multilayered hypodermis may preserve photosynthetic capacity in *P. dolabriformis* while water loss is reduced. Growing conditions affect hypodermal development, thereby substantially altering leaf anatomy. Leaf thickness in *Codonanthe paula* and *Peperomia camptotricha* varies several fold, depending on the season. Shifts in gross shoot form that affect photon interception can be pronounced in Bromeliaceae (Figs. 2.11, 2.12). Additional study is needed to confirm the beneficial effects of these properties on photosynthesis. Consideration of other leaf parameters could be equally rewarding.

Co-occurring taxa often possess leaves with differing forms and life histories, surely a signal that there are alternate solutions to common problems of carbon balance. For instance, secondary hemiepiphyhtic *Peperomia tropaeolifolia* (Fig. 2.8) and juvenile *Marcgravia coriacea* climbing side by side in wet Ecuadoran forest possess foliage that is thick and deep green in the former but pale and delicate in the latter. Longevity also differs. Robust *Peperomia tropaeolifolia* leaves are more durable, a pattern consistent with a life spent no more than 2 m above the dark forest floor. Mature *Marcgravia coriacea* eventually grows into tree crowns where its broader, thicker leaves presumably outperform its less productive, older, shade-adapted foliage below. Use of a portable gas exchange system could reveal much more about how epiphytes with dissimilar foliage share the same type of microsite.

Leaf variegation and shoot architecture

Leafy rosulate shoots unavoidably self-shade all but the youngest foliage. A number of epiphytes, especially tank bromeliads, seem to respond to this type of archtecture with, among other qualities, elaborate ornamentation (Benzing 1986). Leaves of several taxa (e.g., *Vriesea fosteriana;* Figs. 2.13, 2.19) are differentiated into regularly shaped and distributed zones, some Chl-rich (shutters) and others Chl-poor (windows or fenestrae). Shutter patterns are often copied by the aforementioned abaxial cyanic mirrors (Fig.

2.13). Assays of *Vriesea fosteriana* showed that shutters and windows on young leaves differed in nutrient content and function, but became more alike with age (Table 2.2). Young shutters photosynthesized more vigorously than did adjacent windows. Later after windows had become equally chlorophyllous, both areas consumed CO_2 at about the same (faster) rate. Basal shutters routinely exhibited lower light saturation points than did those in older regions of the same blade (Fig. 2.6).

Clearly, fenestration has a physiological concomitant, but what are its benefits? Is division of labor or visual effect most important? Or are there other factors involved – economy for one? The high cost of photosynthetic machinery is evenly distributed throughout a concolorous leaf, but perhaps fenestration allows production of a larger leaf with amounts of N and P similar to those required by a smaller concolorous blade. Large size is important, as it offers room for impoundment of nutrients and moisture and also serves to scatter resource-rich patches among less desirable ones, thereby discouraging herbivory.

Another possible economy comes to mind. Perhaps less incident PAR is wasted by a compact, multilayered shoot if its congested leaves possess both translucent and opaque photosynthetic zones. Even if the presence of windows diminished the quantum efficiency of upper leaves below that of uniformly green surfaces, a larger amount of the total incident energy might still be harvested by the plant as a whole. *Vriesea fosteriana* and its kind might be best suited to operate in moderate shade where sunflecks are common, or in prolonged high exposures where the upper layers of the shoot are routinely light-saturated. Figure 2.7 illustrates how much sunlight passes through shutters and windows of *V. fosteriana* foliage.

Visual effect may have been the prime reason for fenestration as well as other patterns of color. Deposits of dark pigments deep in a rosulate shoot (Fig. 2.9) are well positioned to hide cryptic wildlife, whose survival is enhanced by a dark backdrop within the tank microcosm. Perhaps more intricate patterns are effective in breaking up the outline of lighter-colored tank occupants. Because tank bromeliads utilize fauna to process intercepted litter, plant features that benefit these mutualists will in turn benefit the epiphyte. The notion that some ornithophilous neotropical gesneriads, whose foliage is adorned with conspicuous red/orange spots, maintain these signals to remind resident pollinators of their presence between flowering seasons cannot pertain to all ornamental bromeliads. Some of the most elaborately pigmented species (e.g., *Vriesea fosteriana* and *V. hieroglyphica*) are predominantly bat-pollinated, flower at night, and are wind-dispersed.

Table 2.2. **Data on the structure and function of foliar windows and shutters in *Vriesea fosteriana***

Position in rosette	Area	Chlorophyll content in leaf surface (mg cm^{-2})	Mean element content (% dry weight)					Mean element content (ppm dry weight)						Mean maximum photosynthetic rate (g CO_2 cm^{-2}min^{-1})
			N	P	K	Ca	Mg	Mn	Fe	B	Cu	Zn	Mo	
Distal portion of upper leaf	window	0.0089	1.13	0.09	1.08	0.22	0.11	14	95	19	7	21	1.00	14.6×10^{-5}
	shutter	0.0442	1.58	0.13	0.32	0.22	0.14	12	86	20	8	22	1.09	41.3×10^{-5}
Distal portion of lower leaf	window	0.0400	1.42	0.08	0.67	0.48	0.17	12	105	16	6	29	1.23	62.2×10^{-5}
	shutter	0.0740	1.35	0.09	0.79	0.39	0.20	11	178	15	15	30	1.21	61.8×10^{-5}

Source: Benzing and Friedman 1981b.

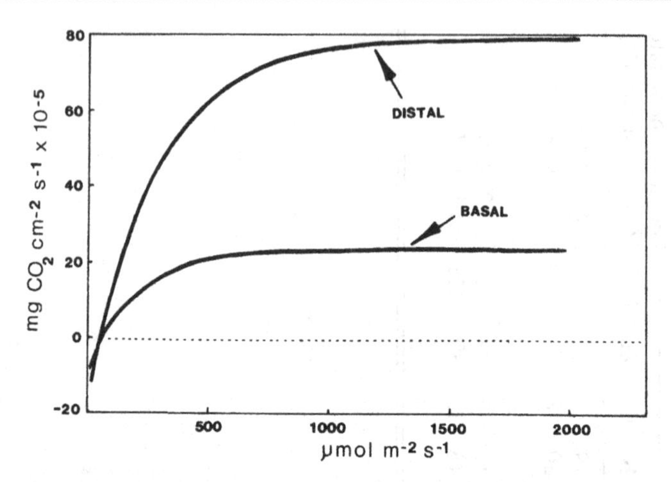

Figure 2.6. Carbon dioxide exchange by segments of *Vriesea fosteriana* leaves representing mature blade tissue and younger samples taken near the rosette center. (After Benzing and Friedman 1981b.)

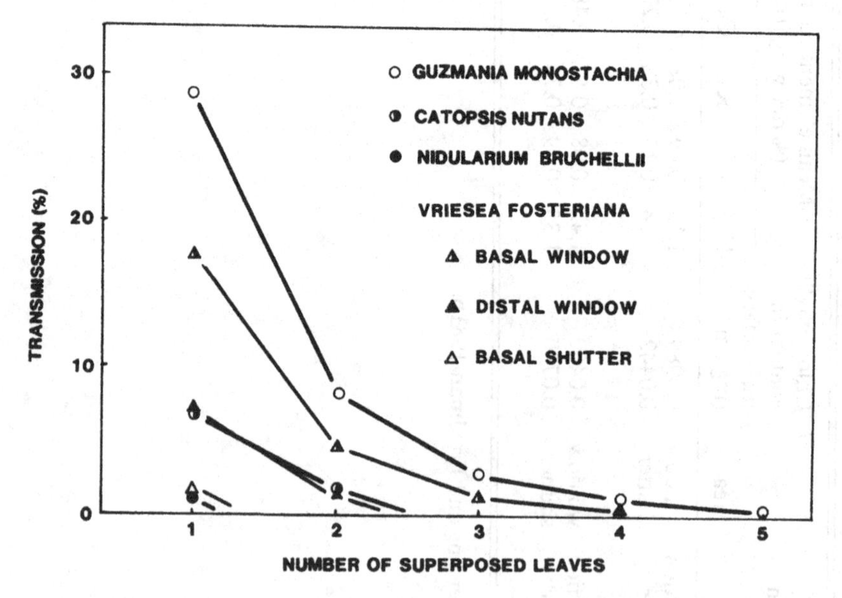

Figure 2.7. The transparency of leaves of three tank bromeliads. (From Benzing and Friedman 1981b.)

Figures 2.8–2.13. (2.8) *Peperomia tropaeolifolia* growing at the base of a tree in rain forest at Río Palenque, Ecuador (×⅕). (2.9) *Vriesea erythrodactylon* (×⅕); note deeply cyanic leaf bases. (2.10) Eroding velamen of the shootless orchid *Campylocentrum pachyrrhizum.* (2.11) *Tillandsia utriculata* growing in deep shade in a South Florida swamp forest (×1/10). (2.12) *Tillandsia utriculata* growing in nearly full sun in South Florida (×1/10). (2.13) The central part of a tank shoot of *Vriesea fosteriana* (×⅕).

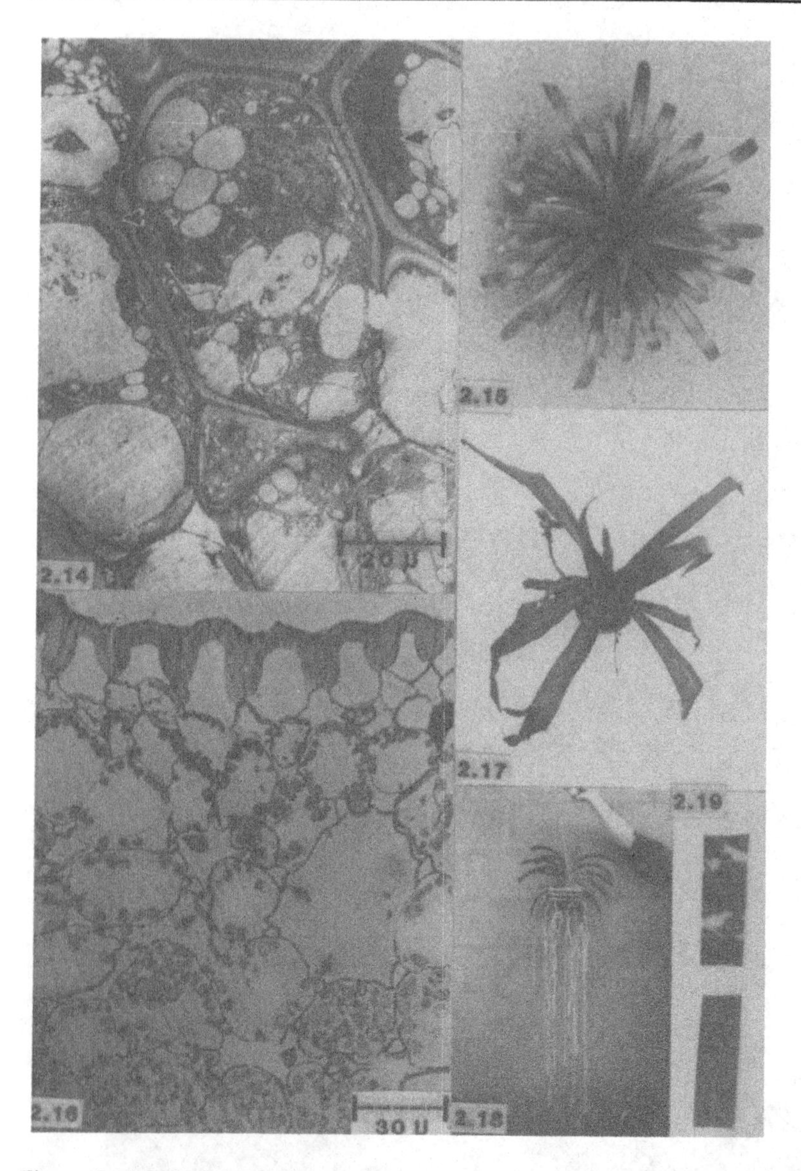

Figures 2.14–2.19. (2.14) Stelar parenchyma cells of shootless *Harrisella porrecta* illustrating presence of chloroplasts (arrows). (2.15) Shoot of *Guzmania monostachia* illustrating the multilayered habit ($\times\frac{1}{5}$). (2.16) Root cortex of shootless *Campylocentrum pachyrrhizum* illustrating the presence of numerous chloroplasts and considerable intercellular space, an eroded velamen, and spirally thickened idioblasts (arrows). (2.17) Monolayered habit of *Nidularium bruchellii* ($\times\frac{1}{12}$). (2.18) A specimen of *Vanda* sp. showing profuse development of an exposed root system ($\times\frac{1}{25}$). (2.19) Young (upper) and mature (lower) leaf tissue segments from *Vriesea fosteriana* ($\times\frac{1}{10}$).

Shoot architecture and ordinary green foliage are combined in ways that again suggest multiple solutions to common problems, including need for reduction of risk. Losing even a leaf or two in a monolayered shoot would seriously reduce photosynthetic – and perhaps impoundment – capacity. Broadly tolerant *Guzmania monostachia* (Fig. 2.15) seems to exemplify a less vulnerable alternative with no corresponding loss of shade tolerance. Its shoots are composed of numerous concolorous but translucent (and presumably inexpensive) leaves, several of which must be superposed to quench a beam of intense sunlight to levels below the individual leaf's light compensation point (Fig. 2.7). As expected, the bicolored foliage of *Nidularium bruchellii* (Fig. 4.25G) and that of similarly monolayered, ecologically restricted, shade-tolerant *Catopsis nutans* are much less translucent and perhaps of necessity better defended against predators.

Extensive research will be required to explore all of these hypotheses and to refine and expand the pigmentation categories. More work in the laboratory, plus field surveys noting habitat specifics, correlated growth forms, and leaf characteristics of resident bromeliads, will be useful. Should the logic presented above to explain the adaptive basis of several pigmentation/leaf combinations be accurate, then the following conditions should apply:

1. Rosettes of red-backed leaves should tend toward a monolayer and be found growing in deep shade. Chlorophyll content should be relatively high and the Chl *a*/*b* ratio, energy-transducing capacity, and dark respiration, low. Texture and defensive chemistry should be designed to promote extended leaf service.
2. Species whose shoots bear fenestrate foliage should be more densely multilayered and occupy moderate to high exposure. A protective role would be indicated by an especially rich fauna in the interfoliar tanks.
3. Plants with dark leaf bases are likely to be those whose shoots are borne singly and exposed rather than as compact clusters or obscured by persistent dead leaves. Unusually rich tank faunas would also be significant here.
4. Species with multilayered shoots routinely encountered in fairly deep shade should have translucent, uniformly green, inexpensive foliage. These species would probably not flourish alongside the most shade-tolerant bromeliads.
5. Anthocyanin development in the adaxial epidermis, where it can act as a sun screen, should occur in specimens exposed to strong illumination.

Vertical stratification

Atmospheric humidity and irradiance vary little throughout the short open canopies of dry forests; resident epiphytes show little or no vertical stratification there. Restriction of species to twigs versus thicker axes, or to knotholes or crotches instead of less absorbent media, probably reflects needs and tolerances for moisture rather than for particular light levels. Another pattern prevails under wetter conditions. Richards (1952) recorded substantial vertical segregation of canopy flora depending on the opacity and stature of some humid forests in Guyana, paralleling Pittendrigh's (1948) findings regarding bromeliad distribution in the cacao plantations and adjacent woodland of northern Trinidad (scheme IV, Chap. 1; Table 2.3). Pittendrigh's "exposure" group occupied upper, well-illuminated perches in dense everwet forest or was distributed throughout drier sparser canopies. Members were either xeromorphic tankless taxa – the so-called atmospherics (Fig. 1.11) – or species with large impoundments relative to overall shoot size (Fig. 4.25A). His "sun" group, most of which had sizable but wider-spreading tank rosettes, inhabited intermediate height in thick forest. Below were his "shade-tolerant" populations, the most hygrophytic of all. Their shallow catchments, fashioned of lax rosulate shoots (Fig. 4.25G), were usually filled with moisture and fallen debris.

There has been extensive follow-up of Pittendrigh's bromeliad survey aimed at detecting the physiological concomitants of his three light-related categories (Benzing and Renfrow 1971b; Griffiths and Smith 1983; Smith et al. 1985, 1986a; Griffiths et al. 1986; Lüttge et al. 1986a). Contrary to Pittendrigh's proposal that the shade-tolerant group is comprised of heliophiles restricted to dark microsites by high humidity requirements, these plants in reality use shade light efficiently (Benzing and Renfrow 1971b). Species adapted to shady microsites are C_3 types, gaining carbon vigorously but subject to steep moisture demands. Over streams where air is nearly saturated, life is possible in full insolation. Exposure and sun forms are often CAM plants (as indicated by $\delta^{13}C$ data) or have accessory moisture sources. The usual association of bromeliad CAM with moderate to high light demands may account for its absence or weak development in most deep-forest species. Furthermore, the extensive foliar indumenta of tillandsioid CAM bromeliads (Figs. 3.6, 3.11, 4.25B–D) prohibit success in overwet sites (Mez 1904; Benzing and Renfrow 1971a). Although these structures allow survival with only periodic wetting, this benefit can be negated by the aforementioned suffocating effect of water-filled shields when humidity remains high. Still, CAM per se does not always accompany drought tolerance. In fact, the obligate CAM bromeliads *Aechmea aripensis, A. downsiana,* and *A. fendleri* all occur only in the wettest forests of Trinidad (Griffiths and Smith 1983). The existence of terrestrial *Bromelia humulis* (Medina, Olivares, and Diaz 1986) in partial shade further obscures the benefits of CAM for some bromeliads. Although CAM succulents have been deemed eco-

Table 2.3. **Trinidad bromeliads: exposure and humidity requirements; habits and photosynthetic pathways**

Category/species	Photosynthetic pathway[a]	Habit[a]	Relative abundance[b] in 7 humidity zones with maximum rainfall (in cm/yr) of:						
			650	500	320	280	230	180	130
I. Exposure group									
Tillandsia didistichoides	C_3	tank	+++						
Catopsis berteroniana	C_3	tank		+++	++				
Vriesea amazonica	C_3	tank		++	+++	+			
Catopsis sessiliflora	C_3	tank		++	+++	++	++	++	
Vriesea procera var. *procera*	C_3	tank		++	+++	+++	+++	++	
Catopsis floribunda	C_3	tank			++	++	++	++	
Tillandsia gardneri	CAM	atmospheric			++	++	++	++	
Tillandsia tenuifolia	CAM	atmospheric					++	++	
Tillandsia stricta var. *stricta*	CAM	atmospheric					++	++	
Tillandsia usneoides	CAM	atmospheric					++	++	
Tillandsia juncea	CAM	atmospheric					++	++	
Tillandsia utriculata	CAM	tank			+	+	++	+++	
Tillandsia elongata var. *elongata*	C_3/CAM	tank (little impoundment)					++	+++	
Tillandsia flexuosa	CAM	tank						++	+++
II. Sun group									
Vriesea johnstonii	C_3	tank	+++						
Glomeropitcairnia erectiflora	C_3	tank	+++						
Vriesea broadwayi	C_3	tank	+++						

Table 2.3. **(cont.)**

Category/species	Photosynthetic pathway[a]	Habit[a]	Relative abundance[b] in 7 humidity zones with maximum rainfall (in cm/yr) of:						
			650	500	320	280	230	180	130
Tillandsia spiculosa var. *micrantha*	C_3	tank	+++						
Aechmea aripensis	C_3	tank	+++						
Vriesea capituligera	C_3	tank	+++	++					
Vriesea rubra	C_3	tank		+++					
Tillandsia complanata	C_3	tank		+++					
Aechmea fendleri	CAM	tank		+++					
Guzmania sanguinea var. *sanguinea*	C_3	tank	+	+++	++				
Vriesea splitgerberi	C_3	tank	+	+++	++				
Vriesea platynema var. *platynema*	C_3	tank		+++	++				
Araeococcus micranthus	CAM	absorptive roots, common ant-nest inhabitant		+++	++				
Aechmea dichlamydea	CAM	tank		++	+++				
Guzmania monostachia var. *monostachia*	C_3/CAM	tank		++	++	+++	+++	++	+
Aechmea aquilega	CAM	tank			++	++	++	+++	
Tillandsia fasciculata var. *fasciculata*	CAM	tank (little impoundment)			++	++	++	+++	
Tillandsia bulbosa	CAM	atmospheric myrmecophyte			+++	+++	+++	+++	

Hohenbergia stellata	CAM	tank		+++	+++	+++		
Aechmea lingulata	CAM	tank		++	+++	+++	++	
Aechmea nudicaulis	CAM	tank	++	++	++	+++	++	++
III. Shade-tolerant group								
Tillandsia anceps	C_3	tank (little impoundment)		+++	+++	+++	+++	
Tillandsia monodelpha	C_3	tank (little impoundment)	+++	+++	+++	+++		
Vriesea simplex	C_3	tank (leaves discolorous)	++	+++				
Vriesea rubra	C_3	tank	+++	++				
Guzmania lingulata var. *lingulata*	C_3	tank	++	+++	+++	++	++	
Vriesea splendens var. *formosa*	C_3	tank	+++	+++	++	++		

Note: +++, abundant; ++, intermediate; +, scarce.
Sources: [a]Griffiths and Smith 1983; [b]Pittendrigh 1948.

nomical water users and heliophilic, this particular population was not favored by full sun; specimens shielded by modest overgrowth showed less moisture stress and more production than did unprotected individuals.

Pittendrigh's (1948) observations showing that neotropical epiphytes partition dense forests along ecoclimatic gradients are generally applicable today; even his proportionalities are representative. He found Trinidad bromeliads to be segregated into groups of 14, 21, and 6 taxa in order of increasing shade tolerance (Table 2.3). Gentry (1982a) working at Río Palenque in west-central Ecuador, named his three exposure levels "canopy-top" (111 species), "mid-canopy" (120 species), and "understory" (17 species). Figure 2.20 gives the taxonomic grouping. Individual more or less free-standing trees provide a parallel: Greater epiphyte diversity occurs in midcrown (Fig. 7.10; Johansson 1975). Kelly (1985) found a higher percentage of sciophytes in a moist montane forest in Jamaica, but the majority were ferns and hence very likely drought-sensitive, as are most pteridophytes. Angiospermous counterparts on the island, a relatively depauperate, generally more heliophilic assemblage, would probably have changed the picture had they been present in the usual proportions.

Wallace (1981) noted vertical distribution of vascular epiphytes relative to carbon fixation pathway in a humid Australian forest (Table 7.2). Ten ferns and 14 mostly orchidaceous angiosperms were surveyed. Species in shady locations on lower trunks and along large branches were all C_3 ferns and woody hemiepiphytes. Those at the upper canopy margin were predominantly CAM species, as determined by $\delta^{13}C$ assays. Among the few highly exposed C_3 types were a woody epiphyte *(Pittosporum undulatum)* and several ferns. Some of these more mesic specimens were either growing in trash-basket *Dendrobium speciosum* impoundments or other substantial humus deposits, or part of their root system had reached the ground. Wallace's findings certainly accord with the view of Griffiths and Smith (1983) that CAM epiphytes usually occupy more xeric microhabitats than do C_3 types (see Chap. 3 for exceptions). Also demonstrated was tapping of different resource bases by diverse canopy-adapted flora at the same level in a forest.

A three-level categorization of forest exposures may fail to describe adequately an epiphyte's position; subdivision is sometimes required. Recall that *Peperomia tropaeolifolia* (Fig. 2.8) almost always grows on tree trunks in deepest shade within a meter or two of the ground. This same well-defined microhabitat is occupied by distinct bryophyte assemblages in other humid woodlands (Pócs 1982; Fig. 3.5), and some epiphytic filmy ferns are found nowhere else. Here again, sciophytism may be dictated more by the desiccating effects of strong irradiance and turbulent air than by some intrin-

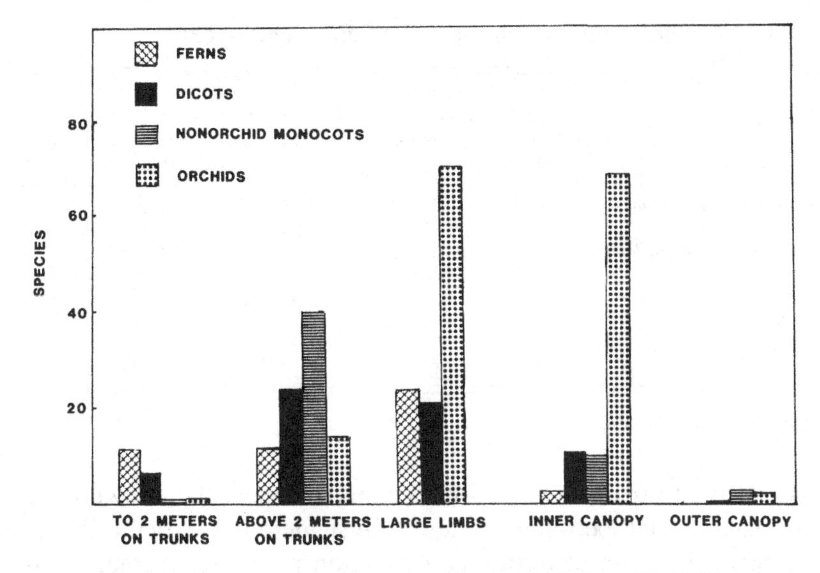

Figure 2.20. Distribution of epiphytes by taxonomic group in wet forest at Río Palenque, Ecuador. (From Gentry and Dodson 1987b.)

sic light requirement. In other microsites, adaptation can be flexible enough to blur an epiphyte's ecological classification; once again, moisture seems to be the decisive factor, although light plays a role. Several Florida tillandsias that root only in tree crowns over much of the state also thrive as terrestrials if substrata are well drained and sufficient irradiance reaches the ground (Fig. 1.13). Conversely, a number of ferns and flowering plants confined to earth at certain locations in tropical America will also root in moist canopy sites. Kelly (1985) observed *Philodendron lacerum* growing as an epiphyte, a secondary hemiepiphyte, and a terrestrial in different Jamaican locations. Sometimes variation in habit seems to be the most decisive factor. Certain *Blakea* and *Topobea* species of neotropical Melastomataceae include lianas as well as terrestrial and epiphytic shrubs (Renner 1986).

Ancestral habitats

Several attempts have been made to identify ancestral habitats by characterizing the light requirements of extant epiphytes (e.g., Pittendrigh 1948; Benzing and Renfrow 1971b; Lüttge et al. 1986a). Lüttge et al. concluded that every tested species, including *Guzmania lingulata* and several ferns

Table 2.4. **Photosynthetic properties of eight epiphytes**

Species	Carbon fixation pathway	Light compensation point (μmol $m^{-2}s^{-1}$)	50% light saturation	Light saturation	Apparent quantum yield
Aglaomorpha heracleum	C_3	5	58 ± 12	160	0.07 ± 0.02
Anthurium hookeri	C_3	5–40	90 ± 25	180–375	0.03 ± 0.01
Drymoglossum piloselloides	CAM	8	28	140 ± 60	0.02
Kalanchoe uniflora	CAM	25–160	70–245	200–650	0.002–0.031
Nepenthes × hookeriana	C_3	5–10	65 ± 25	150–225	0.04 ± 0.01
Platycerium grande	C_3	20	—	>520	0.015 ± 0.004
Pyrrosia lanceolata	C_3	10	90	300	0.02 ± 0.01
Pyrrosia longifolia	CAM	8	45	100–150	0.03

Source: Lüttge et al. 1986a.

also native to dense humid forest, either matched the known qualities of sun plants or fell between sun and true shade types. Specifically, no taxon exhibited light compensation or saturation points comparable to those of obligate understory terrestrials (Table 2.4). Additionally, none were photoinhibited by high levels of radiation as are the sensitive shade-demanding forms. Finally, quantum yields in the linear portion of the PAR response curve, when compared with those recorded for shade-restricted plants, fell short, whereas the rate of light-saturated photosynthesis was greater.

The applicability of current data to questions about the ecology of ancestral stock is debatable in any case. Needed now is a look at the more fundamental and evolutionarily conservative parameters of carbon balance, including finer details of the energy-capturing and -transducing apparatus. Certainly, no sweeping judgments are possible at this time, but there can be little doubt that culture requirements have shifted during phylogeny, and that canopy-adapted vegetation, although broadly divergent, is also convergent with regard to light use – that is, a light-use category can be comprised of both distantly related and closely allied species. Future data are likely to reveal that earthbound precursors of the modern epiphytic flora occupied various habitats, including deserts, wet savannas, and the lower reaches of dark forests. It seems that two or more confamilial lineages can have had different types of ancestral habitats, as appears to be true for Bromeliaceae (Benzing et al. 1985). One point is now indisputable, however: Not all epiphytes are heliophytes. Their foliage, like the high-exposure and shade-adapted phorophyte leaves produced in a single dense tree crown, is suited for just about any level of PAR that exists in a forest canopy.

Evergreenness/sclerophylly

This final section concerns differences in leaf structure and life-span and why these features reveal an important distinction between the carbon budgets of trees and the plants they support. Epiphyte foliage tends to be tougher, more damage-resistant, and longer-lived. Exceptional species with more mesomorphic foliage either are deciduous (e.g., *Catasetum,*, some ferns) or anchor only in wet forests (e.g., many other ferns, *Begonia,* some gesneriads). An economic analogy focusing on cost-effective photosynthesis provides a useful perspective. Fundamental to the model is the assumption that energy return on investment in biomass determines fitness. Committed resources are water, N, and phosphorus (P); the payback currency is fixed carbon. Leaf cost mounts as drought requires increased mechanical strength to resist wear and tear, tighter epidermal barriers to reduce transpiration, and more defensive chemistry to counter predators. Investment is recovered and profit realized through terms of service dictated by leaf cost and amortization rate – a full season or more in the case of the most stress-adapted leaf but much less for the cheaper more porous one with its comparably higher A_{max}. The model also postulates that nutrient insufficiency drives plant evolution in the same direction as does drought. The type of foliage that combats aridity is also needed to keep scarce N and P in prolonged service; these costly commodities require that returns be maximized. Because drought and infertility often co-occur and promote the same plant characteristics, sorting out the effect of either one on a plant is difficult.

Although dissimilar leaf texture and longevity indicate dissimilar productivity and stress tolerance in epiphyte versus phorophyte, the magnitude of these differences and implication for resource use efficiency remain obscure. Data are needed to understand how epiphytes influence mineral cycling, economy, and energetics of hosting forests. Almost certainly, supports are more continuously supplied with moisture than are most of their crown residents and thus would be expected to – and largely do – utilize shorter-lived foliage for speedy but water-expensive CO_2 fixation. Nutritional distinctions between the two floras are not as clear-cut as is water balance; an abundance of nutrients may actually be more available to tank bromeliads and myrmecotrophs than to some supports, and more information will be brought to bear on these subjects shortly.

3 Water balance

Aspects of foliage reveal the fact that stress, particularly drought, is a powerful selective force on plants anchored in tree crowns. Whereas phorophytes are usually characterized by mesomorphic leaves and C_3 photosynthesis, their epiphytes tend to xeromorphism and unusual mechanisms for procuring, as well as greater capacity for storing, water. Therefore, if accommodations to aridity can be identified, so will many of the epiphytes' most distinguishing features. In this chapter, that challenge is met by first laying some groundwork and then examining the nature of water balance mechanisms and their influence on overall epiphyte biology.

Water use and conservation: defense against drought

All land plants must expend water in order to create biomass. As stomata open for CO_2 influx, water vapor exits at a much higher rate. Xerophytic forms manage this unavoidable trade-off with suprising success: Their transpiration ratios (TRs) are in the neighborhood of 100:1, and exceptional performers do considerably better (Table 3.1). But water economy always has its price; productivity is slowed as leaf conductance falls – the lower the TR, the slower it is. In contrast, species native to humid habitats or those arid-land dwellers whose active phase coincides with the rainy season expend as much as 1000 g of water for each gram of dry matter they create. These drought-sensitive taxa serve notice that parsimonious water use is not always the best mechanism and can be decidedly disadvantageous when moisture is plentiful.

No single pattern of gas exchange is beneficial everywhere. Profligate water use, for instance, is obligatory for dominant vegetation in crowded habitats; without it, growth would be too slow to provide competitive advantage. Alternatively, resource-deficient sites mandate a very different kind of performance: As environmental rigor mounts, stress tolerance becomes the major arbiter of success, exceeding the capacity to produce more roots or shade out neighbors. Moisture is probably the resource in shortest supply for most epiphytes; thus aridity constitutes the most formidable barrier to their survival. Griffiths et al. (1986; Table 3.1) noted TR

Table 3.1. **Parameters of local climate versus plant performance (uptake and internal cycling of CO_2, leaf titratable acidity, transpiration, and TR) of six Trinidad bromeliads**

Species	Local climate	CO_2 uptake from 1800 to 0600 h [mmol (kg fr wt)$^{-1}$]	Dawn–dusk ΔH^+ (mol m^{-3})	Internal CO_2 recycling as % of total CO_2 fixed	Transpiration from 1800 to 0600 h (mmol HOH m^{-2})	Transpiration ratio for dark-period CO_2 uptake (HOH/CO_2, w/w)
Aechmea aquilega	dry	6.1	113	89	0.78	49
	wet	36.2	393	83	0.85	10
	wet	49.8	237	61	1.17	10
Aechmea nudicaulis	dry	46.4	301	71	3.08	27
	wet	16.0	239	87	9.13	233
	wet	44.2	233	65	3.29	34
Aechmea fendleri	wet	37.7	230	47	14.62	159
	wet	45.5	332	56	4.24	38
Tillandsia elongata	dry	15.9	271	82	2.80	72
Tillandsia utriculata	wet	3.9	251	95	2.28	239
Guzmania monostachia	wet	0.6	70	94	2.27	422

Source: Griffiths et al. 1986.

ratios below 50:1 for several CAM bromeliads in Trinidad – lower than those quoted for many CAM terrestrials (Osmond 1978). The most efficient species expended only about 10 g of water for each gram of CO_2 consumed during dark acidification (phase I of CAM, normally the period of greatest foliar conductance). Performance varied within and between species, however; there were no consistent relationships among magnitude of CO_2 fixation, WUE, and habitat type.

Structural adaptations to drought

Vascular plants maintain serviceable water economy by utilizing a variety of coordinated structural and dynamic measures. The uncommon poikilohydrous epiphyte with its lightly insulated foliage loses water quickly; desiccation is not fatal, however. The less tolerant homoiohydrous species must deploy special anatomical features (Fig. 3.1) as a first line of defense against lethal drying; xerophytes possess stout-walled epidermal cells covered by a thick evaporation-retarding cuticle. Mesophyll is commonly differentiated into hypodermis and chlorenchyma. A largely achlorophyllous, multilayered, adaxial epidermis occupies up to 80% of total leaf volume in some epiphytic *Peperomia* (Kaul 1977). Much of the leaf interior assumes a similar water-storage role in numerous dry-growing Bromeliaceae, Gesneriaceae, and Orchidaceae.

Succulence and CAM usually coincide, but some epiphytes are exceptions. A sizable hypodermis is indeed a poor indicator of CAM in Orchidaceae, just as Guralnick et al. (1986) and Nishio and Ting (1987) discovered was the case for certain Gesneriaceae and *Peperomia.* Capacity for hypodermal development but not for nocturnal CO_2 fixation increased with altitude among surveyed orchids in Papua, New Guinea (Earnshaw et al. 1987). Only two of over 30 lowland rain forest inhabitants possessed hypodermal layers, yet in most instances $\delta^{13}C$ values indicated considerable CAM activity. Populations farther up the transect (i.e., at 2600–3150 m) tended toward differentiated mesophyll; they were primarily C_3 species, the most negative $\delta^{13}C$ value being −24.2‰. Orchids at the highest sites (at 3400–3600 m) were also C_3 types and grew mostly in earth or on rocks. Although succulent, few possessed a foliar hypodermis. Earnshaw et al. cited seasonal montane rainfall as the impetus for high water-storage capacity; low daytime temperatures purportedly prevented CAM.

Additional less conspicuous xeromorphic features promoting water retention include recessed stomata, elaborate indumenta, and reflective surfaces. But these devices may be more common in terrestrial than in arboreal veg-

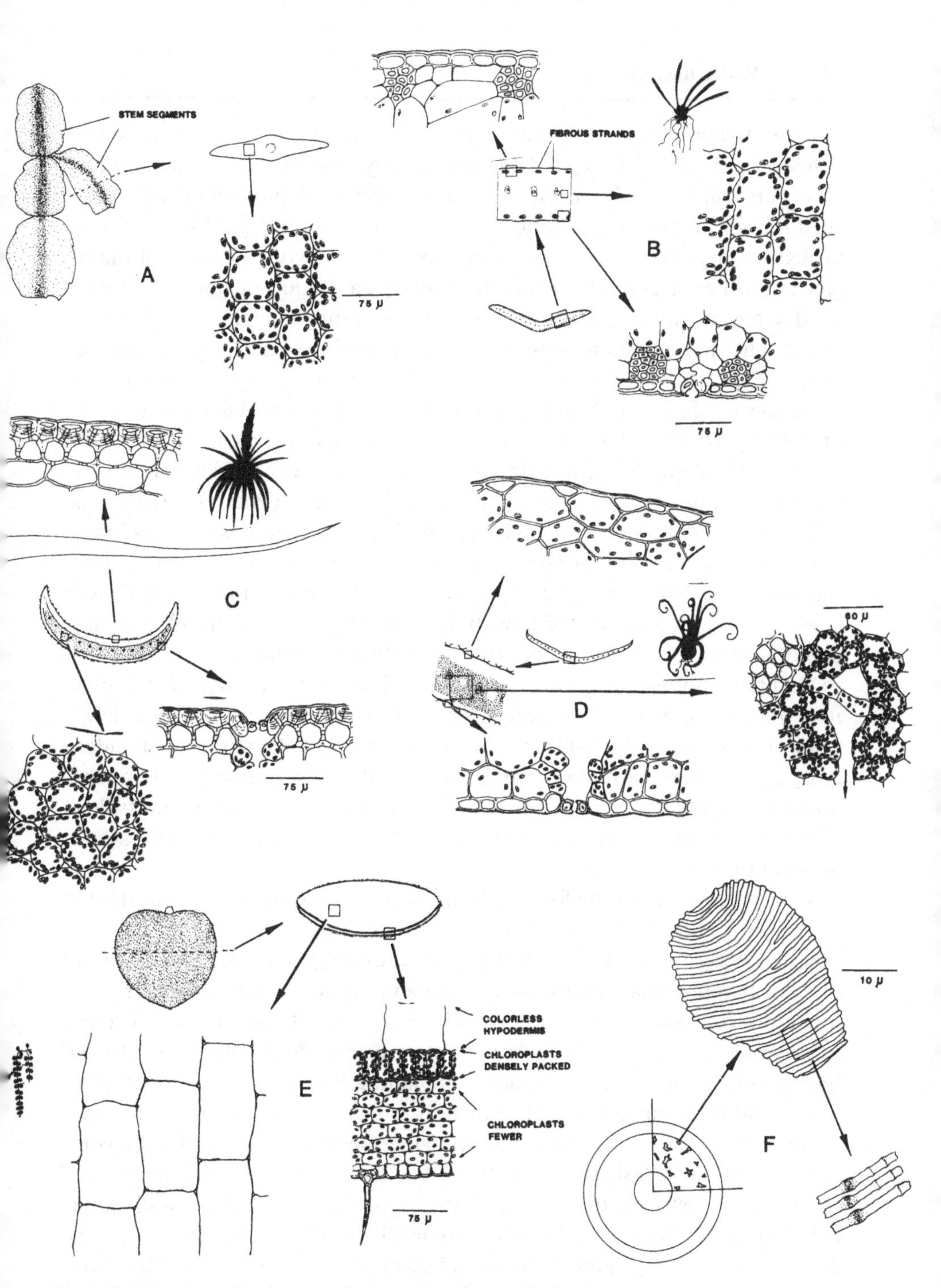

Figure 3.1. Xeromorphic features of selected epiphytes: (A) *Zygocactus* sp. cladophyll in transverse section (T.S.) (×¼); (B) *Encyclia tampensis* leaf in T.S. (×⅛); (C) *Tillandsia concolor* leaf in T.S. (×½); (D) *Tillandsia streptophylla* leaf in T.S. (×⅓); (E) *Codonanthe paula* leaf in T.S. (×½). (F) Spirally thickened idioblast from an orchid root.

etation. Xylem conduits, locally narrowed to ensure that mounting water deficits would first kill expendable organs (e.g., leaves vs. meristems; Zimmerman 1983) probably exist, and may underlie phenomena such as the shedding of unspecialized rather than ant-inhabited foliage (Fig. 4.24c) by droughted myrmecotrophic *Dischidia* species. Likewise, narrow cell diameter and other structural features that retard cavitation of tracheary tissue on dry ground no doubt characterize many epiphytes. One would expect primary hemiepiphytes to present a special case if indeed the quality of their moisture supply improved later in life. But, although terrestrial roots of these species do in fact bear a superficial resemblance to liana stems, those of *Ficus pertusa* and *F. trigonata* descending *Copernicia tectorum* palm trunks in the llanos of Venezuela do not contain broader vessels than do the thinner axes of younger, wholly epiphytic specimens (Putz and Holbrook 1987). There is, however, the apparent shift in water use described below.

Better understanding of epiphyte–water relations requires more data on the mechanical properties of key tissues and organs. Schmidt and Kaiser (1987) have already observed how differentiated elastic moduli seem to help coordinate water-storing hypodermis and chlorenchyma function in order to stabilize photosynthesis during drought. Leaves of succulent *Peperomia magnoliaefolia,* after an induced 50% water loss, exhibited 75–85% dehydration of colorless adaxial layers versus only 15–25% dehydration of underlying green layers. Osmolarity and ion concentrations remained similar in both tissues, indicating that weaker, collapsible cell walls allowed mass flow of contents from a relatively inert water-storage tissue to the more active desiccation-sensitive one.

A rather novel cell type for improving water storage is widespread in epiphytes by virtue of its presence in so many orchids. Spirally thickened idioblasts (Pridgeon 1981; Fig. 3.1F) of various shapes are scattered within diverse parenchyma tissues. Most frequent locations are chlorenchyma, including root and stem cortices, and the colorless hypodermis of leaves. Compared to the more conventional, weaker-walled, collapsible, living hypodermal cells of many succulents (including *Peperomia magnoliaefolia*), these rigid cells can shunt their entire store of moisture to nearby desiccation-sensitive tissue without compromising mechanical support. The contents of an idioblast are exported after bulk-leaf water potential (Ψ_L) becomes low enough to "air-seed" the filled interior through wall pores between spiral thickenings (Zimmerman 1983). These tiny openings, formerly occupied by plasmodesmata, are so small that intruding bubbles will simply cavitate rather than fill the lumina with air. Thickenings opposite intercellular spaces (Benzing et al. 1983; Fig. 4.10) probably prevent massive air embolisms that would be difficult or impossible to dissolve when opportunity for refilling returns.

Dynamic defenses against drought

Leaf anatomy and type of photosynthetic pathway alone cannot maintain acceptable water use without additional fine tuning of the transpiration stream. Of three major interrelated dynamic phenomena that help plants avoid desiccation and maximize use of photosynthetic capacity – that is, adjustment of diffusive conductance, osmotic regulation, and CO_2 fixation via β-carboxylation – the first is the most pervasive. (Information on the dynamics of elastic moduli is too scanty to warrant discussion of any potential benefit to epiphytes.)

1. Adjustment of conductance

Transpiration is limited more by plant activity than is photosynthesis because stomatal aperture often reflects CO_2 demand. As photosynthetic capacity increases in response to higher light intensity for instance, diffusive conductance mounts, mediated by changes in guard cell Ψ_S (solute potential); should PAR level diminish, CO_2 consumption decreases and stomatal aperture narrows. The C_i needed to saturate existing carboxylation capacity is somehow translated into altered guard cell turgor pressure (Ψ_P) according to the optimization hypothesis (Farquhar and Sharkey 1982). But at other times, stomatal closure suppresses water loss, irrespective of photosynthetic capacity or plant water status. High temperature and low relative humidity (RH), both of which increase transpiration without promoting additional photosynthesis, induce narrowing of stomata. Generally, however, guard cell movements minimize transpiration more than they impede photosynthetic output, and WUE remains relatively constant (Fig. 3.18) despite fluctuating conditions.

A few studies have examined the role of stomata as humidity sensors in such epiphytes as the deciduous C_3 orchid *Catasetum integerrimum* and some bromeliads, including the CAM xerophytes *Tillandsia usneoides* and *T. recurvata.* Carbon dioxide consumption by the *Catasetum* plummeted under steady high-level PAR immediately after RH was abruptly depressed from about 70% to 40% (Benzing et al. 1982a). When RH again approached 100% in the sample cuvette, the orchid's gas exchange returned to pretreatment rates within an hour or two. *Tillandsia recurvata* (Lange and Medina 1979) responded to drier night air in much the same fashion as did *T. usneoides* (Martin and Siedow 1981; Fig. 3.2). Griffiths et al. (1986) noted

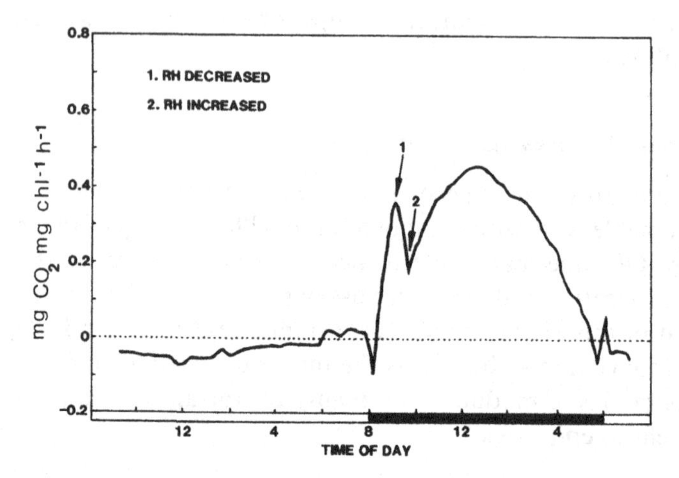

Figure 3.2. Effects of relative humidity (RH) on CO_2 exchange in *Tillandsia usneoides*. Daytime RH was 55%, nightime was 92% up to arrow #1, at which time RH was decreased to 72%. At arrow #2, RH was elevated to 95% for the remainder of the night. (After Martin and Siedow 1981.)

declining leaf conductance in the CAM bromeliads *Aechmea nudicaulis* and *A. aquilega* as dry high-pressure weather cells passed over Trinidad forests. Other epiphytes almost certainly possess similar sensitivities; details will probably vary with the taxon and nature of its moisture supply. Also of interest is the influence on an epiphyte's foliar conductance of conditions in rooting media; stomata of some terrestrial species apparently respond to hormones originating from roots that sense moisture status in earth. Pulse-supplied epiphytes would need an unusually prompt signal–response system to keep pace with changing moisture supplies. Conditions in substrata are not, of course, important to the bromeliad with absorptive foliage.

2. Osmotic adjustment

The effects of mild moisture stress are often mitigated by a plant's ability to regulate osmotic potential. Were appropriate tissues unable to alter Ψ_S so as to maintain turgor under these circumstances, chloroplasts would soon outstrip their CO_2 supply, and yield would suffer. Two Malaysian epiphytic ferns examined by Sinclair (1983b) exhibited somewhat more negative Ψ_S than expected as relative water content (RWC, expressed as percent saturation) decreased. Concentration effects of tissue shrinkage alone seem unlikely to have caused drops of -0.8 to -2.75 megapascals (MPa) for *Pyrrosia adnascens* when droughted for 15 days and approximately

-0.30 to -1.4 MPa after 30 days for *P. angustata.* Data points were too scattered to justify a definitive statement about adjustment, however. Parallel experiments with the epiphytic orchids *Dendrobium tortile* and *Eria velutina* definitely indicated no changes in Ψ_S as desiccation progressed.

The most common organic molecules utilized for osmotic adjustment by vascular plants are proline, betaines, and other low-molecular-weight, usually N-containing, compounds, but epiphytes have not been reported as users. Osmotica among those terrestrials studied so far vary with cost to the plant and suitability in particular habitats (Raven 1985). Quantity and timing of the moisture supply and scarcity of N may have narrowed the options for some epiphytes. Perhaps this question is less critical in tree-crown habitats, however; compared to much other vegetation, epiphytes maintain dilute cell sap, possibly to heighten stomatal sensitivity.

3. Carbon dioxide fixation via β-carboxylation.

The third dynamic response to aridity involves water-conserving CAM, many details of which have already been covered. Important effects on acquisition and storage of water went unmentioned, however. Plants that fix CO_2 at night reach peak Ψ_S in early morning precisely when dew-point temperatures occur. Epiphytes positioned for greatest benefit are those lacking access to continuous supplies of moisture and those equipped with absorptive foliage. Fully hydrated *Tillandsia recurvata* increased its water content via moistened shoots (Lüttge 1987) by 8.4% as malic acid accumulated over the night. Abundant citric acid, synthesized nocturnally by hemiepiphytic *Clusia rosea* (Popp et al. 1987), in addition to considerable malic acid, may accumulate primarily to improve water balance (Lüttge 1988). Although unsuitable as a vehicle for storage of carbon via CAM, citric acid is the cheaper osmoticum, as compared to the malic acid derived from polysaccharide reserves. Citric acid production also provides the less costly pathway to recapture of dark-respired CO_2.

Water balance categories

Poikilohydry

Best known of the poikilohydrous epiphytes is *Polypodium polypodioides* (Fig. 1.6), a small rhizomatous fern whose range extends from the east-central United States into tropical America. Fronds are erect and green during humid periods but soon curl and turn brown with drought (Fig. 3.3). Reputedly absorptive peltate hairs resembling those in canopy-adapted Bra-

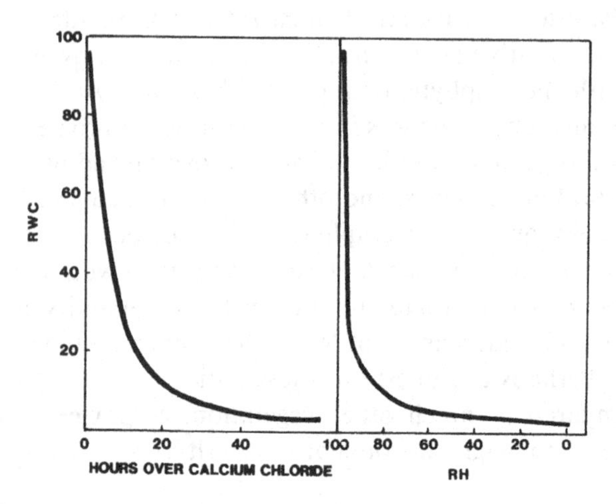

Figure 3.3. Desiccation of *Polypodium polypodioides* over calcium chloride at different RHs. (After Stuart 1969.)

zilian *Pleopeltis angusta* and two *Polypodium* species (Müller, Starnecker, and Winkler 1981) cover both foliar surfaces, but their role here is unclear. Stuart (1969) believed that they promoted uptake in *Polypodium polypodioides,* but that xylem supply was also important because detached fronds rehydrated more slowly than did intact plants. Whether refilling occurred from wetting or from saturated air, moisture content and metabolic activity were tightly coupled. When specimens, dehydrated to 2–5% of RWC, were placed in water, they recovered photosynthetic capacity and full hydrature within hours (Fig. 3.4). Cytoplasmic integrity was conserved in part by vacuolar sap that solidified on drying, thus preventing complete collapse of the protoplast.

Numerous bark-dwelling and epiphyllous algae, bryophytes (Fig. 7.6), and lichens are poikilohydrous; timing probably explains why *Polypodium polypodioides,* a few other pteridophytes, and possibly a couple of gesneriads are the only resurrection epiphytes. Alpert and Oechel (1985) found that the lithophilous resurrection moss *Grimmia laevigata* failed to maintain a positive carbon balance in overly arid sites because its thalli were driest by day, a condition more conducive to respiration than to photosynthesis. Options for poikilohydrous species diminish according to wet/dry cycling and degree of repair needed for recovery, both of which in turn vary according to phylogenetic status. Whereas typical rehydrating bryophytes recover full photosynthetic capacity within hours, desiccated resurrectable pteridophytes need a full day or so, and the few poikilohydrous angiosperms (grasses at

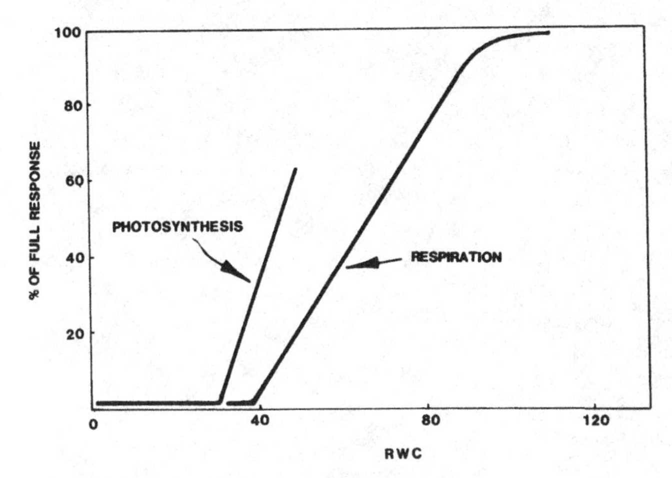

Figure 3.4. Recovery of photosynthesis and respiration in rehydrating fronds of *Polypodium polypodioides.* (From Stuart 1969.)

least) must rebuild a damaged photosynthetic apparatus, a process that requires even more time.

Homoiohydry

There are four categories of homoiohydrous epiphytes.

1. Hygrophytes

Delicate structure and presumably C_3 photosynthesis characterize hygrophytes, and desiccation is rapid in bright sun or in any but the highest humidity. Habitats are pluvial to wet forests or such sites as moist, cool, shaded tree bases in somewhat drier communities (Fig. 3.5). Humus suspended on bark supports their roots. Some taxa (among which ferns, especially Hymenophyllaceae, dominate) may be more desiccation-tolerant than assignment to this category would warrant. Certain South American *Hymenophyllum* survive water losses exceeding 90% (Gessner 1956; Walter 1971), but neither they nor their semipoikilohydrous relatives show evidence of tolerating repeated dehydration.

2. Mesophytes

Mesophytes are evergreen and vulnerable to all but modest drought. Their foliage is somewhat thicker and more resistant to desiccation than that of the hygrophytes. Moderate vigor is supported by C_3 photosynthesis,

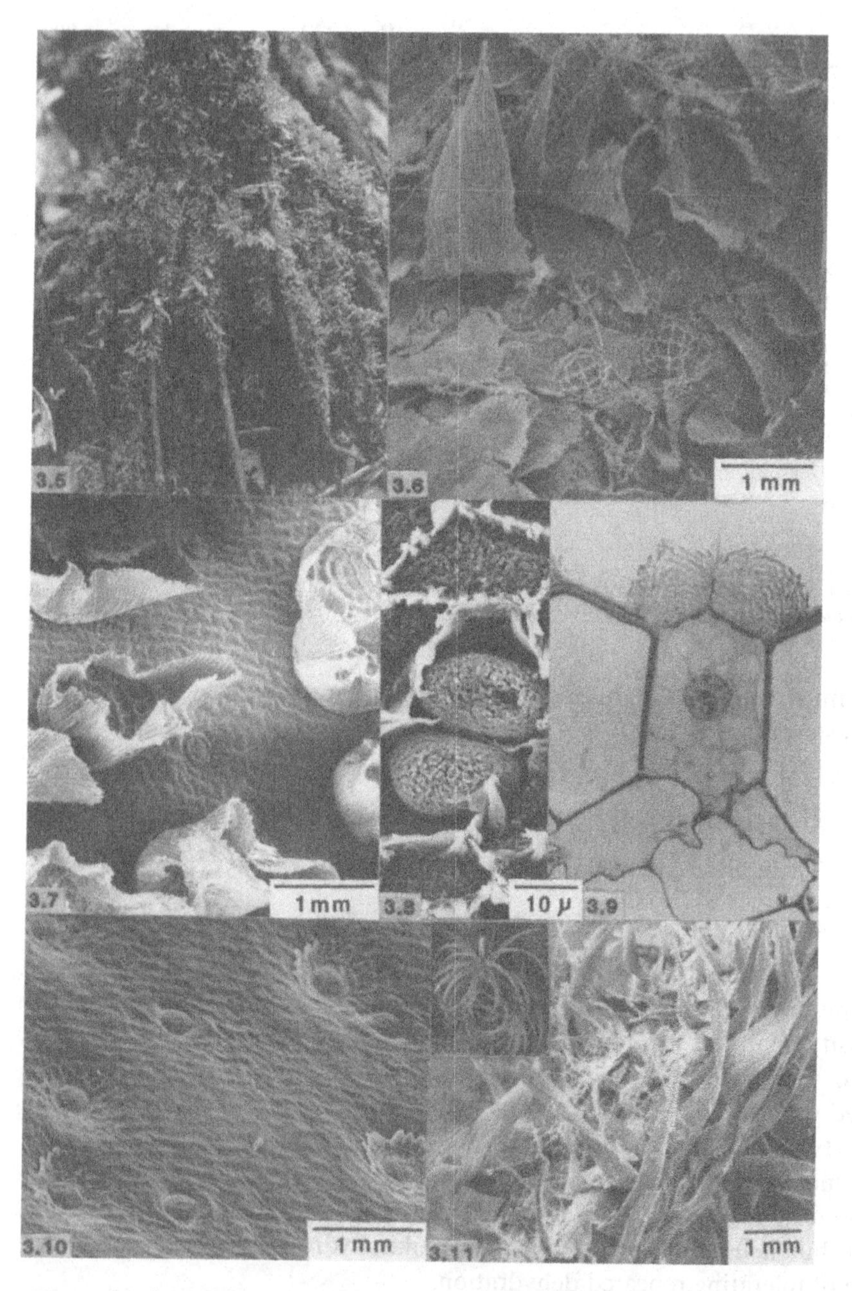

Figures 3.5–3.11. (3.5) A mixed colony of bryophytes and filmy ferns growing on the base of a tree in primary Amazonian rain forest. (3.6) Absorbing trichomes with attenuated shields covering the leaves of the atmospheric bromeliad *Tillandsia tectorum* (×120). (3.7) Absorbing trichomes on the abaxial leaf surface of the atmospheric bromeliad *Tillandsia ionantha* (×175). (3.8) Paradermal section through the root of the orchid *Sobralia macrantha* illustrating the tilosomes (×250). (3.9) Passage cell and tilosome of *Sobralia macrantha* (×275). (3.10) Abaxial leaf surface of the tank-forming bromeliad *Catopsis nutans* showing small, scattered absorbing trichomes (×120). (3.11) Shoot (inset) and portion of a leaf of the atmospheric bromeliad *Tillandsia tectorum* (×50).

perhaps assisted by CAM-cycling and even CAM-idling under stress (e.g., some Gesneriaceae and *Peperomia*). There is no precise demarcation among the meosphytes, hygrophytes, and xerophytes; thus it is not possible to determine the exact size of any of the groups. Many of the sturdier ferns are mesophytic, as are broadleaf tillandsioid bromeliads with shallow tanks (e.g., *Tillandsia imperialis* (Fig. 4.2A), *Guzmania lingulata*). Soft-leaf aroids, gesneriads, and others with only moderately effective mechanical barriers to water loss also belong here.

3. Xerophytes

This subset contains two types of epiphytes: those equipped to avoid, and those adapted to endure, prolonged drought. In reality, drought avoiders (Fig. 1.15) are not xerophytes at all; they are seasonal mesophytes that restrict most of their transpiration, hence vegetative activity, to humid periods. As the substratum dries and heavy water use is no longer supportable, poorly insulated but protective (i.e., mesomorphic) foliage is shed. Senescence is cued by photoperiod and possibly other climatic fluctuations. Less vulnerable perenniating organs lapse into dormancy during dry seasons, storing such reserves as carbohydrates and generating the meristems needed to renew growth when favorable weather returns. Orchidaceae and Polypodiaceae, and less frequently a few other families, contain drought-deciduous epiphytes.

Drought endurers are by far the most numerous and best known of the two "xerophytic" groups. Conspicuous hallmarks are leathery to stiff, thick, long-lived foliage and stout, durable roots. Pseudobulbous shoots of a *Laeliocattleya* cultivar, for instance, have functions for up to seven years although the average is probably three to four seasons for dry-growing evergreen orchids (Poole and Sheehan 1982). Root systems of a *Cattleya* hybrid have remained absorptive for at least four years. Occasional taxa are aphyllous stem (*Vanilla;* Fig. 4.23F) or root (*Polyradicion;* Fig. 4.23C) succulents. Judged on growth in culture, these plants, like earth-rooted counterparts, mature slowly. Habitats range from exposed sites at the tops of trees in humid communities to various locations in the driest microphyllous forests, savannas, and deserts.

Features responsible for the exceptional performance of drought-enduring epiphytes, especially their ability to photosynthesize while moisture is scarce, are numerous and varied. Xeromorphy is routine, but there are no consistent patterns; there are even a few surprises. Those atmospheric bromeliads that exhibited such delicate leaf boundaries (Fig. 3.1D) that Tomlinson (1969) labeled them "hydromorphic" were also found to feature other foliar specializations related to bromeliad epiphytism/xerophytism. Leaves of drought-enduring epiphytes are simple, and margins are entire or less often armed; mesophyll is usually differentiated into water storage and chlorenchyma tissue (Fig. 3.1E). The sine qua non of this group was once thought to be CAM, but recent data like that just described for Papuan Orchidaceae and the work by Ting and others have revealed a more complicated situation.

Sinclair's (1983a,b; 1984) examination of two ferns and three orchids in Malaysia has provided one of the most comprehensive work-ups to date on epiphyte water relations. All five subjects were evergreen and comfortably assigned to the xerophyte category even though their native habitat is aseasonal forest where uncommon dry spells rarely last more than two weeks. These plants were definitely not among the most drought-challenged of the evergreen epiphytes; nevertheless, they all bore moderately thick leaves (0.44–1.81 mm) and, in all but one instance, exhibited characteristics that usually signal high WUE. The orchid leaves showed distinct CAM-type acid fluctuations; data for *Pyrrosia adnascens* gave a flatter curve but one still highest at night, especially in droughted material. Whether watered well or partially desiccated, *P. angustata* appeared to be a C_3 plant.

Under induced moisture stress, both ferns quickly exhibited decreased leaf conductance and lower Ψ_L (Fig. 3.12). The orchids kept opening stomata far longer than did *Pyrrosia,* in part due to a fourfold slower water loss. Substantial leaf conductance in *Eria velutina* was noted day and night until desiccation eliminated all but nocturnal gas exchange. After 25 days, RWC still exceeded 65%. Apparently, all three orchids were able to fix some atmospheric CO_2 throughout the run, whereas conductance in *Pyrrosia angustata* nearly disappeared after the first few days. Cuticular transpiration continued at a diminished rate, however, and RWC fell to 13% by day 30. Prestress levels of Ψ_L, RWC, and conductance were regained within two to three days following irrigation. Fern rehydration was 95% complete after just 24 h.

Wong and Hew (1976) demonstrated CAM-like acid metabolism in *Pyrrosia longifolia* and closely related *Drymoglossum piloselloides,* an unexpected discovery considering the drought performance of *Pyrrosia angustata* and *P. adnascens.* Resurrection-like behavior in Sinclair's two fern subjects, "shriveled fronds" in droughted *Drymoglossum* (Hew 1984), and a photosynthetic system that conserves moisture in so many xerophytic angiosperms – all suggest that at least some epiphytic pteridophytes possess an interesting and perhaps unique array of water balance mechanisms. These findings surely leave no doubt that different water relations sustain different epiphytes facing the same climatic stress; *Pyrrosia* and *Dendro-*

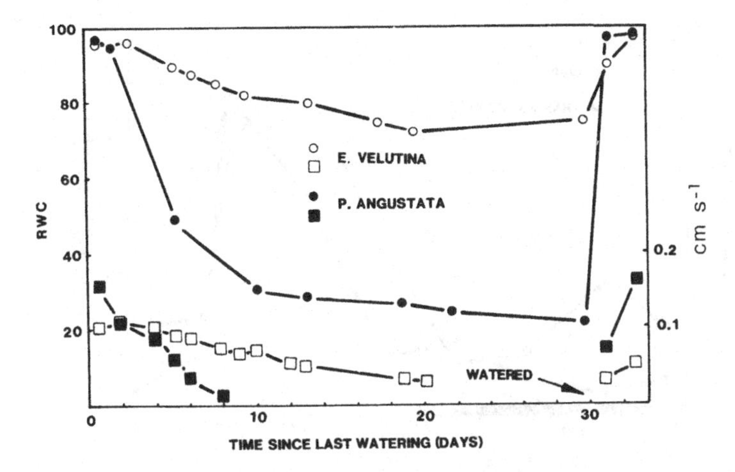

Figure 3.12. Changes in relative water content and diffusive conductance while water was first withheld from, and then provided to, the orchid *Eria velutina* and the fern *Pyrrosia angustata*. (After Sinclair 1983a.)

bium often form mixed colonies and indeed grew intertwined during Sinclair's drying runs.

Differences in WUE and related physiology and ecology of drought-enduring and -avoiding epiphytes were illustrated by comparing two neotropical orchids, *Catasetum integerrimum* (Fig. 1.15) and *Encyclia tampensis* (Benzing et al. 1982a; Fig. 1.10). The latter is a small to medium-size, drought-enduring bark epiphyte whose adult stages consist of numerous, closely attached, determinate shoots, the youngest of which bear one or two coriaceous, slender, elongate one- to four-year-old leaves. Xeromorphy is pronounced. *Catasetum integerrimum* is also sympodial. Each elongate, multinoded, fleshy shoot supports 6–12 thin plicate leaves through the first seven to nine months of life. During this leafy phase, foliated stems mature to full thickness and generate inflorescences bearing staminate or pistillate flowers from basal axils; an extensive velamentous root system matures along with the subtending shoot. Leaves eventually senesce over a few weeks, then disarticulate at preformed abscission zones just above sheathing bases. Roots die at about the same time but remain in place, adding to the absorptive medium available for young ramets. New shoots appear each growing season and persist for three to four more years, gradually shrinking as they, too, support successive attached ramets. Much of the root system is embedded in rotting wood or organic debris, nearly cut off from light or air – a habit which determines this species' status as a humus epiphyte. Where

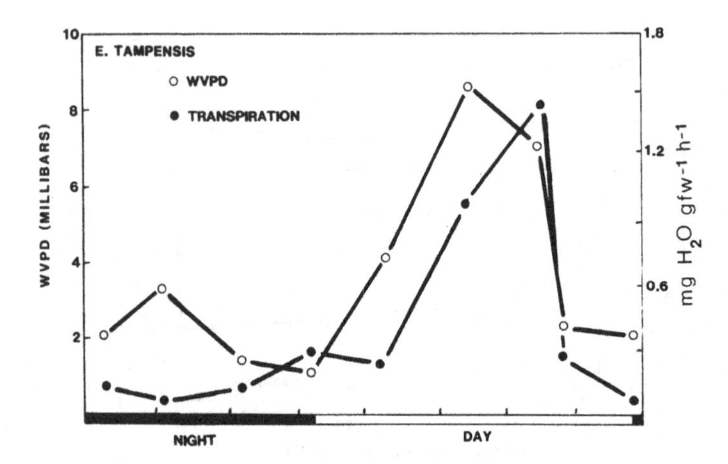

Figure 3.13. Water loss over a full day by the orchid *Encyclia tampensis.* (After Benzing et al. 1982a.)

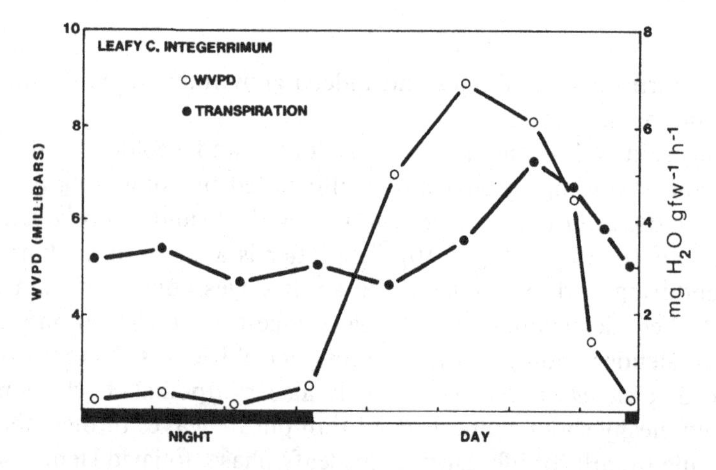

Figure 3.14. Water loss over a day by the orchid *Catasetum integerrimum* in the leafy condition. (After Benzing et al. 1982a.)

the substratum is less penetrable, an exposed mass of upward-growing, determinate roots produces a trash-basket impoundment.

Encyclia tampensis proved to be a CAM species (Fig. 3.16). Transpiration is slow day and night (Fig. 3.13). *Catasetum integerrimum* exhibits typical C_3 photosynthesis while leafy (Fig.3.16); daily carbon gain greatly exceeds that of *Encyclia,* but water use is heavy (Fig. 3.14). Dormant *Catasetum* specimens, reduced to naked pseudobulbs bearing no living roots, lose little CO_2 or moisture (Figs. 3.15, 3.16). Assimilation rate clearly demonstrates

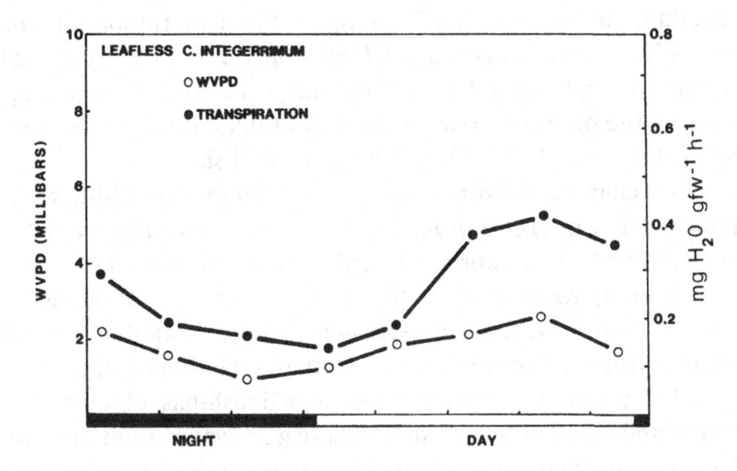

Figure 3.15. Water loss over a day by the orchid *Catasetum integerrimum* while dormant. (After Benzing et al. 1982a.)

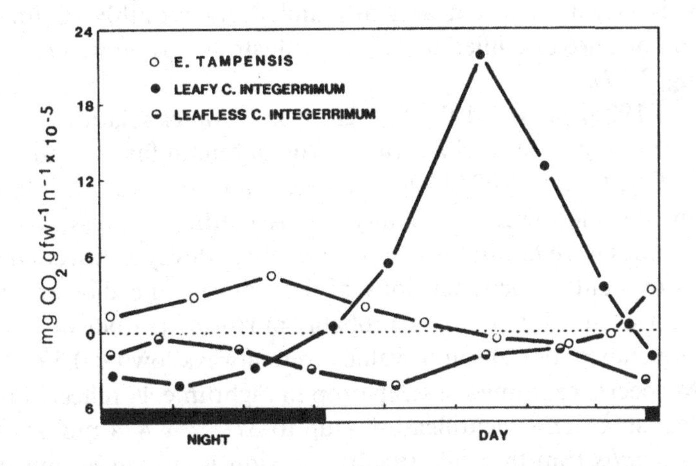

Figure 3.16. CO_2 exchange by *Encyclia tampensis* and leafy and dormant *Catasetum integerrimum* specimens over the course of a full day. (After Benzing et al. 1982a.)

why this species, although leafless for several months each year, matures and grows faster than does drought-enduring *Encyclia tampensis.* Poor WUE explains why activity is seasonal and deep humus the required substratum.

4. Impounding epiphytes

As a rule, leaves provide impoundment capacity; roots do so less often. Bromeliaceae illustrate the tank in its highest refinement; about 1000 species, roughly half of the family, produce rosulate tank shoots comprised of tight, overlapping leaves with inflated bases (Fig. 1.2). Impoundment volume can be substantial; a single large *Glomeropitcairnia erectiflora* has been said to contain several liters of fluid (Pittendrigh 1948), but the average for all species is on the order of a few hundred milliliters. Reasons for variation in tank shape (Fig. 4.25A,B,E–G) will be addressed shortly.

Looser leaves characterize other tank species that intercept little water but considerable litter (e.g., *Anthurium,* Fig. 1.5; *Asplenium*). *Platycerium* (Fig. 4.19) and *Drynaria* feature appressed sterile fronds that form a plant-to-substratum catchment; roots perfuse this cavity, tapping impounded litter moistened with stemflow. Root masses such as those created by several velamentous anthuriums, *Catasetum,* and similar orchid genera, and certain ferns (Fig. 1.19), provide spongelike reservoirs. Trash-basket formation can be facultative and favored by dry substrata (e.g., *Cyrtopodium punctatum*). The impounding mechanism, effective as it may be, is not without limitation, however; tank bromeliads give way to atmospherics as rainfall becomes too scarce. Impoundment types in Ecuador generally occur where precipitation is at least 200 cm annually and 8 cm monthly (Gilmartin 1983). The importance of a filled tank is well illustrated by adult *Tillandsia deppeana* (Fig. 3.17).

Smith et al. (1985) provided the first glimpse of water relations in tank epiphytes by studying 11 bromeliads in a Trinidad humid forest (Table 2.3). Assays of C_3, CAM, and C_3–CAM forms were made at 4- to 6-h intervals over two rainy days followed by a sunny one. According to pressure bomb readings, Ψ_L values were faithful to photosynthetic pathway and secondarily to local environment. Xylem tensions peaked during the day or night, respectively, for C_3 and CAM plants, probably mirroring periods of greatest diffusive conductance, but absolute values were always low ($<$0.59 MPa). Among CAM species examined, a steep drop in nighttime Ψ_S reflected near-record (for leaf succulents) acidification – up to ΔH^+_{max} = 474 mol m^{-3} for *Aechmea nudicaulis* (Smith et al. 1986b). Tension was reduced and acid rhythms were diminished in CAM forms on the stormy days compared to the one clear day, illustrating how quickly energy reserves are affected by changing irradiance.

Guzmania monostachia (Fig. 2.15), the only C_3–CAM intermediate included in the study, performed according to the aridity of its surroundings. In a deeply shaded lowland forest, C_3-type fluctuations were recorded. The same species in a drier forest gap not far away behaved more like a CAM plant. Clearly, some measure of this bromeliad's wide range is based on its flexible water relations and variable photosynthetic mechanisms. Smith et al. (1985) failed to demonstrate why CAM and C_3 epiphytes are so

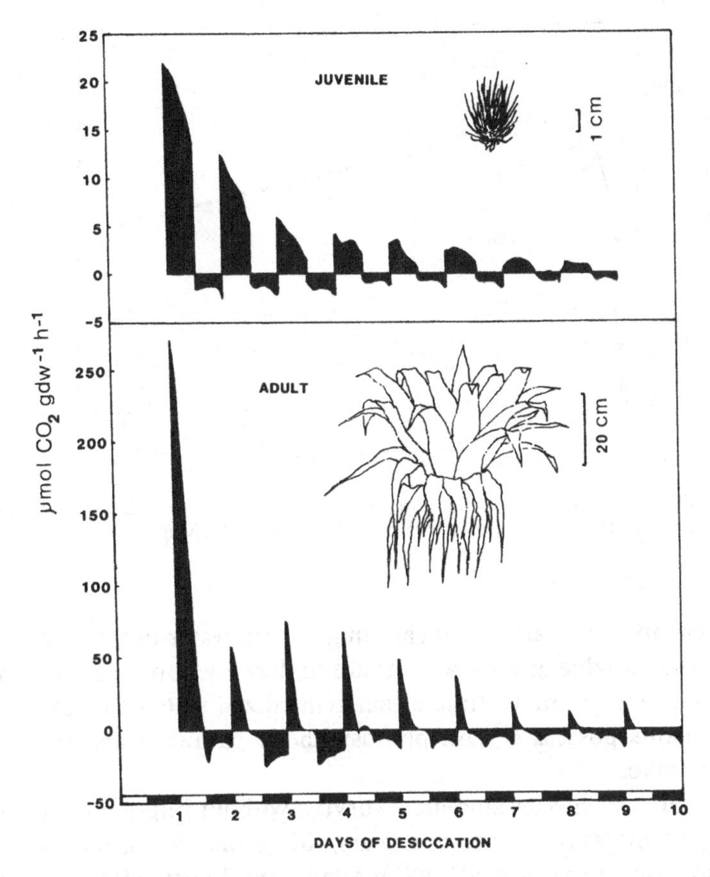

Figure 3.17. Diurnal patterns of CO_2 exchange in juvenile and adult *Tillandsia deppeana* over eight or nine days without watering. Irradiance = 650–900 μmol $m^{-2}s^{-1}$ PAR: T = 25°C/18°C day/night; RH = 50–60%/70–75% day/night. (After Adams and Martin 1986b.)

often vertically segregated in humid forests, as they were here. Values for Ψ_L indicated that none of the monitored species were under markedly greater moisture stress than any others despite their distinct carbon fixation pathways. Filled tanks must have been partly responsible. Subsequent examination of similar materials during the dry season complicated the question even more (Griffiths et al. 1986). At that time, TRs of the C_3 bromeliads approximated those of nearby CAM forms during dark (phase I) fixation and late afternoon (phase IV) activity. Broad intraspecific variation (e.g., TRs between 10 and 49 for different *Aechmea aquilega* populations; Table 3.1) confuses comparison of species. Under at least come conditions, C_3 and CAM bromeliads appear to be equally impressive drought performers in

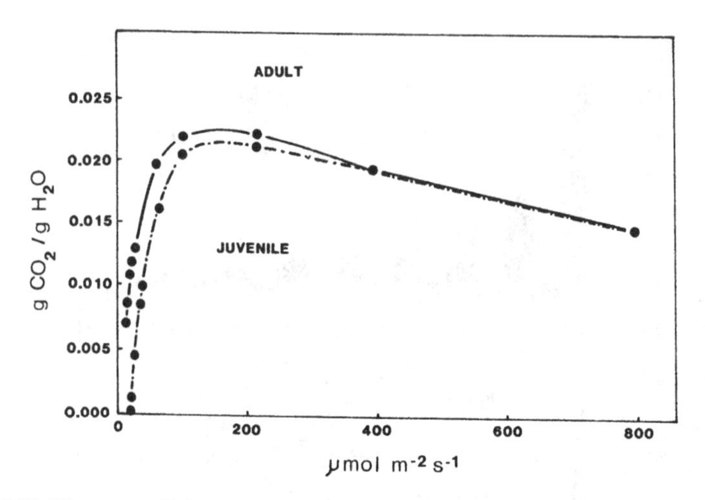

Figure 3.18. Water use efficiency of adult and juvenile *Tillandsia deppeana* under constant leaf temperature and RH (55%), varying PPFD. (After Adams and Martin 1986b.)

spite of their different patterns of carbon gain and water use. However, gas exchange is so variable and its benefits are so dependent on plant condition and rapidly changing microclimate that generalized statements about why particular plants possess certain photosynthetic pathways are not always very informative.

Seedlings of tank species somehow survive without impoundments until adult foliage appears (in one to five years); up to that point, they act more like atmospherics (Benzing et al. 1985). Adams and Martin (1986a) reported decreasing xeromorphy as heterophyllous *Tillandsia deppeana* developed. Juvenile foliage was succulent and densely trichomed, whereas that of mature individuals was broad, thin, and more glabrous. Neither juvenile nor mature *T. deppeana* engaged in CAM, but each possessed distinct water economies to match their dissimilar habits and modes of moisture procurement (Adams and Martin 1986b). Both stages fixed CO_2 (Fig. 3.17) with equally impressive WUEs (Fig. 3.18), comparable to those achieved by some C_4 and CAM species. Gas exchange data revealed additional atmospheric qualities in seedlings and indicated why growth rate increases with maturity. Maximum diffusive conductance, transpiration, and A_{max} were all up to a full order of magnitude greater when impoundment foliage was in place. Response to rewatering after a droughting period was equally instructive. During the first day of desiccation in a growth chamber, net photosynthesis by juveniles dropped to 40% of maximum (Fig. 3.17). Subsequent days brought further diminution, but carbon gain never ceased. Photosynthesis fell abruptly when tanks of adults were emptied, nearly stopping before the end of the first photoperiod. Net CO_2 exchange had become negative by about noon on almost every day beyond the second in a nine-day run. Mature plants drained of tank contents, lightly desiccated, and then rewatered, did not recover full photosynthesis until the next day. Stressed juveniles, on the other hand, regained pretreatment photosynthesis shortly after wet foliar surfaces dried – a capacity probably attributable to a denser trichome cover.

Great functional variety was discovered in seasonal low montane forest supporting tank forms and other epiphytes as well at Rancho Grande in northern Venezuela (Table 3.2). Subjects anchored side by side in a thin dry bryophyte–lichen mat covering the trunk of a large unidentified tree were drought-deciduous *Microgramma lycopodioides,* poikilohydrous *Pleopeltis astrolepis, Stelis* sp. (a presumed CAM orchid), an equally succulent *Peperomia* of unknown photosynthetic pathway, and *Vriesea platynema,* a C_3 tank-forming bromeliad. The bromeliad tanks were empty, and fronds of the *Pleopeltis* had lost enough moisture to curl; CO_2 exchange rates were monitored with a portable infrared gas analyzer. Of the five epiphytes, only *Microgramma* remained sufficiently hydrated to effect net carbon gain, despite signs of foliar senescence. Recovery was also uneven. On the day following a thorough irrigation of all the epiphytes and their rooting mats, *Pleopeltis* had regained substantial photosynthetic capacity. It took another full day for *Vriesea,* either as adults with filled tanks or as nonimpounding juveniles, to respond similarly; *Peperomia* and *Stelis* continued to evolve CO_2 by day and night throughout the 72-h experiment. Foliage on neighboring Guava trees underscored the advantage, as a water source during dry seasons, of ground over a canopy mat or tanks; it consistently achieved the highest of all recorded photosynthetic rates.

Moisture procurement

Epiphytes native to all but the wettest forests or anchored in deep humus must either create sizable impoundments with their own bodies or meet all needs during occasional contact with canopy fluids. These latter PS species are poorly known in terms of moisture procurement. The exceptions – several atmospheric bromeliads, a few orchids, and some ferns – possess analogous absorptive devices. Although different organs and tissues are involved, a single type of architecture is common to all. In each case, mini-impoundment is based upon a biphasic system: a superficial, nonliving,

Table 3.2. **Net photosynthesis and tissue composition of leaves of six epiphytes growing on a single dead tree trunk at Rancho Grande, Venezuela, January 8–10, 1988**

Species (photosynthetic pathway)	Unirrigated	A [net photosynthesis (μmol CO_2 $m^{-2}s^{-1}$)] on days following irrigation: 1	2	3	Leaf composition ($g\ m^{-2}$) N	P
Psidium guajava (C_3)	27.3 ± 9.6	29.9 ± 12.5	—	—	0.64	0.079
Vriesea platynema (C_3)	0	0	10.02 ± 2.60	6.96 ± 2.11	0.26	0.027
Stelis sp.[a] (CAM)	0	0	0	0	—	—
Peperomia sp.[a] (?)	0	0	0	0	—	—
Microgramma lycopodioides (drought-deciduous fern)	3.70 ± 1.7	—	—	—	0.142	0.0214
Pleopeltis astrolepis (resurrection fern)	0	7.15 ± 0.82	—	6.53 ± 0.98	0.104	0.0056

[a]Gas exchange monitored day and night.
Note: The dry mat of suspended humus supporting these plants was soaked with water after the first measurements.
Source: Unpublished data.

imbibing tissue overlying a deeper absorptive one containing transfer cells which traverse a vapor barrier. The two best-known absorptive aerial organs are the velamentous orchid root and the trichomed bromeliad leaf.

Velamentous roots

Several families contain members equipped with velamentous roots; an example is the aroid genus *Anthurium*. But this character is particularly refined in xeric Orchidaceae (Barthlott and Capesius 1975), both drought endurers and avoiders (Fig. 3.19A–G); thus it serves the majority of the epiphyte flora. The specialized nonliving rhizodermis (velamen) forms an insulating but permeable mantle, 1–24 cells thick, around a living core comprised of an often chlorophyllous cortex (Fig. 2.16) and central conductive stele. Cell walls within the velamen are elaborately pitted and sculptured, and sometimes differentiated into several anatomically distinct zones. Pit membranes rupture throughout the system, allowing extensive infiltration by microbes (Fig. 4.14). An epivelamen (Fig. 3.19B,D), if present, is delicate and, except at the pneumathodes, may disintegrate with age (Fig. 2.10). Although somewhat variable in structural detail, velamina are not very expandable or compressible; saturated or dry, a root has about the same diameter.

Upon contact with fluids, the velamen becomes engorged instantaneously by capillary flow. Only the pneumathodes remain air-filled, repelling moisture by some as yet undetermined mechanism. Fluids in a saturated velamen lie against the outmost layer of the cortex – a uniseriate, suberized exodermis, impermeable to water except through its transfer cells (Figs. 3.9, 3.19G). After moisture moves through the cortical parenchyma, another tissue is encountered – a stoutish endodermis which also contains passage cells. From there, the distance to conductive tissue is no more than the dimension of a few stelar parenchyma cells. Velamentous aroid roots are organized along the same lines, but nothing is known about function.

A fair number of epiphytic and a few terrestrial orchids produce tilosomes (Figs. 3.8, 3.9) consisting of numerous lamellate or fibrillar wall protrusions from the velamen just above passage cells (Benzing, Ott, and Friedman 1982b; Pridgeon, Stern, and Benzing 1983). Those comprised of intermeshing branches may act as one-way valves, readily admitting fluids but blocking vapor loss when dry and compacted. Coarser tilosomes (Pridgeon et al. 1983) seem less likely to modulate moisture exchange in this fashion, but perhaps they impede passage of pathogens through what would otherwise be the most easily breached point along the exodermis.

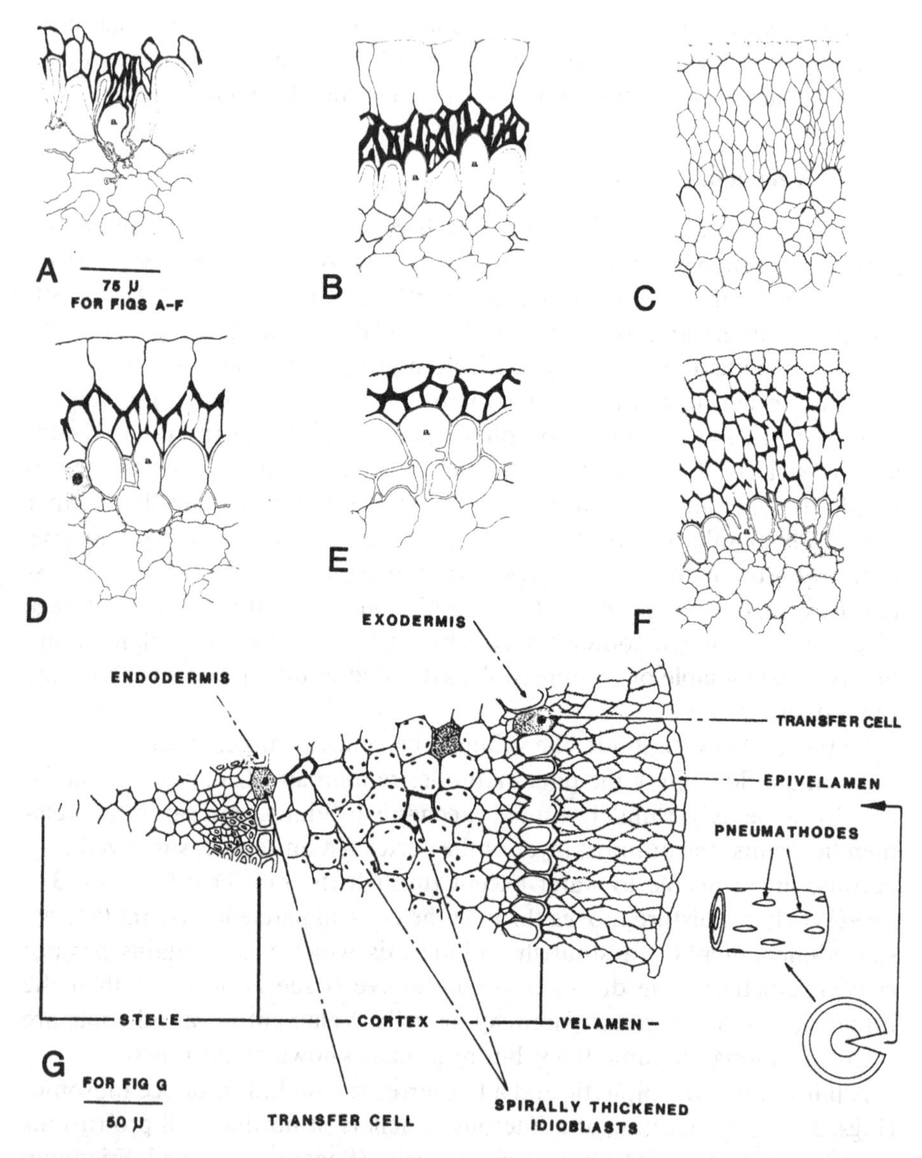

Figure 3.19. The velamentous orchid root in partial transverse section (T.S.): (A) *Kingidium taeniale;* (B) *Phalaenopsis amabilis;* (C) *Vanda parishii;* (D) *Campylocentrum sellowii;* (E) *Harrisella porrecta;* (F) *Encyclia tampensis.* (G) View of the entire root tissue series (a = aeration cells of the exodermis). Figures are not drawn to a common scale.

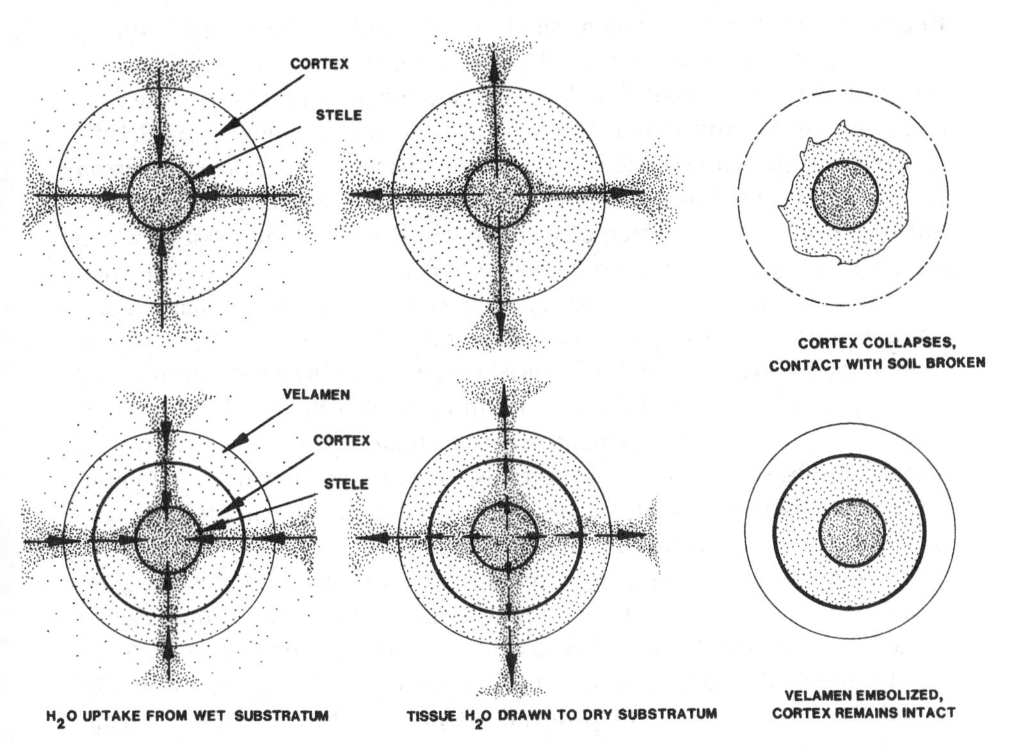

Figure 3.20. A schematic representation of the desiccating effect of dry media adjacent to roots without (upper) and with (lower) a velamen. Dehydration of the nonvelamentous root exceeds that of the velamentous organ.

Embolized (empty) velamina provide protection against water loss to hydrophilic substrata (matric forces generated by dry bark greatly exceed that of the driest air by many megapascals; Fig. 3.20). Jordan and Nobel (1984) observed that two nonvelamentous desert perennials, *Agave deserti* and *Ferrocactus acanthodes,* broke away from dry earth, collapsing the cortex and killing the roots. Regrowth and restoration of contact between plant and rewetted earth was affordable because less than 15% of total plant biomass had been sacrificed. Velamentous species do not need this mechanism, nor could those epiphytes facing drought shortly after every rainstorm bear the cost (e.g., bark orchids with small shoot/root ratios).

Although most authorities agree that aerial function is promoted by a velamen (but see Dycus and Knudson 1957), moisture retention varies, depending on specific anatomical details and overall form. The root's surface-to-volume (S/V) ratio appears to be one of the most important factors: Among 10 species, S/V ratios predicted drying rate better than did velamen thickness, number of velamen cell layers, cortical tissue compaction, or exodermal cell wall development (Benzing et al. 1983). Quite surprising was the discovery that roots of shootless *Polyradicion lindenii,* whose exodermal aeration cells are eroded like those of *Campylocentrum sellowii* (Fig. 3.19D) and lead directly into sizable intercellular airspaces, dried more slowly over $CaCl_2$ than did the other nine species. Roots of this orchid are quite stout, but they bear a thin velamen of just two cell layers, the outmost of which largely disintegrates at maturity. Exodermal U cells (Fig. 4.12) and inner velamen cell walls, however, are very robust and must be unusually vapor-tight. Sanford and Adanlawo (1973) noted that epiphytic West African orchids native to arid locations featured deeper velamina and thicker-walled exodermal barriers than did those on mesic sites. Velamen depth may be more closely tied to procurement problems than to desiccation resistance. Should rain be scanty over long intervals, or roots hang free or cling to non-absorbent bark in wetter sites, a large velamen dead space that prolongs contact with precipitation could be crucial for survival.

Existing information on aerial orchid roots poses many interesting questions. Why, for instance, is there so much variety? Perhaps different aspects of root structure can be modified to achieve equivalent performance. Studies of whole-body integration and consistency within plant groups would be useful. Are roots of all deciduous forms sacrificed at the end of a wet season? Are they so constructed as to minimize anoxia that could accompany long-term engorgement of velamina in deep wet humus? What functional trade-offs exist between requirements for absorption or photosynthesis on the one hand and drought resistance on the other, and what limits are thereby set on the physical dimensions of a given orchid's root system? Finally, how do nonvelamentous epiphytes in droughty microsites avoid root desiccation without impairing future absorptive function? Some clues on this last question may already exist in the anatomical literature. For instance, Wilder (1986) noted a more persistent epidermis and greater sclerenchyma development in first-order roots of predominantly epiphytic compared to terrestrial Cyclanthaceae.

Foliar trichomes

The second absorptive organ mentioned earlier is the trichomed leaf. Although several epiphytic ferns, *Astelia,* and perhaps some orchids bear absorptive foliar hairs, the trichome of Bromeliaceae, particularly of subfamily Tillandsioideae, exhibits the most elaborate structure. Scales (another term for these peltate hairs) feature shields made up of empty cells

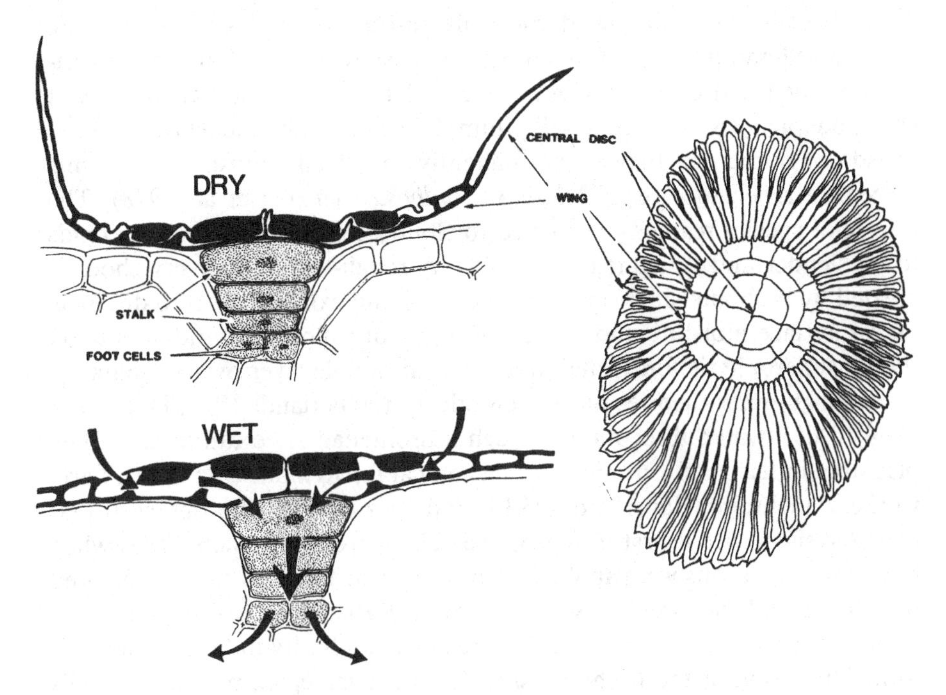

Figure 3.21. An absorbing bromeliad trichome in the dry and wet configuration. Arrows depict the route of water and solute movement into the mesophyll in the wet specimen.

aligned in orderly fashion (Figs. 3.6, 3.21). Centermost, just over the point where the dome cell subtends the shield, are four equal-sized, thick-walled, empty cells. Several additional rings of dead cells, each made up of twice as many cells as the one before, may be present. The outermost portion of the shield forms the wings, containing over twice as many, and now more elongated, cells as the previous ring. Ring cells have thin zones in outer walls that allow wing flexure. Thus a trichomed bromeliad leaf is, like the velamentous root, bounded by an absorbent, nonliving tissue which captures flowing moisture for slower absorption into living tissue.

Most tillandsioid trichome stalks are composed of three to five transfer cells supported by two foot cells. The dome cell is the largest of the chain and contains a dense protoplast with a prominent nucleus. Dolzmann (1964, 1965) noted that it is equipped with an elaborated plasmalemma plus other membrane systems and numerous mitochondria which he theorized are needed for absorption. Walls of both shield and stalk cells are modified in a pattern consistent with that critical function. All transverse junctions between cells – disc–dome, dome–stalk, intrastalk, and stalk–foot – are uncutinized and penetrated by numerous plasmodesmata. Shields are variously shaped and oriented (Figs. 3.6, 3.7, 3.11). Those on leaf surfaces of shade-adapted species are small, immobile, and scattered (Fig. 3.10); if broad and numerous, they are permanently appressed against the epidermis to reduce reflectance (e.g., *Tillandsia bulbosa;* Benzing et al. 1978). The foliage of species regularly subjected to clouds, fog, or high humidity tends to be slender and elongated (Fig. 3.11). A rough-textured silvery shoot is created by trichome shields that are elevated and extended as if to dissipate excess surface moisture and avoid the threat of suffocation. Indumenta on the most xerophytic forms maintain a flat immobile layer over stomata.

Although other theories have been offered (Haberlandt 1914; Dolzmann 1964, 1965), water absorption through a bromeliad scale can be described plausibly in terms of osmotic and mechanical forces alone (Fig. 3.21). Early workers, including Haberlandt (1914) and Mez (1904), who observed that plasmolysis occurred first in mesophyll cells surrounding stalk bases when hypertonic solutions were applied to intact leaf surfaces, correctly evaluated the tillandsioid trichome's absorptive power. Rather than beading up, drops of moisture falling on a leaf quickly spread by capillarity to form a thin uniform film between the trichome shields and foliar epidermis. Shield cells readily imbibe fluid; as filling proceeds, the upper walls of the central disc are forced upward while the wing flattens against the leaf surface. By the time emptying is completed, shield cell lumina have pinched shut. Once again, a thick plug is interposed between dome cell and bulk air, and water is prevented from leaving the plant by its route of entry. In effect, the atmospheric bromeliad's trichome serves as both a one-way valve and an energy dissipator, alternately hydrating the plant and insulating it against water loss and intense insolation. Functional analogy between this sytem and the velamen/exodermis is obvious, although the latter is less dynamic.

The absorptive role of trichomes on leaf sheaths of tank bromelioids has not been investigated adequately. Also, claims that foliar hairs of several polypodiaceous ferns can supplement root function are based solely on capacity to take up eosine (Müller et al. 1981). Pleurothallid orchid leaves bearing hairs with similar qualities showed very little rehydration capacity. Atmospheric bromeliads behaved quite differently. Leaves of these taxa, if droughted to 20–30% deficits, could still rehydrate in a matter of hours after wetting. They did not respond, however, to water-saturated air, contrary to other experimental findings (Fig. 3.22; DeSanto, Alfani, and DeLuca 1976). Garth (1964) demonstrated that *Tillandsia usneoides* required contact with water to survive in a Georgia forest. Terrestrial *Tillandsia purpurea* and *T.*

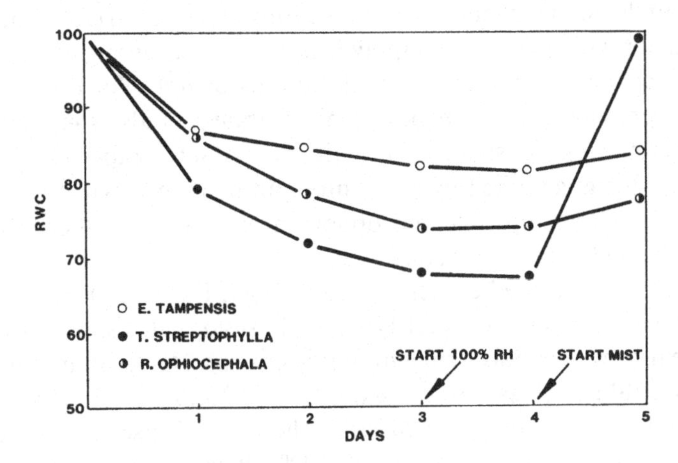

Figure 3.22. Rehydration of partially desiccated leaf sections of two orchids and the atmospheric bromeliad *Tillandsia streptophylla* during exposure to water-saturated air and liquid moisture. (From Benzing and Pridgeon 1983.)

latifolia flourish in Peruvian coastal deserts where fog rather than rain serves as the moisture source except during the pluvial El Niño years. Water vapor as opposed to rain or fog is probably much more important to epiphytes as a transpiration retardant than as a direct source of tissue moisture.

Adaptive uniqueness

Some of the requirements for epiphytism can be rather straightforward – namely, aerial dispersal and holdfast. Other capacities – maintenance of adequate hydrature and nutrition, for instance – are more complex, and margins for error differ. Whereas acceptable mineral nutrition is no less imperative for success in tree crowns than elsewhere, maintenance of a favorable water balance requires quicker adjustment to more numerous and abrupt environmental challenges. Each time an epiphyte's moisture source dries out (a process sometimes requiring only an hour or two), extended drought may follow, and an inappropriate or delayed reaction could have disastrous consequences. Chances are good that much epiphyte novelty centers on water relations. And because carbon gain and transpiration are effected through the same diffusive pathway, unusual water balance is likely to be accompanied by unconventional photosynthesis.

Adaptation to drought in forest canopies, as elsewhere, requires coordinated structure, physiology, and phenology. The CS group has greater opportunity to deploy mechanisms incorporating all three variables than do PS epiphytes, for which moisture supply is never reliable enough to support the demands of mesomorphic foliage, deciduous or not. Nor does classic resurrection behavior appear to be a viable strategy. Pulse-supplied epiphytes survive through possession of high water-storage capacity in long-lived organs, often augmented by mini-impoundment and high WUE. This last phenomenon offers an excellent opportunity to delve further into the more unique aspects of epiphytism.

The work of Sinclair cited earlier (1983a,b; 1984; Fig. 3.12) hints at what may be a widespread and unusual type of physiological response by epiphytes to the moisture supply problem in tree crowns. His ferns and orchids both possessed dilute cell sap: Ψ_S ranged from -0.33 to -0.69 MPa for the orchids and equaled about -0.55 MPa for the ferns. Consequently, turgor pressure (Ψ_P) was low even at RWC = 100%. The Ψ_L at which stomata apparently close (Ψ_{crit}) was also unusually high – only -0.50 to -0.75 MPa for the ferns. Clear-cut orchid data were not obtained, but at effectively zero foliar conductance, RWC for the ferns had decreased only to about 94% – a smaller loss than that accompanying stomata closure in much terrestrial vegetation. Mistletoes hold the opposing record for canopy-adapted forms (Table 3.3), but here very low Ψ_S is needed to divert the host's transpiration stream.

Analyses reported by Putz and Holbrook (1987) for five strangler figs native to the llanos of Venezuela are also of interest. Data were collected for young specimens with wholly aerial roots growing on palm-leaf bases (stage 1) and for older ones after their roots were embedded in earth (stage 2). Compared to the stage 2 hemiepiphytes, the more drought-stressed stage 1 figs were relatively xeromorphic and drought-sensitive; they often produced tubers. Their leaves showed lower stomatal density, higher trichome density, lower bulk tissue elasticity, greater leaf water content, and much reduced foliar conductance by day, at least during the dry season. Moreover, Ψ_S at full turgor was higher on average than that in leaves of ground-rooted conspecifics. Putz and Holbrook concluded that, for these stranglers in open habitats, hemiepiphytism is most beneficial for the access it eventually provides to water rather than to nutrient ions or light.

Clusia rosea, another strangler, further illustrated how water use shifts with access to land soil. Compared to adults, epiphytic juveniles utilized CAM more heavily over the entire year and curtailed phase IV C_3 activity almost entirely during the dry season (Sternberg, Ting, Price, and Hann 1987). *Clusia minor* proved to be a facultative CAM plant, but it also shifted its performance with life stage (Ting et al. 1987). Epiphytic and terrestrial forms alike switched from CAM to C_3 photosynthesis after daily rains had begun; however, capacity to maintain steady yields throughout the year varied with rooting medium. Maximum rates of CO_2 fixation in earth-rooted specimens were about double those reached when CAM prevailed. Productivity of the epiphytic stage was more vulnerable to drought; highest dark fixation rates were only one fourth of those achieved when moisture was plentiful. Hemiepiphytic *Ficus* subjects and a *Coussapoa* species (Moraceae) in the same region survived dry seasons without CAM, once again probably aided by unusually sensitive stomata.

Early surveys of Ψ_S in Jamaican and South Florida epiphytes by Harris and Lawrence (1916) and Harris (1918), Spanner's (1939) data from myrmecophytic *Myrmecodia* and *Hydnophytum,* and Walter's (1971) findings on diverse canopy-based ferns and several phanerophytes anticipated the discoveries of Sinclair, Putz and Holbrook, and Ting et al. In addition, Harris found that osmotica were fully two to three times less concentrated in epiphytic foliage than in that of phorophytes and understory herbs. The data of Table 3.3 are especially enlightening, concerning as they do structurally diverse epiphytes belonging to four families – pretty good evidence that Ψ_S is generally high in canopy-adapted species. But why does dilute cell sap seemingly have special utility for an epiphyte?

At least one hypothesis is plausible. Unlike trees, epiphytes need not support tall water columns or overcome associated flow resistance. There is, then, no particular reason for maintenance of concentrated osmotica which might, in fact, pose greater risk. Dilute cell sap suggests that many epiphytes may be especially well equipped to anticipate drought and minimize the chance of injurious water deficits. High Ψ_S translates into low maximum Ψ_P and potentially sensitive control of leaf conductance (high Ψ_{crit}), assuming of course that osmotic adjustment is negligible. Supersensitive stomata provide similar benefit to other vegetation which appears outwardly to have little need for it. Gas exchange was curtailed at elevated Ψ_L by six evergreen trees in a New Zealand cloud forest (Jane and Green 1985). Deep root penetration, which in these species is inhibited by almost continuously saturated earth, obliges minimal transpiration when the uncommon drought drastically reduces moisture in upper ground layers.

Where growing conditions are extreme, stomata can be quite specialized, as in paleotropical *Paphiopedilum,* a predominantly forest-dwelling, often epiphytic, C_3 genus of slipper orchids. The achlorophylly of guard cells seen in these plants but unknown elsewhere long puzzled physiologists. *Paphiopedilum* apparently grows so consistently in deeply shaded, infertile habitats that relatively low conductance serves full photosynthetic capacity. Suffi-

Table 3.3. **Solute potentials (Ψ_S) of vascular epiphytes**

Family	Species	Ψ_S (MPa)	Location
Bromeliaceae	*Catopsis berteroniana*	−0.56	Florida
	Guzmania monostachia	−0.56	Florida
	Guzmania sintensii	−0.38	Jamaica
	Tillandsia recurvata	−0.58	Florida
	Tillandsia usneoides	−0.90	Florida
	Tillandsia utriculata	−0.52	Florida
Gesneriaceae	*Columnea hirsuta*	−0.43	Jamaica
Polypodiaceae	*Pyrrosia angustata*	−0.60	Malaysia
Orchidaceae	*Dendrobium crumenatum*	−0.49	Malaysia
	Encyclia tampensis	−0.58	Florida
	Epidendrum imbricatum	−0.36	Jamaica
	Eria velutina	−0.33	Malaysia
	Lepanthes ovalis	−0.42	Jamaica
	Macradenia lutescens	−0.61	Florida
	Octadesmia montana	−0.53	Jamaica
	Pleurothallis racemiflora	−0.26	Jamaica
	Polystachya minuta	−0.60	Florida
	Stelis micrantha	−0.26	Jamaica
	Vanilla eggersii	−0.29	Florida
Piperaceae	*Peperomia basellifolia*	−0.42	Jamaica
	Peperomia septemnervia	−0.37	Jamaica
Mistletoes	*Dendrophthora cupressoides*	−1.42	Jamaica
	Phoradendron flavescens	−1.58	Jamaica
	Phthirusa parvifolia	−1.62	Jamaica
	Phthirusa pauciflora	−1.99	Jamaica

Sources: After Harris and Lawrence 1916; Harris 1918; Sinclair 1983a.

cient energy to open stomata is available without the photosystems. There are probably other cases where extremes in the environment have fostered unconventional gas exchange in canopy-adapted flora. Those PS epiphytes from the most stressful exposures grow more slowly than does *Paphiopedilum* (at least in culture), perhaps because they have evolved under even greater deprivation. Atypical epidermal features such as those of some atmospheric bromeliads (Tomlinson 1969) are all the more interesting, given the associated slow maturation. Claims that *Tillandsia usneoides* lacks mobile guard cells while even more diminutive *T. bryoides* has no stomata at all have proved incorrect (Martin and Peters 1984). Findings from

T. usneoides should not discourage further research, however, because this is one of the most vigorous of the atmospheric species. In fact, there is precedent for lack of mechanical control of gas exchange in land plants. Shootless orchids gain carbon in tree crowns by astomatous roots alone (Winter et al., 1985; Fig. 2.5); CAM cannot improve water economy here except perhaps by promoting absorption of dew. Were it not for the close stoichiometry achieved during daytime phase 3 deacidification and refixation (Cockburn et al. 1985), much regenerated CO_2 would be lost. Investigators of stress effects on water and carbon balance in plants would be well advised to examine PS epiphytes, especially those with unusual structure.

4 Mineral nutrition

Plant nutritionists have paid little attention to epiphytes, concentrating instead on plants rooted in earth soil. Consequently, consideration of mineral cycling in tropical woodlands has seldom taken into account the potentially major impact of arboreal vegetation. Reports from a scanty but developing literature on epiphytes, together with relevant aspects of plant physiology in other groups, provide a basis for the following chapter. Epiphytes will be portrayed as plants that not only tap a variety of nutrient sources (at times with novel absorptive organs) but also are significant players in the nutrient and energy economies of many tropical forests.

Nutritional categories

All higher plants require at least six macronutrients and seven trace elements for growth (Table 4.1). Some taxa supplement these basic 13 with others that support out-of-the-ordinary functions. For example, many grasses produce silicon-containing granules that reduce palatability to vertebrates; selenium (Se) can act as a sulfur (S) analog, helping to ward off herbivores; some halophytes and all C_4 and CAM taxa require sodium (Na); and so on. Descriptive epithets are applied in certain cases: halophytes versus glycophytes, to describe occurrence on hyperosmotic media; calcifuges versus calcicoles, to specify calcium (Ca) content of native soil; eutrophs and oligotrophs, to distinguish quantities of key macronutrients needed.

Eutrophs are characterized by several features. Critical concentrations of foliar N and P (those levels required to maximize growth) tend to be elevated (Table 4.1). Shoot/root ratios are generally high, life cycles and longevity of leaves, brief. Mature size and vigor are tied closely to nutrient supply. Adaptation for life on impoverished media is equally multidimensional. Scarcity of macronutrients rarely, if ever, causes serious deficiency for the slow-growing oligotroph, even though concentrations of these elements may be depressed in tissues compared to values for better-nourished conspecifics (e.g., *Tillandsia paucifolia;* Table 4.1). Shoot/root ratios tend to be low, foliage is evergreen, and life cycles are extended. Propagative mode

Table 4.1. **Mineral nutrients present in foliar tissues of a typical eutroph and of *Tillandsia paucifolia* on nutrient-stressed and well-nourished cypress in Florida**

Nutrient	Generalized minimum requirement for eutrophic vegetation		Conc. in vegetative body of *T. paucifolia* growing on nutrient-stressed Florida cypress		Conc. in vegetative body of *T. paucifolia* growing on relatively vigorous Florida cypress	
	% dry wt	ppm	% dry wt	ppm	% dry wt	ppm
N	1.50		0.36		0.35	
P	0.10		0.072		0.085	
K	1.00		0.33		0.54	
Ca	0.50		0.66		0.98	
Mg	0.20		0.17		0.23	
S	0.10		0.05		0.097	
Mn		50		27.5		22.5
Fe		100		154.8		195.8
B		20		15.2		18.3
Cl		100		—		—
Cu		6		9.2		10.0
Zn		20		35.5		41.8
Mo		0.10		1.43		1.60

Sources: Epstein 1972; Benzing and Davidson 1979.

is also characteristic; several seasons pass before adulthood, and the first of usually several polycarpic fruiting cycles occurs. Factors responsible for the oligotrophic nature of many epiphytes include (1) shallow substrata; (2) aridity, which inhibits use of fine roots and beneficial microbes; and (3) flowing canopy fluids, which wash away those root secretions that, for some terrestrials, aid absorption.

Epiphytes are nutritionally diverse, although more regularly confined to acid organic substrata than is terrestrial flora. Eutrophism is probably more common in canopy residents than would seem warranted, considering their lack of contact with the ground. At least one group of shade-tolerant neotropical species requires fertile habitats for maximum development (Gentry and Emmons 1987). Pulse-supplied forms exhibit modest nutritional demands, in keeping with their generally stress-adapted nature. So far, halophytism has not been reported, but salt tolerance has (Benzing and Davidson 1979). *Tillandsia paucifolia* growing on *Rhizophora mangle* in South Florida contained quantities of Na up to several percent of shoot dry weight. Molar concentration of foliar Na greatly exceeded those of K in *Tillandsia flexuosa* and *Schomburgkia humboldtiana* native to a small island off the Venezuelan coast (Griffiths et al. 1989). Orchidaceae are present in numerous Australasian mangals (forests of mangrove species), and other monocots and a number of dicots, ferns, and lycopods also tolerate the high salinity of these habitats (Tomlinson 1986). Sodium may benefit certain epiphytes far from the ocean as it does cabbage, pea, and other natrophilic terrestrials; inland populations of *Tillandsia paucifolia* contained much less Na, although K/Na ratios seldom exceeded 2 (Benzing and Davidson 1979). However, co-occurring *Encyclia tampensis* contained substantially more K than Na (Benzing and Renfrow 1974b), perhaps because roots rather than leaves govern ion uptake in orchids. Broader surveys of atmospheric *Tillandsia* species could determine whether Na moderates demand for K where supplies of the latter are particularly dilute (e.g., telephone wires, treetops). More information about bromeliad trichomes is necessary to understand how a high Na status develops.

Very little is known about epiphyte response to certain unusual soil conditions. Tropical forests located over ultrabasic rocks rich in exchangeable magnesium (Mg), chromium (Cr), cobalt (Co), and nickel (Ni) contain tolerant trees, some of which further concentrate potentially phytotoxic ions. Canopy flora supported by accumulators could experience a type of "ecological magnification," but if so, there are no outward symptoms of extreme exposure. Morat, Veillon, and Mackee (1984) concluded from floristic surveys of New Caledonia forest on a variety of soil types that epiphytes were little affected by local geochemistry. Proctor et al. (1988) encountered similar evidence in North Borneo while studying rain forest growing on a low ultrabasic mountain. One or more epiphytes were anchored to the boles of 3.2–100% of all trees of 10 or more centimeters in diameter at breast height, depending on altitude. In neither instance, however, were phorophytes or their crowns analyzed to reveal possible movements of unusual ions in these ecosystems.

Factors influencing nutrient supply, demand, and use

A plant's ability to secure sufficient nutrients depends on supply and demand, parameters that are neither easily measured nor uniform for all

Table 4.2. **Chemical characteristics of mineral soil and suspended substrata (one sample each) in wet forest at Río Palenque, Ecuador**

				Meq/100g					ppm		
Description of material	pH	% base saturation	Cation exchange capacity	K	Ca	Mg	H	Na	N	P	K
Outer bark of large *Theobroma* branches with associated debris and nonvascular plants	6.2	79.1	123.5	20.0	49.7	25.5	25.8	2.6	3.0	0.34	0.67
Outer bark of *Theobroma* twigs	6.7	85.8	137.4	18.7	67.4	31.5	19.5	0.3	2.2	0.22	0.71
Rotten wood of *Theobroma*	7.1	90.2	163.3	4.6	112.3	30.1	16.0	0.3	1.5	0.09	0.18
Fern root ball	5.2	56.4	135.1	7.5	57.3	11.1	58.9	0.4	1.8	0.10	0.25
Carton of ant nest-garden	6.3	78.4	115.3	20.1	56.2	12.2	24.9	1.9	2.9	0.39	0.79
Earth soil	6.3	55.3	31.1	0.5	14.0	2.5	13.9	0.2	0.3	—	—

Source: Unpublished data.

species, even among members of the same community. Diverse epiphytes in a single tree crown, for example, draw on different nutrient sources (Tables 4.2–4.4) and employ key elements with characteristic mineral use efficiency (MUE). Plant chemistry as an index of nutritional sufficiency, MUE, or habitat fertility is less reliable for comparisons between than within species. For instance, robust bromeliads whose vigor results from abundant resources in tanks or ground soil may concentrate foliar N (but less so P and K) no more than do some of the slower-growing, oligotrophic atmospheric *Tillandsia* species (Table 4.4). Differences in relative concentrations of available nutrients in widely divergent substrata far exceed those in the tissues of plants drawing on those sources.

Field and Mooney (1986) proposed an informative measure of nutritional status that should be applied to epiphytes. Their index – potential photosynthetic N-use efficiency (PPNUE) – recognizes the tight correspondence between invested N and A_{max} in foliage. Emphasis on N is warranted because supplies of this element, more than any other, influence plant productivity. Expressed as instantaneous photosynthetic yield relative to N content,

Table 4.3. **Chemical composition of rainwater, throughfall, and stemflow in the forest canopies of central Amazonia, eastern Panama, Haiti, and South Florida**

	Central Amazonia[a]			Eastern Panama[b]			
Nutrient	Rain-water	Through-fall	Stem-flow	Rainwater	Throughfall	Haiti[b] Stemflow	South Florida[b] Stemflow
Na	0.12	0.27	2.11				
K	0.10	1.24	6.58	0.2–1.4	2.4–5.6	3.00	0.25–1.30
Ca	0.07	0.25	1.72	0.4–3.6	2.0–2.8	1.00	3.04–9.60
Mg	0.02	0.19	0.97	0.04–0.64	0.44–0.80	4.30	0.52–0.61
Mn				0.003–0.68	0.010–0.036	trace	
$N(NH_4^+)$	0.169	0.05	9.20			1.23	
$N(NO_3^-)$	0.11	0.56	0.27				
$N(NO_2^-)$	0.002	0.01	0.02				
N (total)	0.41						0.40–0.76
$P(PO_4^{3-})$	0.003	0.151	0.095	0.033–0.075	0.024–0.068	0.15	0.017
S						0.17	
Fe				0.053–0.340	0.036–0.416	0.41	
Zn				0.017–0.079	0.029–0.062		

Note: Values are expressed in milligrams per liter.
Sources: [a]Junk and Furch 1985; [b]Benzing and Renfrow 1980.

PPNUE is lowest in species with the lowest A_{max}, and both parameters increase together. In contrast, leaf longevity and A_{max} are inversely related. Construction of durable leaves doubtless mandates proportionally larger investment of N in nonphotosynthetic substances such as alkaloids rather than in, for instance, RuBPc/o, which represents up to half of the soluble protein in C_3 leaves. Presumably PPNUE also shifts within populations according to nutritional sufficiency – that is, the individual member's success in obtaining N.

Because photosynthesis and hence leaf amortization are slowed by drought, nutrient scarcity, and shade, PPNUE should be modest and foliage relatively long-lived in many canopy-adapted plants. Indeed, preliminary data obtained using a portable photosynthesis analyzer in wet premontane forest at Río Palenque, Ecuador and in northern Venezuelan cloud forest confirmed this prediction. When A_{max} and N were expressed on a leaf area basis, the relationship between the two for Ecuadoran C_3 *Anthurium trinerve* and the bromeliad *Guzmania monostachia* fell closer to values for evergreen shrubs and trees than for deciduous shrubs and annuals (Field and Mooney 1986). Data for two epiphytic ferns, a co-occurring bromeliad, and a pho-

Table 4.4. **Mineral element composition of leaf blade tissues (% dry wt) representing the three nutritional modes in Bromeliaceae**

		Mineral element					
Species	Habit/Habitat	Ca	K	Mg	N	Na	P
Bromelia karatas	terrestrial/Jamaica	1.19	2.48	0.14	0.67	0.064	0.068
Pitcairnia bromeliafolia	terrestrial/Jamaica	1.06	1.02	0.14	1.37	0.036	0.084
Catopsis floribunda	tank epiphyte/Jamaica	1.15	1.59	0.18	1.36	0.32	0.091
Aechmea nudicaulis	tank epiphyte/Jamaica	0.38	1.94	0.20	0.057	0.18	0.057
Guzmania linqulata	tank epiphyte/Trinidad	0.44	1.59	0.25	0.88	0.19	0.063
Tillandsia balbisiana	atmospheric/Florida	0.83	0.34	0.14	0.36	0.41	0.035
Tillandsia usneoides	atmospheric/Florida	0.67	0.50	0.29	0.82	0.55	0.012

Source: After Benzing and Renfrow 1974a.

rophyte are provided in Table 3.2. No doubt, PPNUE values would vary more in a larger sample; future inquiries into the influence of canopy anchorage on N use could produce some surprises.

Nutrient sources in forest canopies

Atmospheric input

Nutrients enter forest canopies from both air and earth. Ions from the ground become accessible to epiphytes following transport upward in the transpiration stream of trees; ease of uptake thereafter depends on the plant's location and capacity to tap leachates, litterfall, and animal products. Delivery from the atmosphere involves several vehicles of varying utility to particular species, the major types being dry deposition (including vapor) and wet fallout. Vapor provides S and N, the latter as NH_4^+, NO_3^-, and NO_2^- ions. Dinitrogen fixation is discussed below. Airborne particulates are chemically varied, originating from land, sea, and animal sources (Clarkson, Kuiper, and Lüttge 1986). Particulates containing many elements were located on shoots of *Tillandsia usneoides* sampled across the southeastern United States, most likely lodged among the overlapping trichome shields (e.g., *T. tectorum,* Fig. 3.11; Shacklette and Connor 1973; Connor and Shacklette 1984). Proportions of aluminum (Al), barium (Ba), gallium (Ga), iron (Fe), and yttrium (Y) indicated a terrestrial origin for these finely divided solids. Rainfall and mist contain significant quantities of all the essential elements (Table 4.3).

Nutrient input from the atmosphere varies. For instance, although enough S often arrives to equal or exceed local plant requirements, rates of N, P, and K deposition are inadequate (Clarkson et al. 1986) except in the low-demand community (e.g., rain-fed peat bogs). Forests on deep, heavily leached, acid soils maintain nutrient capital primarily by tight recycling of atmospheric inputs (Jordan 1985). In numerous tropical woodlands, existing pools of several critical elements, much contained in vegetation, can represent many years of accumulation.

Nutrient flux through a forest canopy is nonuniform in time and space. More than half the annual wet deposition of several nutrient ions fell on a relatively dry Honduran forest within one to ten rainy days (Kellman, Hudson, and Sanmugadas 1982; Figs. 4.1, 4.2). Finer sampling would almost certainly have turned up variation even during individual events because solute concentration drops quickly after rain has scrubbed particulates and vapor from the atmosphere. Ion concentrations in stemflow and throughfall are further altered by passage over leaf surfaces where ions in dry deposition and those originally in leaf interiors have accumulated since the last storm. Moreover, enrichment at one level may be accompanied by depletion at another as uptake by resident biota and/or host foliage deprives plants farther downstream. Ions of Ca and K are routinely more concentrated in these canopy flushes than in rainwater (Table 4.3), but others, including nitrogenous ions, may be more or less so. Jordan and Golley (1980) recorded considerable aboveground nutrient scavenging in a wet Venezuelan forest, but the responsible agents were not identified. These authors also noted that P sometimes occurred in stemflow and throughfall at concentrations equal to or surpassing those in many ground solutions. Brasell and Sinclair (1983) recorded the presence of more K in annual throughfall than of total exchangeable K in the top 30 cm of terra firma!

Data collected by Nadkarni (1984, 1986) are especially applicable to this discussion because quantities are fairly representative, the habitat is epiphyte-rich, and the time of delivery and vehicle are identified. Precipitation (Fig. 4.2) tended to be more nutritive in dry (N: 0.28 ppm, P: 0.95 ppm) than in wet (N: 0.05 ppm, P: 0.11 ppm) periods. But the greater volume of wet-season rainfall mitigated dry-season influence on community fertility. Clouds and mist are also significant ion sources in upland and montane areas (Clarkson et al. 1986). Effects of specific carriers on local epiphytes are impossible to predict. Nevertheless, failure to intercept entering ions should not preclude subsequent capture during recycling by supports. By whatever

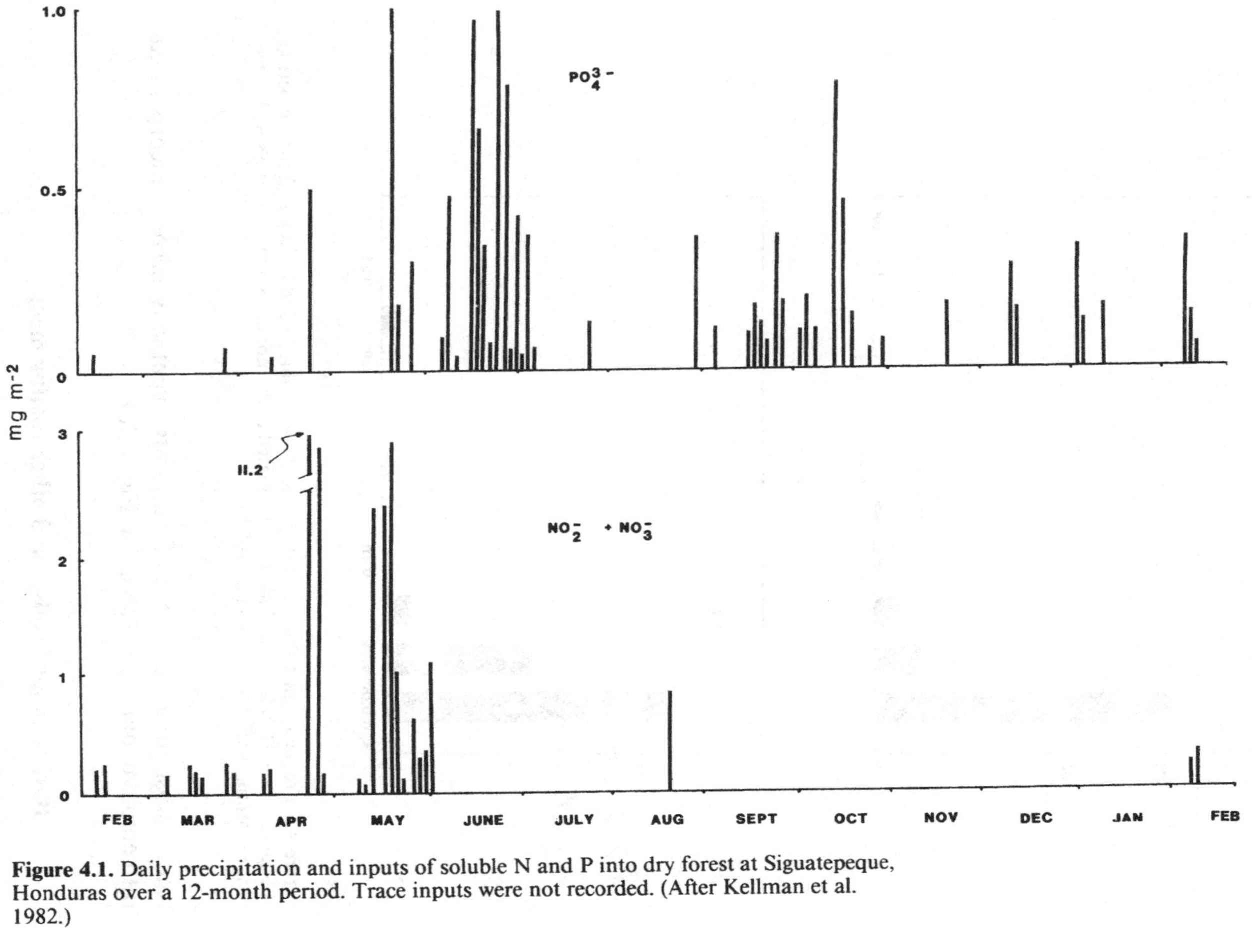

Figure 4.1. Daily precipitation and inputs of soluble N and P into dry forest at Siguatepeque, Honduras over a 12-month period. Trace inputs were not recorded. (After Kellman et al. 1982.)

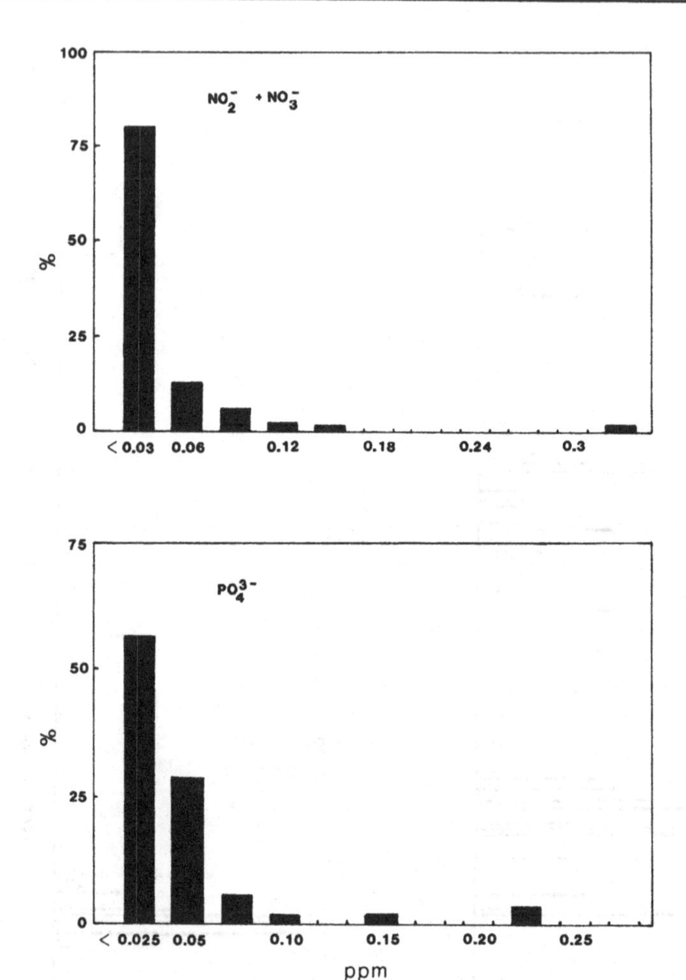

Figure 4.2. Frequency distribution of daily inputs of soluble N and P over a 12-month period at Siguatepeque, Honduras. Included are estimated daily inputs, excluded are days with trace concentrations. (After Kellman et al. 1982.)

route, substantial quantities of nutrients eventually become incorporated into epiphyte biomass (Table 7.8, Fig. 7.26).

Bark, canopy "soils," and other rooting media

So far, emphasis has centered on the epiphyte's access to ions in airborne inputs and canopy washes. Nutrition of epiphytes, except for the atmospheric bromeliads and a few others, must also be affected by characteristics of supporting bark and the nature of diverse additional materials such as ant nest-gardens, suspended humus, rotting wood, animal by-products, bodies of captured animals, and the like. At this point, too little is known about the chemistry of these substances, or the functional qualities of the organs positioned to exploit them, to generalize with confidence.

A few data compare tree crowns with earth as resource pools. The bark of depauperate South Florida cypress yielded modest quantities of N, P, and K upon extraction with dilute hydrochloric acid (Table 4.5); in better nourished trees, K and P values were roughly five- to sevenfold higher, N about threefold higher. Exchange capacity, an important aspect of land soil quality, was not measured. Canopy soil which had accumulated under epiphytes on two phorophytes in Singapore contained appreciable silt and sand-size mineral particles (Johnson and Awan 1972). The large organic fraction had a low ash content and a pH between 5.8 and 6.2. Material removed from *Swietenia macrophylla* harbored 12 different cyanobacteria, at least two with potential for N_2 fixation. Diverse rooting media supporting epiphytic aroids, bromeliads, gesneriads, and ferns in an abandoned *Theobroma cacao* plantation at Río Palenque, Ecuador revealed considerable disparity in important chemical characteristics (Table 4.2). Several points are worth noting. Compared to the earth below, all five canopy media had superior capacity for cation exchange, contained more nutritive ions, and a preponderance of N over P and K. Neutral to moderately acid pH was recorded, but conditions at Río Palenque may be atypical: Suspended humus collected in pluvial forest in northwestern Ecuador yielded readings down to pH 3.3.

Sampling on a more intensive scale, Nadkarni (1984) discovered that large fractions of several essential ions were tied up in suspended humus in a Costa Rican cloud forest (Fig. 7.26). These substances and those in the bark itself would presumably have become available for plant reuse when bark decayed, but perhaps only over many years of slow release. Using a cellulose disc assay, Nadkarni (1986) did indeed observe slower decomposition in canopy debris than she had found in ground litter at the same site. After eight weeks, 23–45% weight loss was recorded for earth-incubated samples; discs placed under suspended humus were essentially unchanged. Humic canopy soils harbored fewer invertebrate processors, had lower water but higher fiber content, had a higher carbon/nitrogen ratio, and were possibly dissipating polyphenols more slowly than did ground litter. Densities of mites, adult beetles, holometabolous insect larvae, Collembola, amphipods, and isopods were on average 2.6 times higher in earth compared to canopy soils. Only ants were equally represented in both microsites. Also demonstrated was substantial N mineralization in suspended and

Table 4.5. **Mineral element content in stemflow and outer bark and foliage of dwarf and vigorous *Taxodium ascendens* hosting *Encyclia tampensis* and *Tillandsia paucifolia* in South Florida**

	Nutritional status of host	Element (mg/5 g dry wt)					
		N	P	K	Ca	Mg	Na
1. Bark							
Water extract	dwarf	0.22	0.004	0.11	0.28	0.02	0.24
	vigorous	0.38	0.004	0.37	6.22	0.08	0.13
HCl extract (0.01 N)	dwarf	0.13	0.014	0.14	33.9	0.33	0.24
	vigorous	0.34	0.096	0.74	107.0	1.11	0.20
Nutrients remaining after both extractions	dwarf	5.31	0.063	0.07	42.7	0.12	0.03
	vigorous	6.46	0.251	0.33	47.6	0.53	0.02
2. Foliage	dwarf	53.0	2.80	21.5	103.5	4.20	1.95
	vigorous	72.0	6.50	33.5	165.5	7.50	3.75
3. Stemflow[a]	dwarf	0.76	0.017	0.25	3.04	0.52	0.60
	vigorous	0.40	0.017	1.30	9.60	0.61	0.64

[a]Stemflow mineral element content is expressed in milligrams per liter.
Source: After Benzing and Renfrow 1974b.

forest-floor humus. Microbial biomasses were similar. Nadkarni (pers. comm.) believes that the histosol-like suspended soil in cloud forest is largely derived from epiphytes (including nonvascular forms) rather than from trees and that it contributes the most to ground litterfall.

Contrary to the pattern reported by Nadkarni, loci of intense litter decomposition have been noted at a somewhat lower cloud forest site in northern Venezuela. There, up to fivefold greater detritivore density (number of animals per unit volume of substratum) was recorded in epiphytic tank bromeliad shoots compared to the forest floor (Fig. 7.12; M. G. Paoletti, B. R. Stinner, and D. H. Stinner pers. comm.). Central watertight tanks surrounded by older, looser, leaf bases create progressively drier humus-based microcosms and living space for diverse biota. Other likely factors contributing to habitat quality include superior cover against predation and enrichment of tanks by resident N_2 fixers.

Single analyses of rooting media (in effect, most of the data available today) provide limited insight on tree crowns as nutrient sources for epiphytes. Information on base saturations and pH should be particularly

Table 4.6. **A comparison of ten paired samples taken from behind persistent leaf bases of *Copernicia tectorum* and from ground soil below hosts supporting strangler figs in the llanos of Venezuela**

		ppm		meq/100 g			
Source	pH	N	P	Ca	Mg	K	Cation exchange capacity
Soil (upper 10 cm)	4.8	0.23	29.4	2.8	3.0	0.5	21.2
Leaf axil deposits	4.3	0.78	17.2	2.8	4.4	3.0	49.9

Source: Putz and Holbrook 1987.

important to arboreal flora, when one considers the organic nature of their rooting media. Histosols possess considerable electronegativity near chemical neutrality, but, unlike mineral colloids, retain little capacity to adsorb nutritive cations under acid conditions. Unless Ca^{2+} or other base-forming ions are more abundant in tropical forest canopies than in many supporting mineral soils, epiphytes may be denied the benefits of important nutrient-mobilizing microbes. Nitrifiers, for instance, require high base saturations for vigorous activity in organic media. Additional data, like that in Tables 4.2–4.4 and the information collected by Putz and Holbrook (1987; Table 4.6) and considered later in a discussion of stranglers, are needed for a better understanding of epiphyte nutrition. Equally critical to this synthesis will be information on mineralization of suspended humus, particularly the activities of N-cycling microbes.

Impoundments

Assessment of input to impounding epiphytes would add perspective on this very important group of canopy residents. Urine sample bottles left for nearly a year to collect fluids and solids from live-oak tree crowns in central Florida intercepted substantial quantities of mineral nutrients (Table 4.7). Materials trapped in leaf axils of *Guzmania monostachia* (Table 4.8) in a South Florida swamp forest also contained significant amounts of the essential mineral elements, in some instances more in the tanks than was present in the biomass of individual plants. Clear water drawn from phy-

Table 4.7. **Quantities of elements intercepted over a 10-month period by sample bottles suspended in live-oak crowns near Tampa, Florida**

	Sample number			
Element (mg/bottle)	1	2	3	4
N	27.2	7.19	15.6	3.20
P	3.23	0.41	1.46	0.28
K	3.54	1.34	1.76	—
Ca	5.36	5.74	11.7	2.00
Mg	0.81	1.09	2.72	0.20
Na	2.43	0.84	3.68	0.30

Source: After Benzing and Renfrow 1974a.

Table 4.8. **Distribution of nutritive elements in the tissues and tanks of two *Guzmania monostachia* specimens growing in a swamp forest in South Florida**

	Mineral content (mg/g dry wt)			
Element	Vegetative body	Mature inflorescence	Tank contents (mg/liter)	Percentage of total plant pool replaceable from tank contents
N	151.2	50.6	197.8	100.2
P	18.8	10.7	11.3	38.3
K	399.5	90.4	17.0	3.5
Ca	189.6	21.1	288.0	136.7
Mg	116.3	18.3	24.0	17.8
Na	48.4	2.56	4.4	8.6
Mn	1.02	0.24	0.45	35.7
Fe	1.41	0.37	2.91	163.5
B	0.27	0.068	0.55	162.7
Cu	0.035	0.017	0.031	59.6
Zn	0.042	0.16	0.28	271.8

Source: After Benzing 1980.

totelmata of nine *Vriesea platynema* specimens during the early dry season at Rancho Grande, Venezuela contained K, Ca, Mg, and Na in concentrations that exceeded those reported for rainfall and canopy washes in Table 4.3 by as much as two orders of magnitude (B. R. Stinner and D. H. Stinner pers. comm.). Biological importance to the bromeliad was not assessed, but would vary depending on whether the ions were in particulates or solutions. Organization and dynamics of the food webs within impoundments influence plant benefits, but again, few details are available. One point is already clear, however: Although data are scarce, the arboreal habitat is emerging as one of great nutritional heterogeneity. Substrata for epiphytes are diverse in origin and chemically varied even across short distances within individual tree crowns.

Nutritional modes

Epiphytes can be categorized according to the substrata utilized. Long-standing labels like nest-garden, tank, humus, trash-basket, bark, and atmospheric describe the obvious quirks of the better-known types. Taxa at one extreme – that is, the hemiepiphytes – differ little from nearby terrestrial vegetation in the way nutrient ions and moisture are acquired during part of the life cycle; roots exhibiting no obvious novel form or function compete at some point with those of hosts and understory vegetation. At the other extreme, certain species possess features that would offer little or no nutritional utility elsewhere. Symbioses that resemble those serving terrestrial vegetation (Figs. 4.3–4.15) have unknown effects on arboreal flora. In the following discussion, the subjects are (1) trophic mutualists; (2) partially earth-adapted forms and creators of substitutes for earth; (3) carnivores; and (4) those remaining taxa without any direct grounding or assistance from obvious trophic symbionts. Categories are not always mutually exclusive. Figure 4.22 proposes a scheme of how various canopy substrata came to be utilized during epiphyte evolution and corresponding degrees of plant specialization. Note that a plant type suited for one rooting medium or for a specific primary nutrient source (e.g., bark/twig dwellers, carnivores) sometimes had multiple antecedents adapted to different conditions. Additionally, the transition from terrestrial to aerial nutrition was probably accomplished by several routes.

Inputs from fungi and diazotrophs

The nutrition of most terrestrial flora is aided by microbes, especially mutualistic fungi. A thin hyphal thread, finer than any rootlet, is the more cost-effective absorptive organ because the quantity of such immobile nutrients as P accessible to a plant is determined by the amount of substratum contacted, whether by roots or hyphae. Reports of epiphyte mycorrhi-

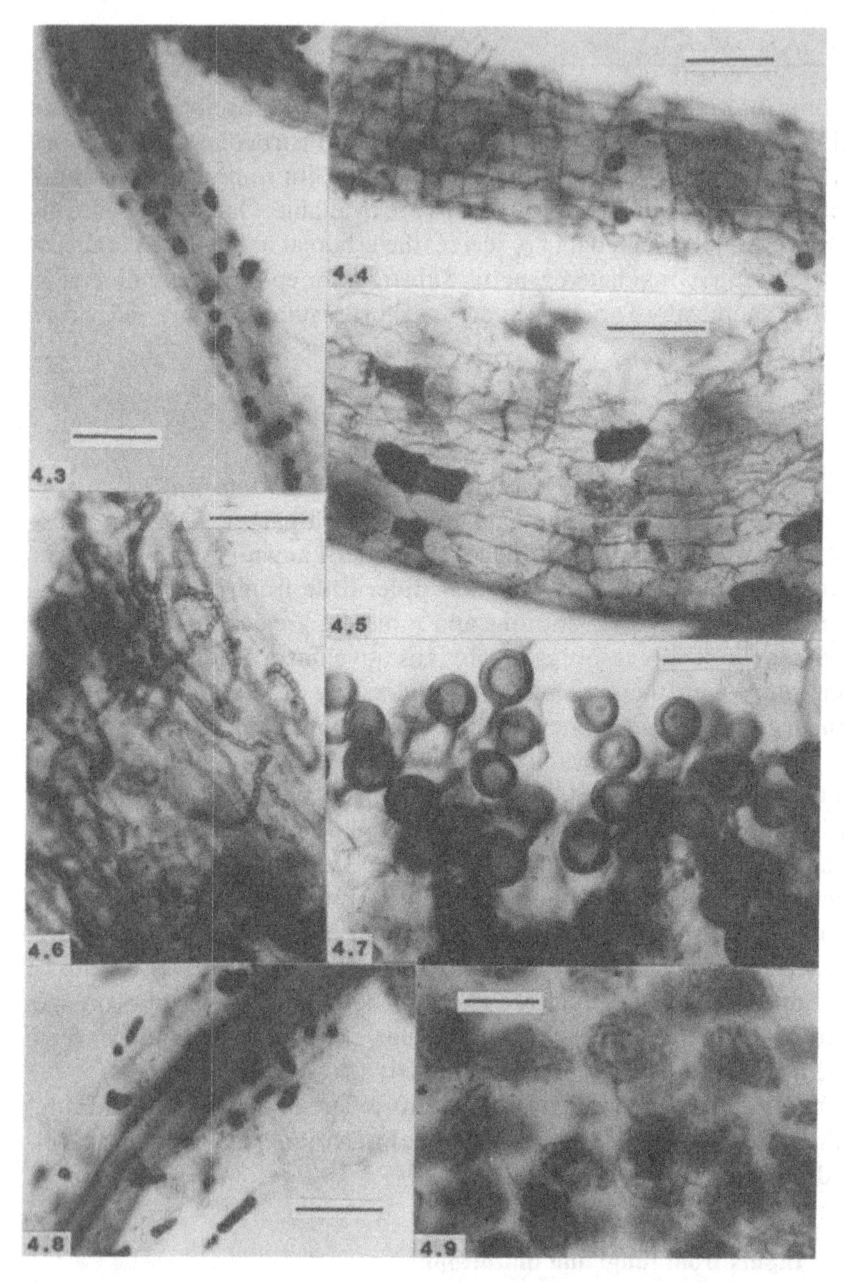

Figures 4.3–4.9. (4.3) Microbiology of the epiphyte rhizosphere: (4.3) multicellular rounded spores associated with root of *Macleania cordifolia* (Ericaceae); (4.4) hyphae and vesicles on surface of root of *Clusia* sp. (Clusiaceae); (4.5) hyphae and fungal inclusions in root of *Topobea* sp. (Melastomataceae); (4.6) cyanobacterium on surface of root of *Peperomia cacaophylla* that had been growing in an ant nest-garden; (4.7) chlamydospores produced by a fungus associated with *Topobea* sp.; (4.8) network of thin, poorly staining hyphae and elongate multicelled spores associated with root of *Macleania cordifolia*; (4.9) fungal coils in a root of the orchid *Campylocentrum fasciola*. Bars = 100μ.

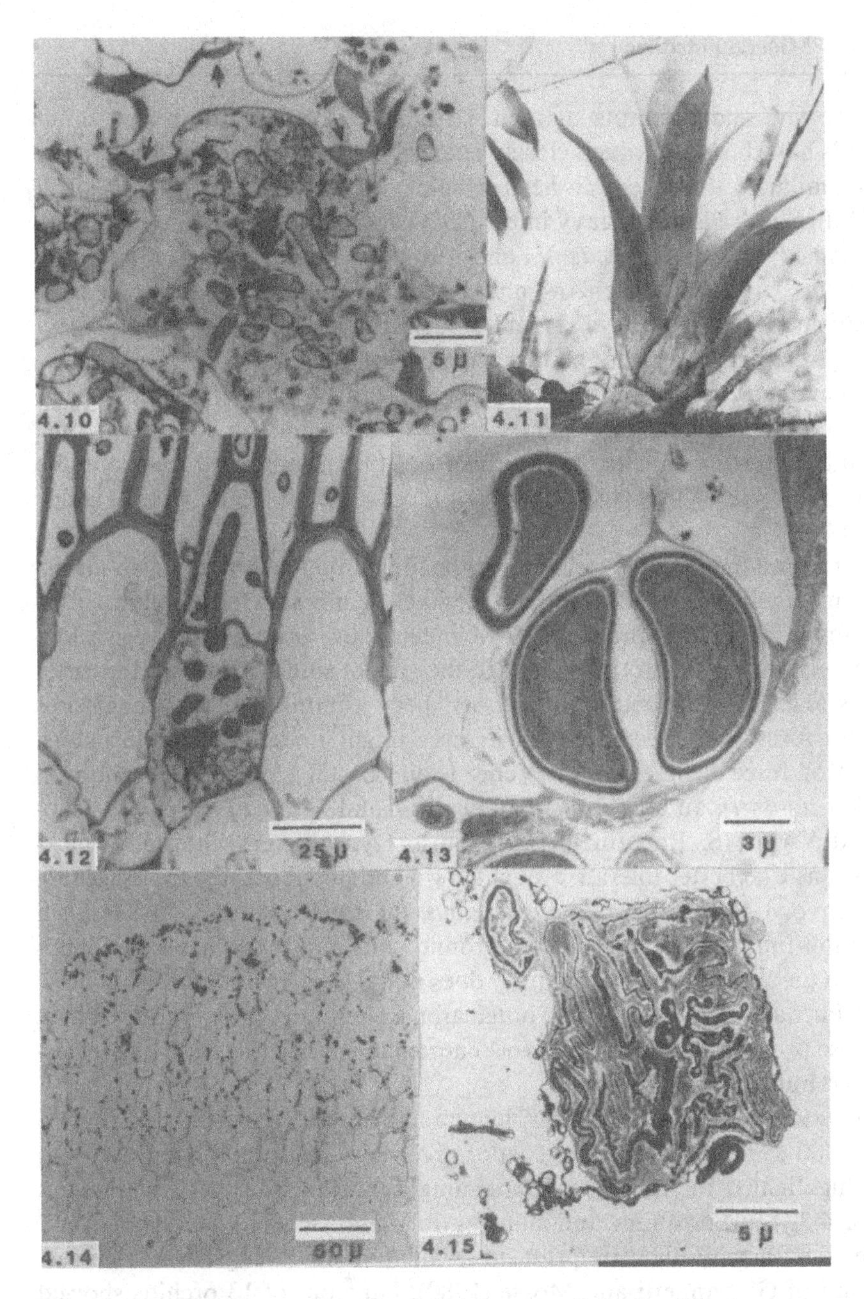

Figures 4.10–4.15. (4.10) Fungal hyphae infecting root cortical cells of the shootless orchid *Polyradicion lindenii.* (4.11) The upright habit and loose cuticular wax on leaf bases of *Catopsis berteroniana.* (4.12) Fungal hyphae infecting a passage cell between thicker-walled U cells in the root exodermis of the shootless orchid *Polyradicion lindenii.* (4.13) Unicellular cyanobacteria located in the velamen of *Sobralia macrantha.* (4.14) Autoradiograph of microbes in velamen cells of the orchid *Epidendrum radicans* after a feeding of [^{3}H]leucine. (4.15) A fungal peloton produced in a root cortex cell of the shootless orchid *Harrisella porrecta.*

zas are few and attempts at identification cursory. A number of adult orchids have been examined closely enough to detect hyphal coils and pelotons similar to those seen in heterotrophic seedlings (Figs. 4.9, 4.10, 4.12, 4.15). In South Florida, heavy infections were found in shootless *Harrisella porrecta, Polyradicion lindenii,* and *Campylocentrum pachyrrhizum* on *Fraxinus caroliniana* (Benzing and Friedman 1981a) and less extensive involvement in leafy encyclias and epidendrums. Hadley and Williamson (1972) were persuaded that neither terrestrial nor epiphytic forms in Malaysia remained significantly mycotrophic beyond early juvenile stages. The question of how long plants remain infected is an important one. Orchid fungi can enhance sorption of P (if not carbon) throughout the entire life cycle of terrestrial *Goodyera repens* (Smith, 1966, 1967; Hadley 1984).

Were wind the sole dispersal mode, the fungi involved in vesicular–arbuscular mycorrhizas (VAMs) would be unlikely symbionts for epiphytes. Certain fossorial rodents promote VAM inoculation of terrestrial vegetation, but the large spores also pass through the guts of some earth-based invertebrates (e.g., "soil isopods"; B. Stinner, pers. comm.). Related fungivores inhabit forest canopies where their feces might infect susceptible plants. Roots of three bromeliads in Venezuelan coastal cloud-forest canopy – *Aechmea lasseri,* an unidentified *Guzmania,* and *Vriesea platynema* – supported VAMs (S. Rabatin, pers. comm.). Hyphae were unusually thick-walled, as if to better tolerate drought. Two and possibly three fungal genera were involved, of which *Glomus tenuous,* the single identified species, is a common inhabitant of P-deficient ground soils and may have been abundant in the understory. Propinquity does not always ensure infection, however. Putz and Holbrook (1987) noted arbuscules in terrestrial roots, but not in those produced by the same two Venezuelan strangler figs growing in suspended humus on palm trunks.

An extensive survey at Río Palenque and in additional humid sites at 800–1800 m in northwestern Ecuador revealed fungi in epiphyte roots and also highlighted the difficulty of assigning significance to the phenomenon (Figs. 4.3–4.9; Bermudes and Benzing in press). Roots were cleared and stained with acid fuchsin prior to evaluation by the gridline intersect method of Giovannetti and Mosse (1980). Nine out of 13 orchids showed pelotons and overall infection rates between 10% and 90% (Fig. 4.9). Vesicles, extensive intracellular hyphae (Figs. 4.4, 4.8), and chlamydospores (Fig. 4.7) characteristic of VAM were noted in a fern, two gesneriads, two *Peperomia* species, a *Topobea* (Melastomataceae), a *Burmeistera* (Campanulaceae), and a *Cyclanthus* species. Infection rates in the finest roots were mostly well below 50%, however. Members of Araceae, Bromeliaceae, and Marcgraviaceae were almost free of fungi. The condition and associations of roots taken from three ericaceous genera (e.g., *Macleania cordifolia;* Figs. 4.3, 4.8) and additional nonorchids were especially difficult to interpret. Hyphae were of several thicknesses, degrees of septation, and affinities for acid fuchsin. Numerous spores, including multicelled forms, were also common (Figs. 4.3, 4.8). Hyphae penetrated hairs and other root cells of epiphytes in quite a few cases, but nowhere could an association be declared unequivocally mutualistic. Atypical mycobionts or unusual morphology reflecting harsh environment require that identification employ more than the usual criteria. Moreover, long-held but incorrect convictions regarding the types and frequencies of mycorrhizas in certain taxa should be abandoned in further surveys (Newman and Reddell 1987).

A casual nutritional mutualism exists between microbes and roots of many earth-bound plants. Algae and bacteria in the velamina of epiphytic orchids and aroids (Figs. 4.13, 4.14) may engage in similar relationships. Conceivably, some prokaryotes reduce N_2 whereas others increase the root's capacity to capture other ions. Acetylene-reduction assays of root segments from *Campylocentrum fasciola* and a *Sobralia* species at Río Palenque yielded negative results, but adjacent media may be more accommodating. There was an uncommon amount of nitrogenase activity in the phyllosphere of certain epiphytic orchids compared to other foliage in an Indian forest (Sengupta et al. 1981). The reason for this phenomenon was not identified, but epiphyte leaves, rather than offering any special chemical advantage, may simply live long enough to encourage denser bacterial colonization. Bentley and Carpenter (1984) examined fronds that were several years old and recorded transfer of newly fixed N from epiphyllae to their host palm *Welfia geogii. Zamia pseudoparasitica,* one of only two epiphytic cycads but a common resident in certain Panamanian wet-forest habitats, possesses characteristic coraloid roots containing cyanobacteria (H. Luther, pers. comm.).

Additional findings suggest that N_2 fixation by free-living photoautotrophs is even more widespread in humid tropical forest canopies. Solids taken from bromeliad tanks and the seepage zones below them, plus encrustations from branches, all reduced acetylene (Bermudes and Benzing unpubl. data). Rates varied with the sample and its state of hydration. Microbes in air-dried scrapings from *Theobroma* branches at Río Palenque fixed N_2 at rates of 5.4–17.7 ng (g sample)$^{-1}$h^{-1}. Parallel sampling at the same sites after rain showers yielded higher values ranging from 8.5 to 110.0 ng (g sample)$^{-1}$h^{-1}. Fritz-Sheridan and Portécop (1987) obtained similar results in cloud forest of La Soufrière, Guadeloupe from cyanophytic epiphyllae and lichen endophytes on the bark of oft-hemiepiphytic *Clusia* and *Norantea* and some free-standing trees near the summit (1467 m). Rates of acetylene reduction by epiphyllae on *Clusia mangle* differed by up to 47 fold on misty versus dry sunny days. Nitrogenase activity in *Norantea spiciflora* increased with leaf age (range = 0.03–2.22 nmol $cm^{-2}h^{-1}$). If precedents elsewhere apply in tropical communities, biological N_2 fixation may be quite significant for epiphyte nutrition. Certain temperate sites are known to be substantially benefited by canopy-based input. The single epiphytic lichen *Lobaria oregana* added 1.5–7.0 kg N $ha^{-1}yr^{-1}$ to a Pacific Northwest forest (Pike 1978), up to half of that received in precipitation at the Costa Rican site surveyed by Nadkarni (1984; Fig. 4.23).

Feeding by ants

Specialized epiphytes engage in at least two nutritionally advantageous ant–plant associations (Madison 1979a; Huxley 1980). Certain members of Araceae, Bromeliaceae, Gesneriaceae, Orchidaceae, and Piperaceae regularly root in nests constructed of carton by arboreal ants (Figs. 1.17, 5.10, 5.11); they also produce myrmecochores, some with conspicuous, edible appendages, that promote dispersal from established to developing ant nest-gardens. These communities merit closer study to determine whether plant requirements match the substratum characteristics of specific ant architects. Carton is a composite material that can be quite nutritive (Table 4.2) if it includes soil or feces; insect species using more inert plant fiber produce inferior seed beds. Adding to the complexity of nest-garden substrata are honeydew and the ants' potentially microbiocidal secretions (Maschwitz and Hölldobler 1970). In addition, myrmecochores could enhance nest quality if their volatiles suppress growth of saprophytes and ant pathogens (Davidson and Epstein in press). However, mycelia of at least one fungus, *Cladosporium myrmecophilum,* regularly permeates cartons with no obvious effect on animal or plant inhabitants.

Ants may play a far greater role in providing nutrients in tree crowns than is currently realized. These insects regularly colonize debris trapped by *Platycerium* (e.g., *Pheidole* in Borneo) and Papuan *Drynaria quercifilia,* a fern whose fronds produce unusually amino acid-rich nectar (Koptur, Smith, and Baker 1982). Ants also often colonize the bases of epiphytes that have not yet been identified as benefactors of ant-borne nutrients (e.g., one or more species of *Aeschynanthus, Araeococcus,* and *Dendrochilum*). Carton galleries regularly crisscross much bark surface in some Amazonian forests (pers. obser.), allowing plants that have only sporadic association with ants to contact beneficial ant products. Longino (1986) has proposed that ant nest-gardens are only the most obvious manifestation of a widespread and general use of carton materials by epiphytes.

Quite intriguing is the near-total restriction of ant-fed ant-house plants (Fig. 4.24) to the forest canopy (Thompson 1981). One or more epiphytes of at least seven families (Asclepiadaceae, Bromeliaceae, Melastomataceae, Nepenthaceae, Orchidaceae, Piperaceae, and Polypodiaceae) regularly harbor ant colonies in domatia (Huxley 1980). Probably other epiphytic taxa also merit inclusion. For instance, *Markea* (Solanaceae) sometimes houses ants in what have generally been considered storage tubers; related *Ectozoma* and *Juanulloa* may do likewise. Most ant-fed ant-house epiphytes offer only transitory trophic rewards (e.g., *Schomburgkia*) or none (e.g., *Tillandsia*) beyond those needed for pollination or seed dispersal. The most important contribution these epiphytes give to their resident fauna is low-cost housing (although they may also provide valuable volatiles, support for aphids which ants can "farm," and occasionally a little nectar and/or edible seed appendages). In return, the epiphyte is definitely "ant-fed," sometimes with a modicum of protection thrown in. This situation is quite unlike that of the terrestrial arborescent ant-house forms that shelter ants in what are probably nonabsorptive thorns and stems (e.g., *Acacia, Cecropia*); here, defense rather than nutrition is the greater benefit to the plant. These ant-guarded ant-house terrestrials feed and house massive and costly populations of aggressive ants–a pattern ideal for defense. But contrary to ant-house ants living in epiphytes, those protecting large ant-house terrestrials rely heavily on host rewards, a practice unlikely to bring in substantial nutrients from beyond the myrmecophyte. Finally, terrestrial ant-house plants are not noted for occurrence on infertile ground where ion supplements would be most beneficial.

At present, demonstration of nutrient flux between ant and ant-house epiphyte is preliminary (Fig. 4.17; Benzing 1970a; Huxley 1978; Rickson 1979). Experiments have shown simply that ant-borne nutrients carried into nesting cavities eventually end up distributed throughout the myrmecophyte; no assessment has been made of nutrient budgets or peculiarities of uptake. Hydnophytinae are reputedly fed by ants who pack domatia with refuse rather than ejecting it, supposedly an unusual practice for arboreal species. But closer study may reveal that ants inhabiting cavities of nonepiphytes do the same. Tank epiphytes also have mutualistic associations with an extensive microflora, diverse invertebrates, and the occasional vertebrate; more about this remarkable phenomenon later.

Hemiepiphytism

Secondary hemiepiphytes early in life, and the primary hemiepiphytes at more advanced stages, root into the forest floor. Shift in nutritional status with transition to or from canopy anchorage seems probable, but in fact need not always occur. Putz and Holbrook (1987) examined decaying debris in persistent leaf bases of the Venezuelan coryphoid palm *Copernicia tectorum* hosting *Ficus pertusa* and *F. trigonata;* 6750 cm^3 of suspended organic soil had accumulated per linear meter of trunk. Nutrients were abundant (Table 4.6), originating in part from numerous termites, ants, rodents, iguanas, birds, and snakes that regularly inhabit trunk crevices. One might even argue that hemiepiphytism is, in this instance, favored by the nutrients accessible to canopy roots. Values for total N, K, magnesium (Mg), and cation exchange capacity all exceeded by substantial margins those in the upper 10 cm of the ground below. Significantly, foliar composition in the epiphytic stages showed no signs of nutritional privation. Colonization of palm trunks after the hemiepiphyte has an alternative also accords with the chemical evidence that canopy soil at least equals ground soil as a source of nutrients. Roots of free-standing strangler figs can penetrate debris on *Copernicia tectorum* trunks up to 9 m away. Wholly terrestrial species offer a parallel in some oligotrophic Amazonian forests (Sanford 1987) where apogeotropic roots of numerous plants ascend trunks up to 12 m, presumably in response to Ca^{2+} gradients created by stemflow. Some trees produce adventitious roots to tap suspended humus in temperate and tropical humid forests (Nadkarni 1981).

Humus-based nutrition

Humus-based epiphytes fall into two classes: Class 1 are nonimpounders that draw mineral ions from rotting wood, ant nests, or, more commonly, layers of suspended humus and the plant life clothing bark; class 2 are impounders of litter or other nutritive solids which are then held against absorptive roots or leaves (e.g., Figs. 4.19, 4.25A). High humidity is required by most but not all humus epiphytes. Tank (phytotelm) bromeliads can inhabit drier sites because of their ability to intercept and hold moisture as well as falling debris. Species requiring a humus mat are, like their substrata, always found in humid locations. Arboreal ant nest-garden flora comprise a small specialized subset of class 1; they are a common feature of moist neotropical woodlands, but rarer in dry and pluvial forests.

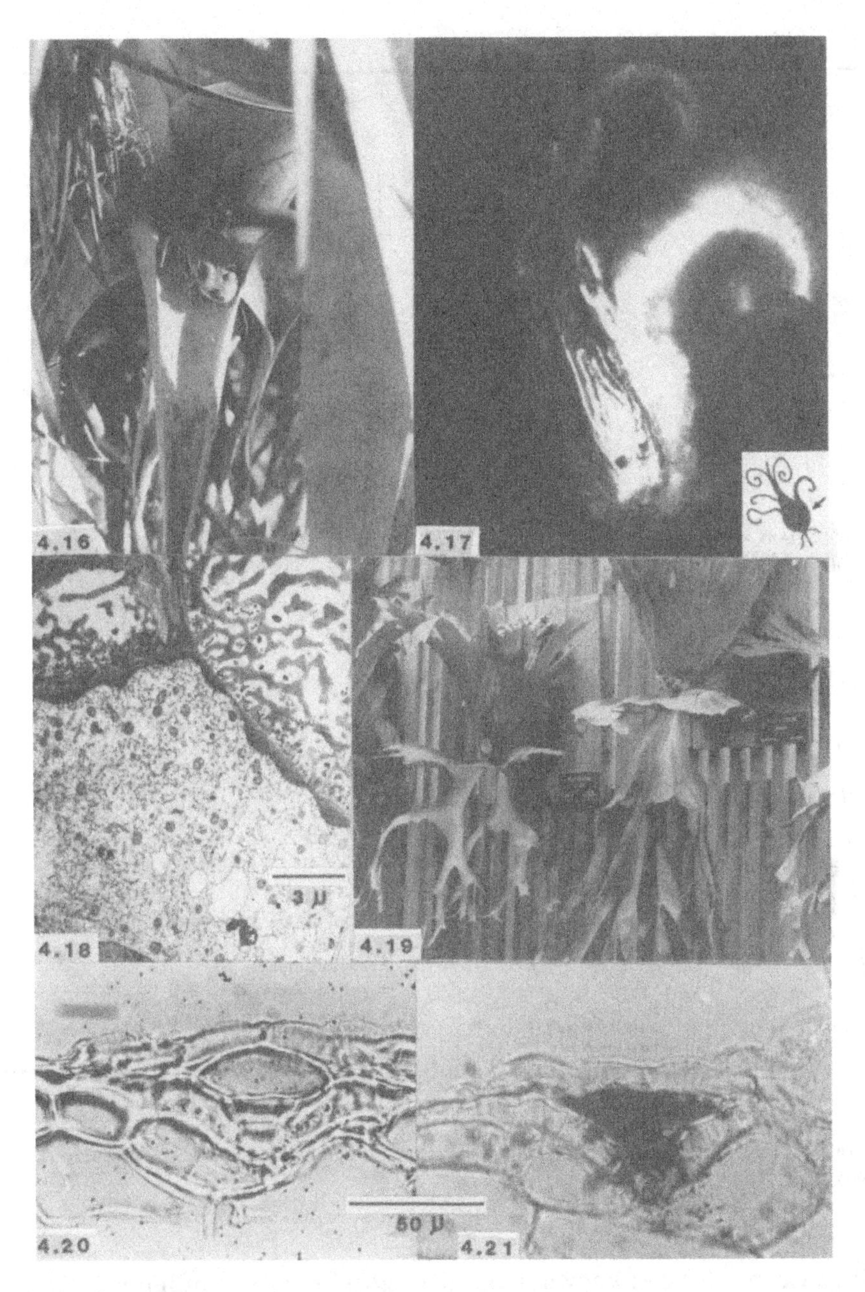

Figures 4.16–4.21. (4.16) Small frog resting in the leaf base of a large *Tillandsia utriculata* specimen in Florida. (4.17) Autoradiograph of myrmecophytic *Tillandsia caput-medusae* three weeks following exposure to a radiocalcium source. Inset illustrates plant habit and point of isotope application. (After Benzing 1970a.) (4.18) Electron micrograph of part of the passage cell of *Sobralia macrantha.* Note the dense organelle-rich cytoplasm and adjacent uneven wall below the adjacent fibrous body. (4.19) *Platycerium* spp. illustrating dimorphic foliage responsible for impoundment. (4.20) Autoradiograph of a single untreated absorbing trichome of *Tillandsia streptophylla.* (4.21) Autoradiograph of a single absorbing trichome after a half hour exposure to [^{3}H]leucine. (See Benzing et al. 1976, for detailed experimental protocol.)

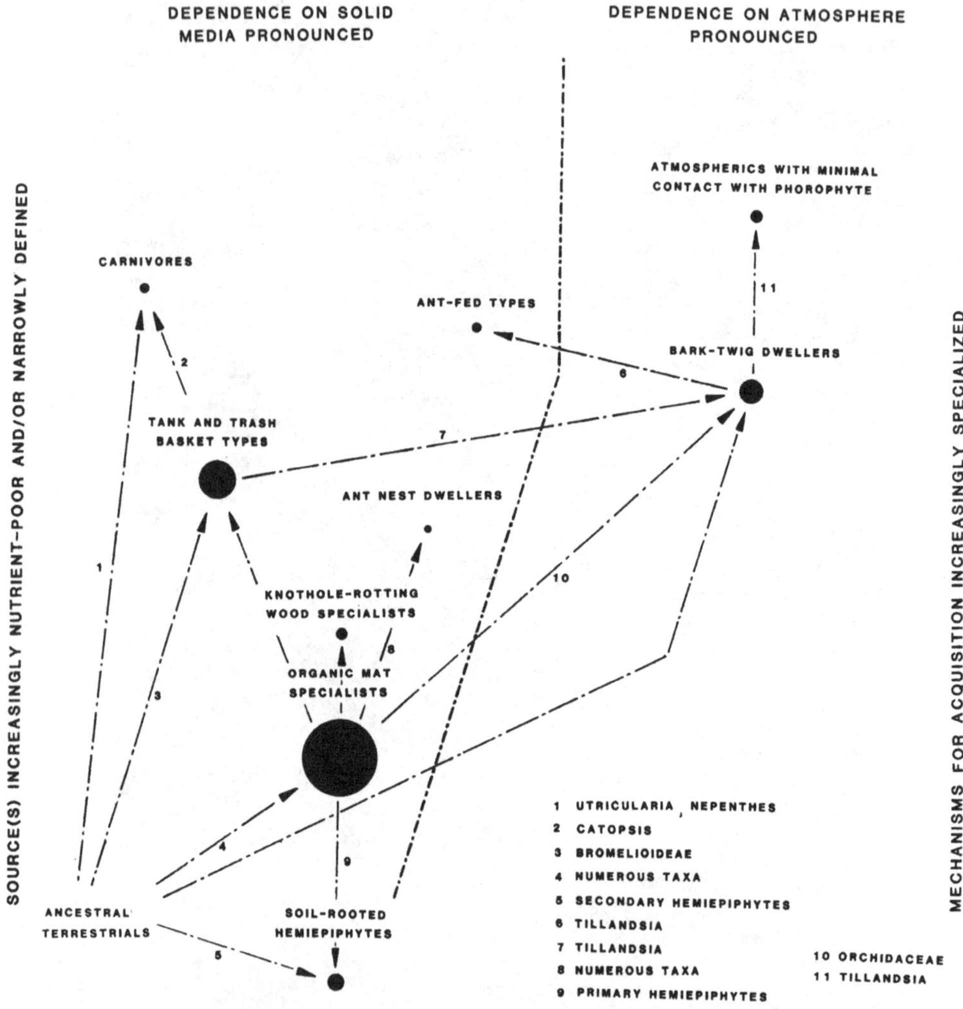

Figure 4.22. Putative evolutionary relationships among the nutritional modes of vascular epiphytes.

Whereas too little is known about the nutritive quality of humus mats, decaying wood, and arboreal ant nest-gardens to warrant much discussion of their utility as mineral ion sources for epiphytes, the impoundment mechanisms and its consequences for plants and associated fauna has attracted considerable attention. Because of the variety of tank types present in Bromeliaceae (Fig. 4.25A,B,E,G) and the family's dominance in so many neo-

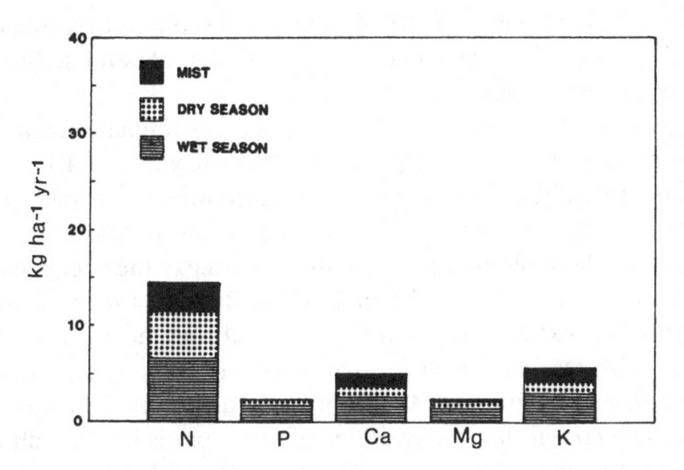

Figure 4.23. Wet and dry season atmospheric inputs of five elements in a cloud forest habitat in central Costa Rica. (After Nadkarni 1984.)

tropical forests, this group will provide the focus for much of the following discussion. Various bromeliads with catchments have been categorized as carnivores, ant plants, and trash-basket forms, but most qualifying taxa lack familiar labels altogether. In fact, nutrition promoted by impoundment is multifaceted, in accordance with the kind of substrates utilized and the manner in which they arrive and are processed. In all instances, animals assist to some degree, but the methods vary, as does the fate of the fauna involved. The use of catchments to capture animal prey rather than to intercept and process litter with mutualists requires the greater plant specialization; thus epiphyte carnivory will be addressed first.

Epiphytic carnivores

Tradition decrees that carnivorous plants possess devices such as traps, secretions, fragrances unequivocally designed to attract, kill, and degrade fauna for trophic gain. Implicit is a second criterion: that supporting substrata must be sufficiently infertile to justify investment in such devices (Givnish, Burkhardt, Happel, and Weintraub 1984; Benzing 1987b). Cost scales upward as plant features originally concerned with capture of energy become more specialized for new functions. Strong selection must account for the elaborate trap leaves of a *Dionaea* or a *Nepenthes,* considering how inefficient these organs are for light harvest. The longevity of a great many, but certainly not all, carnivorous traps suggests that extended service is required to justify expensive construction and reduced photosynthetic output. Pitfall *Nepenthes* features some of the most durable and self-shading leaves among the carnivores.

Additional external factors beyond nutrient scarcity influence plant economy and determine the utility of particular kinds (e.g., Figs. 4.16, 4.24B, 4.25H) of faunal use. The familiar botanical carnivores – sundews, pitcher plants, bladderworts, and other active trappers – are relegated to moist, exposed habitats where photosynthetic output can repay the energy costs of carnivory (Thompson 1981; Givnish et al. 1984; Benzing 1987b). Protocarnivores (plants that extract nutrients from animal prey in relatively lesser amounts and with less expensive devices vis-à-vis carnivores) should be more flexible in their requirements for moisture and light. Just how often conditions in tree crowns favor any degree of carnivory is unclear, although the existence of so few prey-utilizing epiphytes suggests that they are exceptional (Thompson 1981).

Carnivorous epiphytes belong to two, possibly three, families: Lentibulariaceae (*Utricularia* and *Pinguicula*), Nepenthaceae (Fig. 4.25H), and perhaps Bromeliaceae. The epiphytic bladderworts seem to differ little from terrestrial relatives and, in fact, grow under much the same conditions; all are essentially aquatic, rooting in moist humus and bryophyte mats or in bromeliad tanks (Fig. 1.14). Sodden moss-covered rocks and fallen logs accommodate similar species and presumably harbor the same small swimming prey. No obvious features differentiate Cuban *Pinguicula lignicola* or *P. casabitoana* from terrestrial congenerics. Canopy-adapted *Nepenthes* contains mostly vining hemiepiphytes that again differ little in habit and nutritional mode from fully earth-based relatives. The case for carnivory in epiphytic Bromeliaceae remains equivocal.

Bromeliad carnivory

Claims that label certain bromeliads as carnivorous are not convincing. William M. Wheeler (1921), in his classical work on neotropical ant plants said of *Tillandsia* with hollow bulbous bases (i.e., ant-house species, Fig. 4.24B) that ants "make fatal incursions into water-containing chambers." Wheeler's trapping sequence could not be corroborated using either Mesoamerican *Tillandsia butzii* or *T. caput-medusae* (Benzing 1970a). Dissected rosettes proved to have dry axils teeming with brood and adult ants, and all attempts to fill intact rosettes with water by immersion or spraying failed. Present, nevertheless, were nutrients and the capacity to incorporate them: Ant-deposited debris contained considerable soluble N; radiocalcium administered to axils of several intact shoots was taken up (presumably by trichomes) and translocated throughout the plant body (Fig. 4.17).

Picado (1913) first suggested that epiphytic Bromeliaceae include carnivorous species; amino acids placed in the phytotelmata of certain tillandsioids in Costa Rica were seemingly absorbed by adjacent leaf surfaces. Digestive enzymes occurred in mucilage released by injured specimens, but no secretory glands were described, nor could Picado prove that the observed proteins were of botanical rather than microbial origin. Two rather typical phytotelm forms, an *Aechmea* and a *Nidularium,* also took up amino acids from tank fluids (Benzing 1970b), but absorption of substances released by degrading tissues, although consistent with litter and prey use, is not proof of carnivory.

Catopsis berteroniana (Fig. 4.11), an epiphytic bromeliad ranging from southern Florida to southeastern Brazil, is a reputed carnivore (Fish 1976). Its rosette is upright, yellower than most, and covered with a loose, whitish, cuticular powder. Leaf bases are coated more heavily than are blades. Exposed sites at the tops of tree crowns support the densest populations, and tanks contain relatively more animal remains and less plant material than is the case with most other phytotelm bromeliads. Fish presumed that prey acquisition by *C. berteroniana* is enhanced by cuticular reflectance of ultraviolet (UV) radiation that confuses flying insects as they orient toward sky light while negotiating canopy obstructions. Colliding animals tumble into tank fluids; escape is prevented by the lubricating effects of copious wax particles on leaves. Prey subsequently drowns and decomposes, releasing nutrient ions that enter the shoot via absorptive trichomes. No digestive secretions were reported.

Catopsis berteroniana captured more prey than did several other tank bromeliads when tested in southern Florida (Fish 1976; Frank and O'Meara 1984), but more definitive experiments are needed to assess the proposed role of UV reflectance in that activity. In addition, quantification of nutrients in plants and tanks, and rates of interception, are essential in order to determine whether prey represents significant nutritional input. If the shoot is unusually attractive to terrestrial insects and enough of them can be trapped to reduce need for litter significantly, then designation as a low-grade carnivore would be justified. *Brocchinia reducta* (Fig. 4.25E) and perhaps closely related *B. hectioides* are the best candidates for bromeliad carnivory (Givnish et al. 1984), but both taxa are strictly terrestrial.

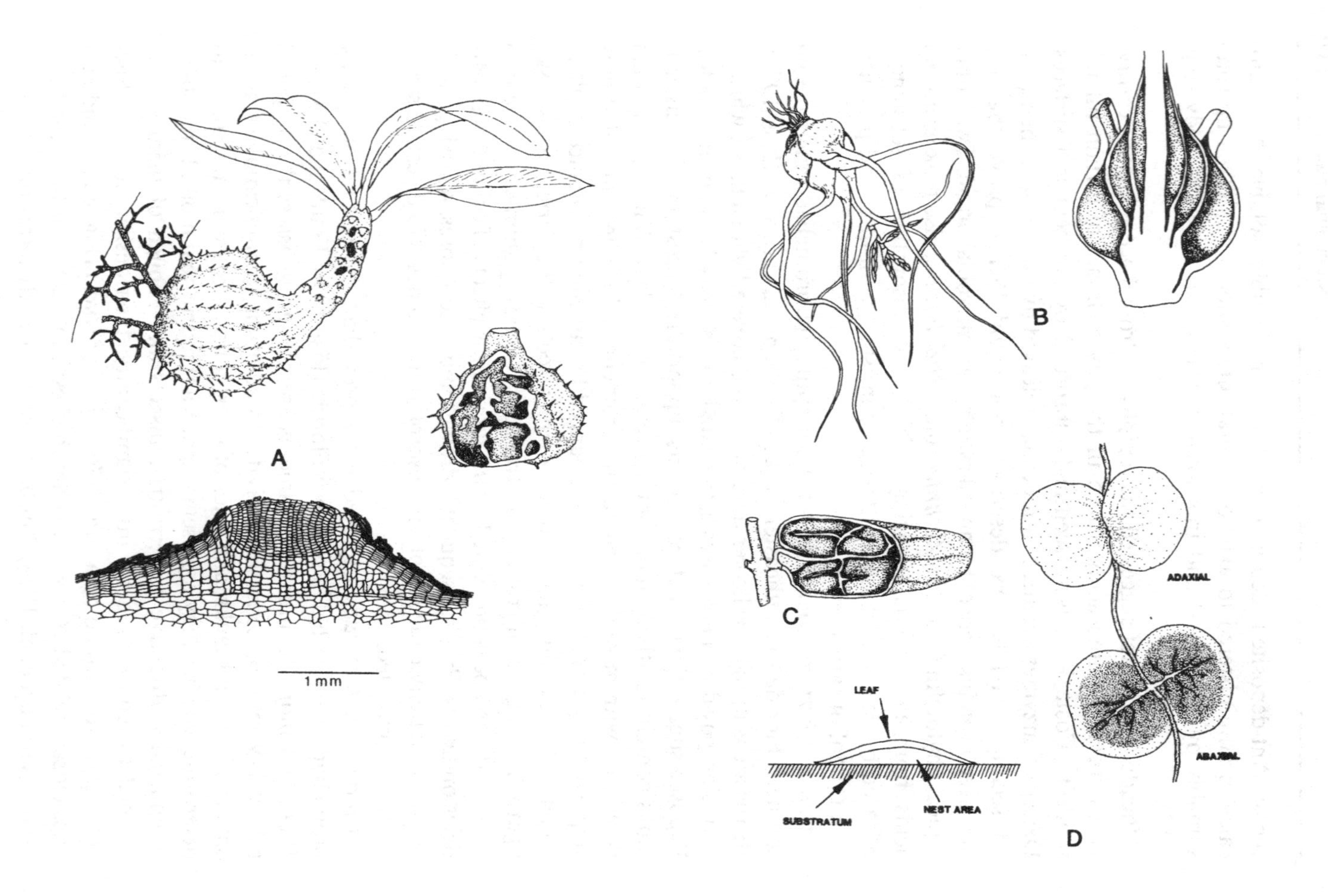

A
1 mm
B
C
LEAF
SUBSTRATUM
NEST AREA
ADAXIAL
ABAXIAL
D

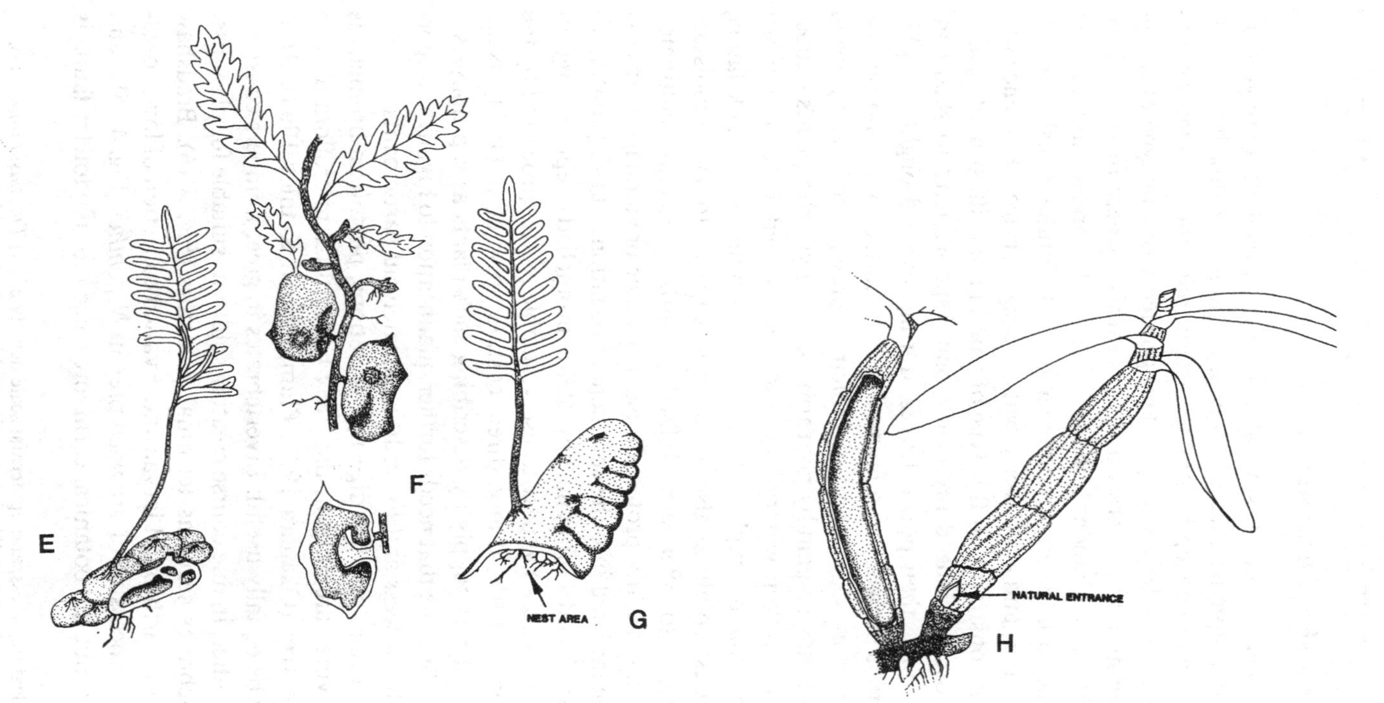

Figure 4.24. Ant-fed ant-house epiphytes. (A) *Myrmecodia tuberosa;* habit, tuber chambers, and transverse section (T.S.) through absorptive papillae. (B) *Tillandsia bulbosa;* habit and bulb. (C) Pouch leaf of *Dischidia rafflesiana.* (D) Convex leaves of *D. collyris,* including illustration of an ant nest chamber between leaf and bark. (E) *Lecanopteris mirabilis.* (F) *Solenopteris bifrons,* including tuber anatomy. (G) *Lecanopteris carnosa.* (H) *Schomburgkia* sp., illustrating hollow pseudobulbs. (B–G redrawn from Huxley 1980.)

Animal assistance to noncarnivores

Carnivory is only one of several mechanisms available to phytotelm and other types of epiphytes that gain ions from fauna. Feeding by live ants takes plant economy one step beyond the use of prey and involves no animal destruction at all; Thompson (1981) suggested that this type of mutualism is more common in tree crowns than is carnivory because of low cost, frequent drought, and abundant shade. Indeed, the organs modified to house ant colonies are usually durable stems that simultaneously provide mechanical, vascular, and even tissue water storage services. Evidence of a plant cost/benefit constraint on this type of animal use is illustrated by *Dischidia* whose most specialized myrmecotrophic species tend to shed conventional leaves first, then pouch leaves, during severe drought. Trophic myrmecophytism is probably also favored by the shortage of ant-house cavities in forest canopies (Davidson and Epstein in press). The lowest price of all for animal use may be paid by noncarnivorous phytotelm plants – those that attract biota into their watery impoundments to feed, hide, or oviposit in safety. The phytotelm epiphyte that invests in housing for tank fauna, unlike the ant-house epiphyte, also increases its capacity to obtain moisture and does not have to offer special food rewards or tolerate cultivated Homoptera. Of course, any protection against other phytophagous insects thus lost must be factored into the cost/benefit equation. At this point, hundreds of terrestrial and epiphytic phytotelm bromeliads representing all three subfamilies (Fig. 2.4) appear to use vegetable matter processed by the biota they nurture at little, if any, direct cost. Givnish et al. (1984) recognized the nature of this supply by describing such plants as saprophytes – an accurate label, but one that needs further qualification to highlight important aspects of the process and to credit adequately all participants.

Whether or not they produce digestive secretions, phytotelm bromeliads and pitfall carnivores are alike in that they both utilize microflora to help process raw resources (Okahara 1932; Plummer and Kethley 1964). Resident fauna may be equally useful: Invertebrates in great numbers consume debris and each other, in due course creating solutions suitable for plant use. Occasional vertebrates serve as terminal predators (Fig. 4.16). Bradshaw (1983) reported that arthropod larvae and a varied collection of lower organisms hasten the processing of drowned prey in *Nepenthes* (Fig. 4.25H) and *Sarracenia* traps. But the botanical carnivore's need for phytotelm fauna is

Figure 4.25. Habit, leaf thickness, and epidermal scale density of (A) *Tillandsia imperialis;* (B) *T. flexuosa;* (C) *T. schiedeana;* (D) *T. usneoides.* Habit of (E) *Brocchinia reducta;* (F) *Billbergia zebrina;* (G) *Nidularium bruchellii;* (H) pitchers of *Nepenthes* sp.

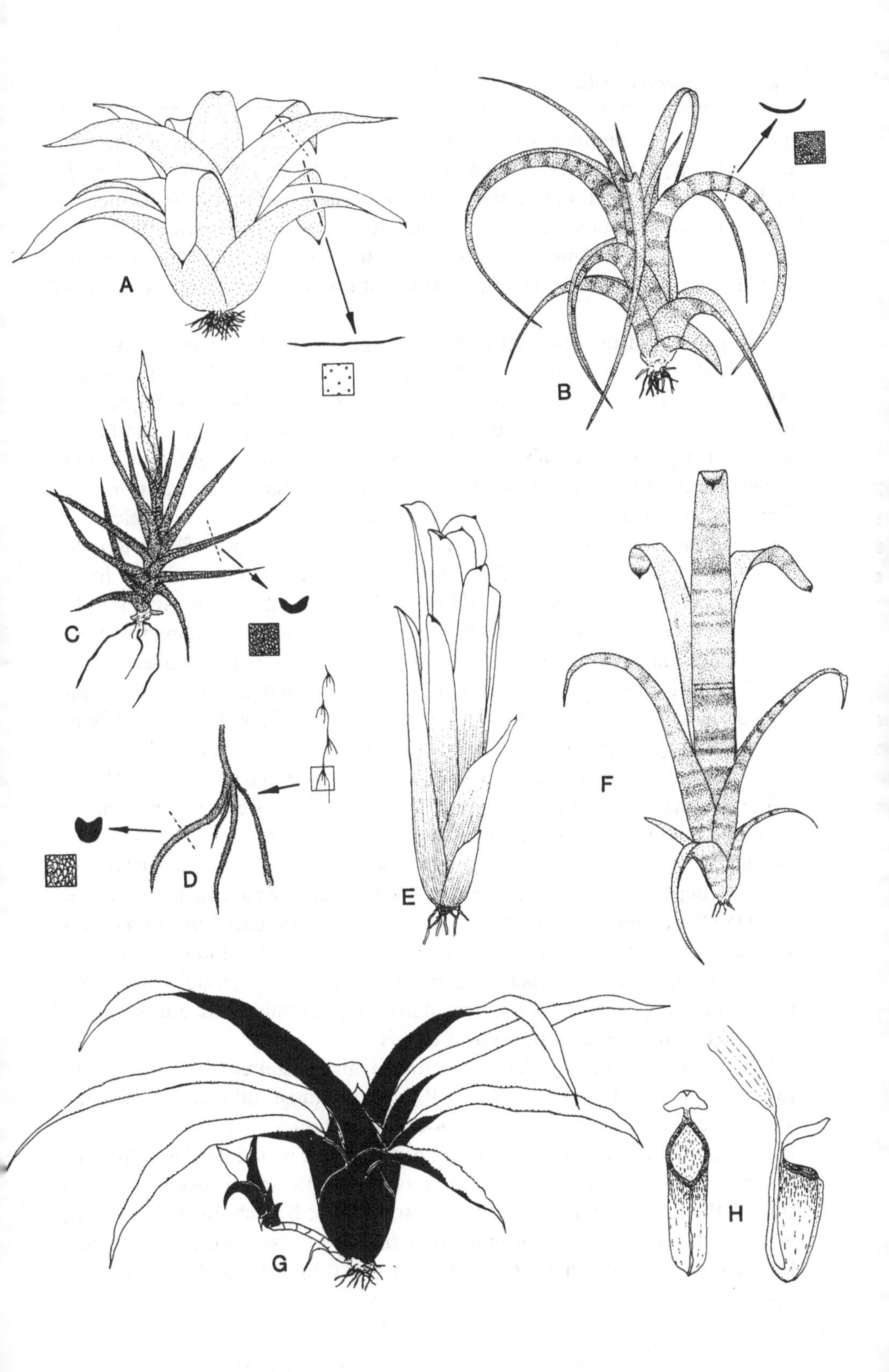
A
B
C
D
E
F
G
H

questionable. A significant difference between the association of animals with pitcher plants, especially the prey-digesting kind, versus that with bromeliads lies in the commensalistic or competitive nature of the former in contrast to the mutualistic nature of the latter (but see Bradshaw 1983). The majority of tank bromeliads might therefore be categorized as animal-assisted saprophytes. Moreover, their nutrition is humus- rather than animal-based.

Comparison of the material captured in pitchers versus most bromeliad tanks further illustrates why few Bromeliaceae are likely to be carnivorous. First of all, the nutrient sources available to forest-dwelling phytotelm plants differ in quality, quantity, and requirements for acquisition and utilization. Litter, although abundant, is less nutritive by weight or volume because of a preponderance of structural polymers that resist digestion; the richer but scarcer animal material is (except for chitinous exoskeletons) more rapidly degraded, probably even without the assistance of invertebrate scavengers. Whereas such mobile elements as K can diffuse quickly from moist solids in both cases, others (N in particular) are doubtless more tightly bound in litter. Just as organic N can be tied up in soil for many years, degradation may be similarly slow in intercepted plant material. Debris collected from leaf axils of adult *Guzmania monostachia* growing in a mixed cypress/hardwood forest in South Florida (Fig. 2.15, Table 4.8) showed N:K ratios ranging from about 5:1 to 39:1 (Benzing and Renfrow 1974a). Because ratios for living, healthy foliage average between about 1:1 and 2:1, either the tank litter was already purged of much of its K by leaching or mobilization prior to abscission, or the N is much more refractory to further biological use. Nutrient release may also vary from one bromeliad to another, depending on the antibiotic character of a particular litter source. Laessle (1961 and Frank (1983) describe broad differences in animal and vegetable contents of tanks, depending on location in the forest; associated nutritional modes were labeled "dendrophilous" or "anemophilous." Primary sources of nutrients for anemophilous types remain obscure but probably include more than wind-blown debris.

Shoot/leaf morphology also determines which resources are most useful to particular phytotelm species. All but the most tubular forms (Fig. 4.25E,F) easily fill leaf axils with debris by maintaining large catchment surfaces (Fig. 4.25A,G). In contrast, pitcher leaves (Fig. 4.25H) have little or no capacity to intercept falling materials; quite often a leaf extension even shields the pitcher orifice. Pitchers are sought out by rich nutrient sources, whereas the humus-based bromeliad, acting more like a kind of botanical filter-feeder, subsists on a diet of randomly settling particulates. It is worth

Table 4.9. **Impoundment capacity of an *Aechmea bracteata* specimen and dilution of an applied dye solution after rain showers**

Tank number by position from the shoot center outward	Capacity of tank (ml)	Percentage of tank capacity filled by 2.6-cm rainfall	Percentage of dye remaining in the tank after 2.1-cm rainfall
Center	93	81.0	39.5
1	61	75.0	16.0
2	43	100	59.5
3	49	89.0	35.0
4	20	100	62.0
5	15	100	66.0
6	46	100	46.5
7	8	63.0	67.0

Source: After Benzing et al. 1971.

noting that the occasional bromeliad that apparently uses degraded animal tissue rather than intercepted humus (e.g., terrestrial *Brocchinia reducta*; Fig. 4.25E) possesses a more funnelform shoot whose architecture resembles that of a single pitcher-plant leaf.

Unlike phytotelm bromeliads, pitcher plants are protected from overfilling by their umbrella-like flaps (Fig. 4.25H). Torrential rains, if funneled in by catchments equal in area to those of most impounding Bromeliaceae, would wash out costly plant enzymes and also allow prey to escape. Those epiphytic bromeliads featuring the common outspread rosette so essential to litter impoundment (Fig. 4.25A) regularly overflow, but turnover is slower than might be expected. Experiments using a nonabsorbent dye in water-filled axils of *Aechmea bracteata* (Fig. 7.5) indicated that loss of tank solutes was substantial but far from complete during a brief shower (Benzing, Derr, and Titus 1971; Table 4.9). Loose, water-logged solids in the same rosette would probably have been even less affected. Various types of tank shoots should be examined for structural features that impede turbulence and turnover of leaf axil contents during rainstorms.

Plant longevity further differentiates carnivorous from humus-based systems by its effects on nutrient processors. The average life-span of a *Sarracenia purpurea* pitcher, and therefore of its aquatic community, is approximately one year (Fish and Hall 1978). Although individual bromeliad tank leaves may also live just a single season (Fish 1983), animals no doubt

Table 4.10. **Percentages of four mineral elements translocated by *Billbergia* sp. following root or tank absorption**

Nutrient element applied	Organ of application	Number of plants treated	Mature leaf blades	Ramet	Rhizome	Mature leaf bases
Ca	roots	4	3.6	48.7	33.9	13.6
	central tank	4	28.3	61.7	9.1	0.9
P	roots	3	17.5	46.1	8.4	27.9
	central tank	1	86.2	11.5	3.0	3.0
Zn	roots	4	2.8	64.2	18.6	14.4
	central tank	2	23.3	50.3	19.2	7.7
Fe	roots	6	7.5	50.5	20.9	21.1
	central tank	6	32.5	24.8	32.0	11.2

Source: Burt and Benzing 1969.

migrate among adjacent congested axils that persist for several years. They may even cross over into the attached ramets of longer-lived colonies where distances are much shorter than those between the mouths of trap leaves in most pitcher plants. The time factor must account in fair measure for the greater complexity of biota in the phytotelm of a humus-based bromeliad as opposed to that of a carnivore.

Finally, detritivores, so crucial to the success of a humus-based phytotelm epiphyte, do not come cost-free; while contributing nutritive ions, they can also carry them off. This price is best viewed as an investment, however; a larva voids more plant nutrients while maturing than it contains after emergence. Use of invertebrate detritivores and saprophytes also obviates the expensive production of glands and digestive secretions; quite likely, this outlay would be prohibitive if plants had to extract needed ions exclusively from vegetable debris without assistance. Moreover, bromeliad trichomes (Fig. 3.21), when compared to secretory organs lining a pitcher, may be more resistant to damage when tanks occasionally dry out. The utility of leaf bases – and presumably the hairs located there – as gleaners of nutrients, even such relatively immobile ones as Ca, has been demonstrated with several species: Nutritive ions were translocated from tank fluids throughout entire plants within a few weeks (Burt and Benzing 1969; Benzing 1970b; Table 4.10).

Bromeliad nutrition revisited

Bromeliad shoot form varies considerably, even including, within a single widely tolerant species, both sun and shade forms (Figs. 2.11, 2.12). Specimens in dark humid environments produce open lax shoots that maximize photon capture (Fig. 2.17); those exposed to higher light levels produce more upright foliage and deeper tanks. Unfortunately, the full benefit of a particular shoot conformation cannot be read reliably from exposure levels alone. For instance, some phytotelm bromeliads (e.g., the funnelform billbergias; Fig. 4.25F) perhaps promote novel and cryptic mutualisms. A member of this group would seem sufficiently well off in that its shoot morphology ensures protection from direct-beam insolation and moisture stress; tanks need only an occasional refill. But an additional benefit sometimes arrives in the form of a frog that tightly occludes a foliar tube with its head in order to secure sanctuary during the hottest, driest part of the day (E. McWilliams, pers. comm.); meanwhile, valuable animal excretions are probably voided for plant use. More will be said later about tank shape and tolerances of species with particular shoot forms.

Trophic mutualism was previously offered as a reasonable explanation for patterned tank foliage; some of these interactions may be fairly intimate. Fauna are often endemic to bromeliad leaf axils, but specificity is not absolute. Certain mosquitoes utilize shoots of a single genus (Istock, Tanner, and Zimmer 1983), but other users exploit more diverse plant species. Several treehole mosquitoes in Florida provide a possible parallel as they cue on chemical and physical traits that relate less to tree identity than to probable duration of the cavity's moisture supply (Bradshaw and Holzapfel 1988). Conceivably, stimuli of the type that guide ovipositing *Wyeomyia smithii* to young *Sarracenia purpurea* leaves (Bradshaw 1983) account for the breeding patterns of some bromeliad inhabitants. If so, analysis might show whether the attractants involved are produced by other larvae, food materials, or the bromeliad itself, in which case the substances would work to the plant's own benefit. The humus-based and carnivorous mechanisms may both incorporate chemical lures, but the botanical carnivore's fragrance often elicits a feeding, rather than egg-laying, response.

Atmospheric nutrition

Neither humus nor appreciable animal products are available to PS species. At the extreme, mist and root/leaf tangle epiphytes of Orchidaceae barely contact host surfaces at all (Fig. 1.7). Certain atmospheric bromeliads possess roots that securely grip the substratum but are essentially nonabsorptive (Fig. 1.11). Those of some other forms (e.g., certain ferns, *Hoya*) often seem too reduced to provide much benefit beyond mechanical support. As noted previously, the specialized shoot epidermis and root mantle alternatively promote hydration and reduce drought stress for many bromeliads and orchids. Similar or analogous features may exist in other taxa.

The velamen of orchids, which proves so efficacious for water balance, may have another impact on mineral nutrition, especially where nutritive fluids arrive sporadically. It is not clear at this point how different thicknesses and arrangements of rhizodermal cells (Fig. 3.19A–G) affect bulk exchange between imbibed canopy fluids and those still flowing along root surfaces. Very likely, diffusion is too slow to allow capture of ions from moving fluids by exodermal transfer cells seated up to 24-cell widths and several millimeters beneath the root surface. And of course, stagnation within the engorged velamen could be advantageous because the first solutions to arrive during a storm are the most heavily charged with nutrients.

Greater insight into involvement of the velamentous root in epiphyte nutrition will come through investigation of the exodermis, particularly its transfer points, under conditions that approximate those in nature. Like transfer cells elsewhere, exodermal passage cells of *Sobralia macrantha* (Figs. 3.9, 4.18) contain dense protoplasts equipped with many mitochondria and membranes (Benzing et al. 1982b). What little information is available offers a tantalizing glimpse of orchid root function. For instance, velamina stripped from *Vanilla planifolia* exhibit unusual capacity to bind cations (Böttger, Soll, and Gasché 1980). Radioisotopes have been used to demonstrate uptake and translocation by aerial roots of several orchids (e.g., Haas 1975). Absorption kinetics have never been measured. Dycus and Knudson (1957) had previously obtained rather puzzling results which suggested that such roots are impenetrable to ions except where velamina have been distorted by growth against bark or some other solid object.

Direct evidence of foliar trichome involvement in bromeliad nutrition was obtained by autoradiography using tritiated amino acids (Benzing, Henderson, Kessel, and Sulak 1976). Trichome stalk cells (Figs. 4.20, 4.21) of all exposed tillandsioids were found to contain substantial quantities of tritium, whereas adjacent epidermal cells were unaffected. In another set of experiments, excised leaf blades of atmospheric *Tillandsia* took up more ^{45}CA, ^{32}P, and ^{35}S in three hours than did comparable samples of sparsely trichomed, tank-forming *Catopsis nutans* (Benzing and Pridgeon 1983). Parallel runs using leaf tissue of nine pleurothallid orchids produced accumulations comparable to or below those of *C. nutans* even though the capitate trichomes of the former have been presumed by some investigators to be quite absorptive (Pridgeon 1981). Nyman et al. (1987) demonstrated metabolic involvement in the accumulation of numerous amino acids through intact leaf surfaces of *Tillandsia paucifolia.* Longer-term studies showed that rootless, otherwise intact, *T. paucifolia* specimens possessed remarkable ability to concentrate inorganic ions (Benzing and Renfrow 1980). Daily half-hour immersion in an enriched nutrient solution brought about a 20-fold increase in P content within 120 days. Levels of N and K also increased but to a much lesser extent. Immersion in equimolar solutions (10^{-5}–10^{-7} M for trace elements) killed *T. paucifolia* within 60–90 days. Post hoc examination revealed copper (Cu) concentrations up to 20 fold above initial levels in damaged subjects; zinc (Zn) and molybdenum (Mo) were also much elevated. Concentrations of boron (B), Fe, and several macronutrients were little affected. Affinity for certain relatively toxic elements such as lead (Pb) and arsenic (As) has been used to control *Tillandsia* infestations in South Florida. Slow growth, lack of contact with soil, and a propensity for accumulating additional technological metals like cadmium (Cd), tin (Sn), and vanadium (V), as well as certain synthetic organics, have all encouraged use of *Tillandsia recurvata* and *T. usneoides* as monitors of air quality (e.g., Shacklette and Connor 1973; Schrimpff 1984; Fig. 4.26).

Vegetative reduction

Unusual epiphyte habits beyond those designed to promote impoundment and nidification seem to be associated with nutrition and resource economy. The most curious examples come from Bromeliaceae and Orchidaceae. Although vegetative reduction is involved in both instances, different organ systems are affected. In *Tillandsia* and closely related *Vriesea* of Bromeliaceae, a near-complete suppression of root growth has occurred. Every intermediate condition possible between profuse and very sparse rooting exists in less specialized relatives (Fig. 4.25A–D). Sarcanthinae (Orchidaceae) exhibits a progression of comparable degree but opposite direction (Fig. 4.27A–D). Shoot development is limited to the short monopodial stem needed to support undiminished root production and an occasional axillary scape. Congeneric, leafy, caulescent, and shootless forms (e.g., *Campylocentrum*) demonstrate how rapidly vegetative form has changed in some clades. Consideration of physical environment, related mortality rates, and the abil-

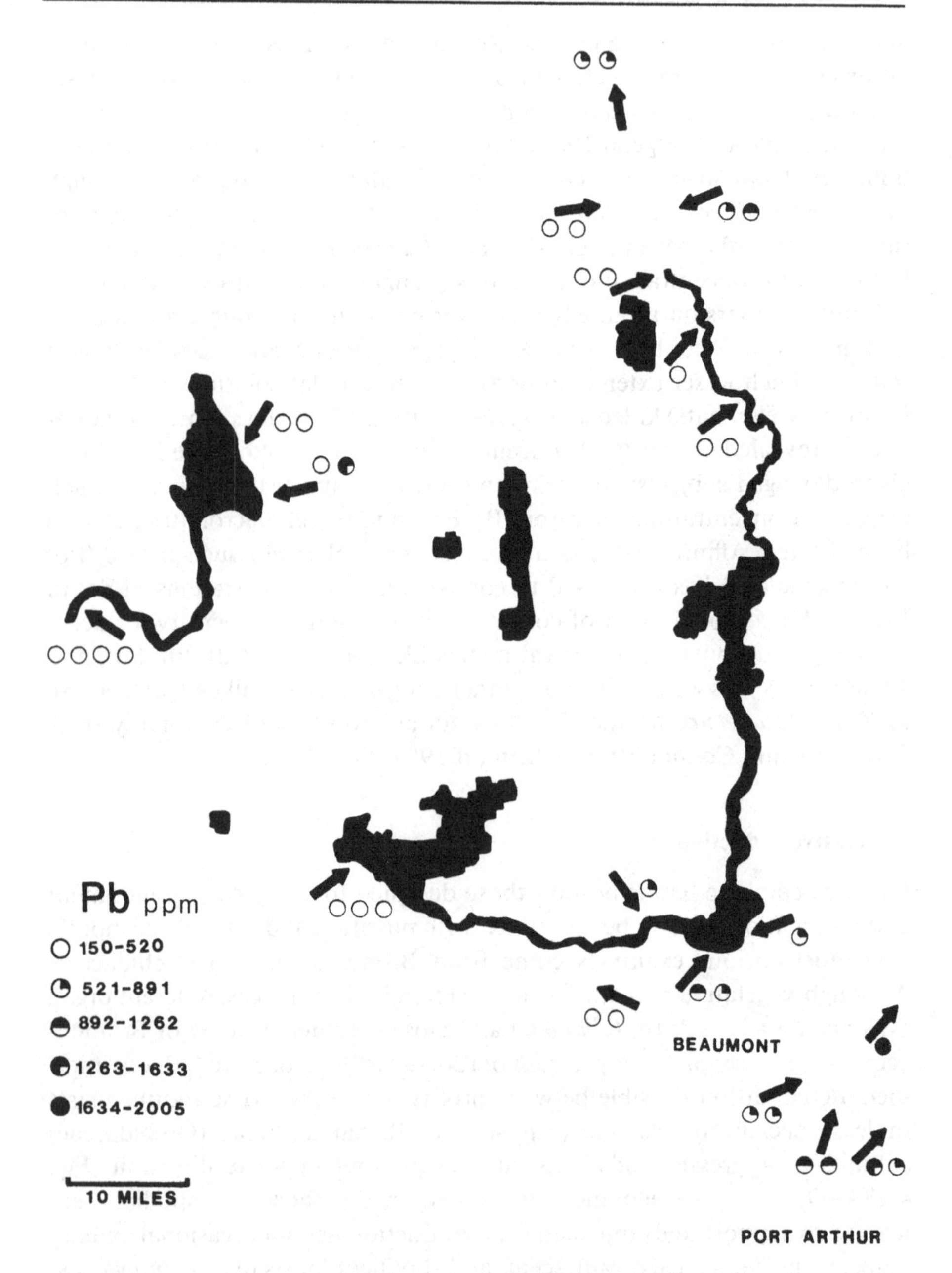

Figure 4.26. Occurrence of lead among samples of *Tillandsia usneoides* collected in the Big Thicket National Preserve, a series of land parcels north and northwest of Port Arthur, Texas, where manufacturing and petrochemical industries are concentrated. Elevated readings remote from Port Arthur represent collections near major highways and the effects of automobile emissions. (Unpublished data.)

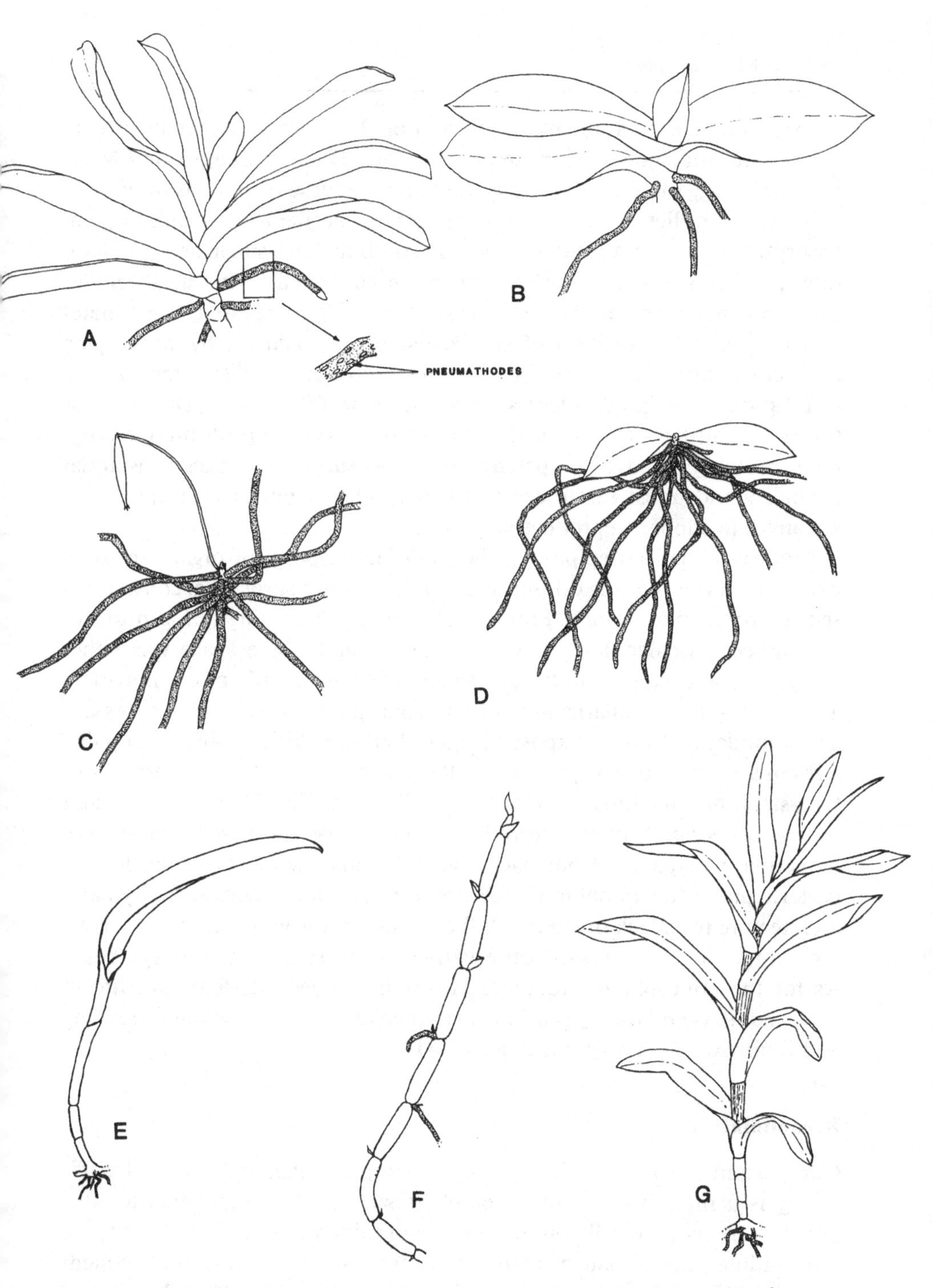

Figure 4.27. Habits representing the gradations between leafy and shootless orchids of subtribe Sarcanthinae and other selected taxa: (A) *Vanda* sp; (B) *Phalaenopsis amabilis;* (C) *Polyradicion lindenii;* (D) *Kingidium taeniale;* (E) *Pleurothallis* sp.; (F) *Vanilla barbellata;* (G) *Epidendrum anceps.*

ity of plants in general to emphasize root or shoot development alternately in order to mitigate specific resource scarcities (Bloom, Chapin, and Mooney 1985) suggest an hypothesis to explain why these reductions occurred.

In both families, all three basic vegetative mechanisms – carbon gain, absorption of moisture, and of ions – are combined in a single organ system, thus providing equal opportunity for similar benefit. This arrangement could enhance fitness in two ways: (1) by diverting energy and scarce material away from production of specialized organs, and (2) by simplifying ontogeny enough to shorten the life cycle (see Chap. 5). Either mechanism could promote epiphyte success. Constraints on PS forms especially favor this restructuring: At the same time that climatic rigor curtails productivity, high mortality imposed by patchy, disturbed substrata obliges substantial fecundity and, most important to this hypothesis, unusual allocations of resources to support reproductive power.

Among specialized epiphytes, the multiple purposes of organs in some cases and corresponding diminution in others is not surprising, considering the environmental context. From strictly aquatic beginnings, progenitors of the earliest vascular flora moved to life on land where space was either energy-rich and exposed to the drying atmosphere, or dark, moist, nutritive, and subterranean; polarization of the sporophyte body was a necessary accommodation. But the exposed twig- or bark-inhabiting epiphyte of today lives in a more uniform space where the benefit – indeed, the necessity – of long-standing functional and structural differentiation between shoot and root systems has become blurred. The advantages of extreme vegetative reduction in epiphytic Bromeliaceae and Orchidaceae must be considerable because major functional trade-offs accompany both progressions. By basic architecture (e.g., lack of stomata), the orchid root would seem to be a poor second to a leaf in terms of conservative water use during photosynthesis. As for the atmospheric bromeliad, foliar trichomes interfere with life in shady habitats and totally preclude existence in overly humid ones (Benzing and Renfrow 1971a; Benzing et al. 1978).

Recapitulation

Circumstantial evidence favors the presumption that nutritional insufficiency is a major constraint on epiphyte success. But arguments for this position are based wholly on the absence of mineral soil in tree crowns, the often dilute and transitory nature of canopy solutions, and the frequent capacity of resident vegetation to utilize alternative nutrient pools, often at what appears to be considerable expense. In fact, epiphyte growth, like that of other vegetation, is to various degrees governed by constraints arising from several quarters. Another look at resource economics will bring this point into clearer focus.

Plants acquire and deploy energy, water, and key nutrients according to patterns that minimize the impact of the scarcest commodity (Chapin, Bloom, Field, and Waring 1987). Abundance in the others allows improved access to those in lesser supply; cost–benefit functions tend to be optimized at many levels. A desert community, for example, traps much less of the abundant PAR than of the meager rainfall by allocating proportionally more biomass to root than to shoot tissue. Similarly, the brief life cycle of an annual weed is made possible by production of a large leaf surface area and only enough root tissue to tap a relatively rich soil. At a still finer level, distribution of N between energy-transducing and carbon fixing machinery in green organs is tailored to match available amounts of radiation or CO_2. Allocation for light and dark reaction mechanisms is emphasized in shade and sun exposures, respectively. In essence, investment of resources in tissue is trimmed to avoid overcapacity that would prove to be superfluous or even hazardous (e.g., excess leaf area where moisture is scarce). On those ion-rich "soils" in humid canopies, and even on rooting media of lesser quality, epiphyte growth may be limited more by frequent cloudiness and shading than by too few essential ions. In arid woodlands, drought may well supersede nutritional deprivation as the primary stress factor. But characteristics of ion supply are never wholly without effect on plant structure and function, and these influences vary with location and epiphyte type.

Bearing in mind that atmospheric as well as soil-borne nutrients occur in numerous (and sometimes little-studied) forms that are variously available to particular epiphytes, one suspects that greater trophic diversity exists among species anchored in tree crowns than is currently recognized. Should uptake of critical ions from vapor or wind-blown particulates prove to be more significant for certain of these species, even distinctions among PS forms may have to be erected. One point is already clear: The colloquial name "air plant," which has long been assigned to epiphytes lacking obvious means of securing moisture and nutrients from fertile solid media, is apt but poorly understood. Utilization of the chemically diverse solids in tree crowns is probably a more varied process than is tapping the atmosphere for essential ions – and not much better defined. Epiphytic vegetation offers considerable challenge to persons wishing to study unusual mechanisms of plant nutrition, but efforts must be broad and integrative. Only comprehensive cost accounting will reveal how ion supply relative to other factors has helped shape form and function among the multiple types of arboreal flora.

5 Reproduction and life history

Constraints peculiar to forest canopy habitats helped shape epiphyte natural history, but the selective forces are not always apparent. Some characters and lineages appear to have been affected more than others; for instance, iteroparity is nearly routine in the group, whereas breeding mechanisms and modes of pollen and seed dispersal are much more diverse. Multiple paths to similar ends complicate the search for a common theme. Enough data are available, however, to offer tentative judgments on several aspects of the plant life cycle that permit success in tree crowns. This chapter deals with such aspects and the factors responsible for their existence.

Breeding systems

Pollination: identity of vectors

If breeding systems differentiate terrestrial from canopy-based pteridophytes, the fact remains unreported. Comparisons of angiosperms are easier, and pollination has been studied in numerous epiphytic flowering plants, especially neotropical Orchidaceae. Pollinators of these taxa tend to be more species-constant and specialized than those serving nearby terrestrials, although sharing of pollinators is sometimes possible. Avians are especially important in northern South America, where nectar-feeding birds and the flora they serve reach unparalleled diversity. Large, heavily ornithophilous families include Bromeliaceae, Ericaceae, Gesneriacese, and Loranthaceae; birds frequently pollinate Cactaceae, Marcgraviaceae *(Norantea),* and Rubiaceae *(Ravnia, Manettia)* as well. The relationship between epiphytism and avian pollination is particularly apparent in families with diverse floral syndromes. Ornithophily in Bignoniaceae is rare except in two epiphytic genera: *Gibsoniothamnus* is entirely pollinated by birds, *Schlegelia* partly so (Gentry and Dodson 1987b). Chiropterophily is well developed in some Central American vrieseas (Bromeliaceae) and *Marcgravia* but is definitely a minor relationship overall. Even less common is the pollination by small rodents of two epiphytic Costa Rican *Blakea* species (Lumer and Schoer 1986).

Insect visitors are extremely diverse. Pleurothallidinae, a neotropical assemblage of about 3,800 mostly epiphytic orchids, along with largely paleotropical *Bulbophyllum* and many small-flowered relatives, are generally pollinated by flies (e.g., *Bradesia, Drosophila,* tachnids). Moths service certain Cactaceae and orchid taxa (e.g., *Epidendrum, Polyradicion*); scarab beetles pollinate Costa Rican aroids (Bawa et al. 1985). Inflorescenses of *Philodendron bipinnatifidum* heat up and emit fruitlike scents sought out by Dynastinae beetles (Gottsberger 1986). The best known pollination liaison is that between male euglossine bees and the Cymbidieae and Maxillarieae – more about this later. Families with inconspicuous reproductive organs are Araliaceae, Moraceae, Piperaceae, Myrsinaceae, and Urticaceae (the last two families are only marginally epiphytic); those of Melastomataceae are similarly drab except for *Blakea* and *Topobea* which produce large bright flowers (Gentry and Dodson 1987b). Anemophily is rare in canopy-based flora, if it exists at all: The best candidates are some *Peperomia,* the two epiphytic grasses, and a couple of mistletoes.

Access to particular pollination syndromes has varied among the epiphytes owing to diverse origins and disparate habitats. Little can be said about phylogenetic constraints, but plant size, population structure, height in the canopy, type of forest, local climate, and the co-occurrence of taxa with the same floral syndromes have clearly influenced reproductive patterns. In humid forest at La Selva, Costa Rica, Bawa et al. (1985) discovered that epiphytic and nonepiphytic vegetation alike shared vectors that were largely restricted to the same canopy level. Surveys of 143 trees were made in order to identify where major groups of pollinators were most active. Precise data were collected for 58 woody species, and floral characteristics were used to infer primary visitors for the rest. In the subcanopy, where epiphytes were most numerous, pollinator diversity was richest. Whereas 44.2% of the canopy overhead was attended by medium-size to large bees (e.g., *Euglossa, Centris*), only 17.6% of the subcanopy was so visited; the rest was serviced mostly by hummingbirds and such insects as small bees, beetles, butterflies, and sphinx moths.

Production of a few showy flowers presenting high caloric, unusually fragrant, or otherwise readily detectable and specialized rewards is common in epiphytes and probably mandated by small plant size and diffuse dispersion (Ackerman 1986). If these plants were utilizing generalized vectors or were less attractive to pollinators, competition with more floriferous trees and lianas could reduce fecundity. In effect, the modest resources available to scattered epiphytes for sexual reproduction have to be invested not only in seeds of appropriate size and mobility, but also in faithful, if not exclusive, pollinators. Suitable animal vectors are wide-ranging trapliners and those foragers susceptible to floral deception. Most canopy-dwelling Bromeliaceae, Cactaceae, Ericaceae, Gesneriaceae, Melastomataceae, Rubiaceae, and many orchids are visited by such animals. Common qualities of these epiphytes are long reproductive periods effected by individual plants that produce each day for weeks or months a few conspicuous, high-reward flowers protected from generalized foragers. Three deception systems also foster reliable visitation; Orchidaceae illustrate all three: Large canopy-based *Epidendrum* is predominantly a food-deception group, and some Araceae and *Dracula* mimic the substrates of fungus gnats. Several slipper orchids offer nesting cues. Mate mimicry, although best known in pseudocopulatory terrestrials, also occurs in epiphytic *Trichoceros* and *Oncidium henekenii.*

Autogamy versus outcrossing

An understanding of breeding systems and their advantages to particular taxa is generally hindered by lack of data on pertinent habitat characteristics and the difficulty of measuring meaningful genetic structure within populations (Loveless and Hamrick 1984). Except for a few cases like that offered below to illustrate the consequence of habitual outcrossing versus autogamy in two *Tillandsia* species, no correlations among epiphytes have been demonstrated between breeding system and habitat type or various life history characteristics. Moreover, even if a single mating pattern were to serve all canopy-adapted flora, genetic structure would vary if for no other reason than the influence of seed dispersal. Not only do seeds move independently of pollen, but embryos carry two rather than one allele for each gene.

Orchidaceae demonstrate how disparate even those breeding mechanisms maintained by related epiphytes can be. Most family members are self-compatible (Dressler 1981), but quite a few regularly outcross, sometimes with floral contrivances featuring almost legendary complexity and specificity. Best known of the allogamous forms in tropical forests are the cymbidioids (e.g., *Catasetum, Stanhopea, Coryanthes*) whose flowers attract pollen vectors via complex combinations of volatiles. Males of a single local euglossine bee population often visit a single plant species – occasionally even just one among two or more intraspecific chemotypes – to forage not for food but for the fragrance that reputedly aids subsequent mate procurement and/or reproduction (Williams and Whitten 1983). Such foragers are attracted to no other orchid although members of other plant families are used as nutrient sources. The euglossines that service the cymbidioids are also notable for long lives and ability to forage over great distances – additional qualities that have promoted orchid and bee speciation and maintained plant hyperdispersion (Benzing in press). Horticulturists have long exploited orchids for their extraordinarily weak reproductive barriers that reflect histories of ethological isolation. Most insect pollinators of Orchidaceae are, however, weaker fliers than the euglossines, and quite a few may not be particularly faithful. Publicity accorded the most spectacular coevolved orchids has obscured the fact that the vast majority engage in more mundane sexual liaisons, and some do not attract pollinators at all.

Autogamy obliged by cleistogamy has been reported in a number of Orchidaceae (Dressler 1981). Occasional flowers may be chasmogamous. Regularly self-pollinating populations tend to be abundant and widely substratum-compatible, often to the point of earning a reputation for weediness (e.g., *Caularthron bilamellatum* and *Spathoglottis plicata*). Both allogamy and autogamy foster profligacy in some other epiphytic taxa. As a group, *Tillandsia* tends to be allogamous, yet some of the most widespread species are regular, if not obligate, selfers. Ball moss *(Tillandsia recurvata)* is a selfer that ranges across most of tropical and subtropical North and South America; in heavily infested tree crowns, its biomass may exceed that of host foliage (Fig. 7.3). By way of contrast, the most broadly distributed of all bromeliads is Spanish moss *(T. usneoides)* whose flowers fail to yield capsules in at least certain southeastern United States populations unless cross-pollinated. The only published study of breeding system influence on isozyme pattern revealed much homozygosity within, but great variation among, local populations of *T. recurvata. Tillandsia ionantha,* a less widely distributed outcrossing Mesoamerican taxon, possessed more alleles per locus but little variation in allele frequency among populations (Soltis, Gilmartin, Rieseberg, and Gardner 1987). Cleistogamous *T. capillaris* must rank among the most inbred of all members of this predominantly epiphytic genus (Gilmartin and Brown 1985). Related taxa probably exhibit less consistent genetic structures. Wide-ranging species such as *T. paucifolia* feature self-compatible populations in some regions (e.g., South Florida) but will not self-fertilize elsewhere (e.g., northern South America). *Catopsis nutans* is dioecious in Central America; South Florida populations produce perfect flowers only.

Marcgraviaceae appear to be mostly autogamous (Gentry and Dodson 1987b). Another more diverse group of purportedly self-fertilizing epiphytes is that associated with ants. Numerous ant nest-garden species and certain ant-fed ant-house epiphytes regularly produce fruits without visitation by pollinators (e.g., *Hydnophytum* spp., *Anthurium gracile, Epiphyllum phyl-*

lanthus, Aechmea mertensii), a practice possibly fostered by their aggressive zoobionts (Madison 1979a). Any resulting homozygosity may pose fewer problems for obligate ant nest-garden species than for plants that must accommodate to more diverse rooting media. Additional factors apparently override myrmecophily in dictating mating capacity in some situations. A Costa Rican *Tillandsia caput-medusae* proved to be self-incompatible in a pollinator-free greenhouse, whereas a less colorful, equally ant-prone collection from southern Mexico fruited heavily in the same setting. Gentry and Dodson (1987b) believe that self-compatibility and autogamy are much more common in epiphytes than in moist, tropical-lowland terrestrials. Ackerman (1986) agrees and considers these features to be complementary, if not obligatory, for epiphytes with hyperdispersed populations or specialized pollinators such as trapliners.

Integration of flowering within the community

A 3.5-year study by Stiles (1978) of a mixed guild of ornithophilous taxa at Finca La Selva, Costa Rica focused on the group's sharing of pollinators in the face of climatic variation. The epiphyte contingent included one species from each of two bromeliad and three gesneriad genera; most of the participating taxa rooted in the ground. Several points are worth noting. Canopy-adapted species flowered most abundantly during or just after the dry season (Fig. 5.1). On average, they were at the peak of flowering for less time (2.90 ± 1.03 months) than were bird-serviced shrubs and small trees (3.29 ± 1.57 months) and herbs (3.44 ± 1.03 months), but longer than vines (1.93 ± 0.45 months). Epiphyte flowering was synchronized with that of other local ornithophilous plants so that food was continuously available except for substantial reduction in November and December. Staggered reproductive schedules could reflect past competition for common pollinator service, although other interpretations are possible. No attempts were made to identify flowering stimuli in this study, but many cultivated bromeliads and orchids reproduce according to day length; seasonal shifts in temperature and humidity have doubtless also been adopted as cues.

Stiles further noted that the guild provided continuous nectar supplies despite significant perturbation – in this case, drought. Response to stress involved alteration of the flowering sequence. For example, the epiphyte *Alloplectus coriaceus* showed a pattern somewhat typical of several other species: If the climate was wet during its fall blooming period (October through December), additional flowering took place from February to March. If October through December was dry, no second flowering

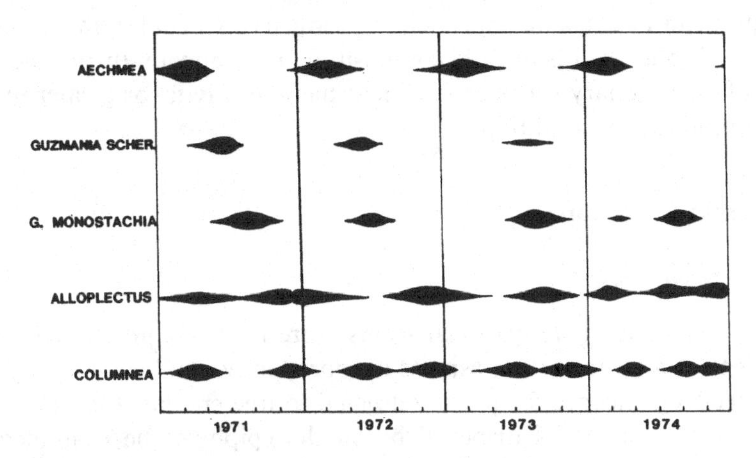

Figure 5.1. Flowering phenologies of five epiphytes comprising part of a guild of ornithophilous plants serving hummingbirds in a Costa Rican humid forest. (After Stiles 1978.)

occurred. The bromeliad *Aechmea nudicaulis,* on the other hand, showed a pattern opposite and complementary (from the standpoint of pollinating hummingbirds): After dry weather in December through January, it bloomed more profusely at its single annual flowering period than it did following wet winter months. These different responses to the same stress may reflect the presence of tank leaves and less need for a continuous supply of external moisture. Whatever the mechanism, it seems that some epiphytes constituting part of a diverse trophic base for hummingbirds in a Costa Rican rain forest exhibit considerable resilience in their capacity to maintain a shared pollinator resource.

A statistical test was used to discriminate random from coordinated flowering schedules in less seasonal understory cloud forest at Monteverde (Murray et al. 1987). Two mixed guilds containing numerous epiphytes and nonepiphytes were identified, one served by long- and the other by short-billed hummingbirds. Except for *Guzmania nicaraguensis* and two terrestrial species, those in the group possessing elongate corollas underwent anthesis on schedules that neither minimized competition for, nor promoted simultaneous use of, the common pollinators. Only soil-rooted *Burmeistera tenuifolia* flowered more in concert with certain members of its short-flowered guild than was predicted by the same null model. Of the numerous possible reasons for the random phenology, a fluctuating environment seemed to be most persuasive. Although plant fecundity is sometimes limited by inadequate pollinator service at Monteverde, observed variations in flower abundance and reproductive schedules, in addition to possibly frequent shifts in guild composition, have apparently precluded the kind of evolutionary sorting that would theoretically favor greater seed set under more stable conditions.

Movement of diaspores

Seed types

Seed and spore dispersal seems more narrowly prescribed for epiphytes than are most other aspects of reproductive biology. Gravity takes on special significance for plants relegated to tree crowns. Madison (1977) reported efficient aerial dispersal by all the epiphytes he examined, but modes of conveyance differed widely. Dust-type propagules (Fig. 5.2B–E) predominated, especially in Pteridophyta and Orchidaceae, but were present in lesser taxa as well (e.g., Bignoniaceae, Lentibulariaceae, Melastomataceae, and Rapateaceae). Winged or plumed diaspores (Fig. 5.2J–M) occurred in 1190 species of 37 genera, mostly in Asclepiadaceae, Tillandsioideae of Bromeliaceae, and parts of Gesneriaceae and Rubiaceae. In all, 84% of epiphyte species possessed wind-borne spores or seeds. The majority of families featured one or more seeds (often with sculptured walls) in fleshy fruit destined for animal consumption (Fig. 5.2F,G,I). Still, only about 4400 species in 191 genera were involved. Unusual types included adhesive forms (*Peperomia;* Fig. 5.2N) and the diverse myrmecochores produced by ant nest-garden and ant-house plants (Fig. 5.2I). Each leathery *Vanilla* capsule liberates many thousands of tiny seeds in a sweet, sticky semifluid which is avidly collected by bees and ants (Fig. 5.2E). Dwarf mistletoes have explosive fruit containing clinging seeds. A few pteridophytes (*Psilotum,* some ferns) utilize small gemmae plus spores. Myrmecophytic *Lecanopteris mirabilis* disperses its spores in groups of 16 held together by filaments. The reason for this novel condition may center on ant dispersal (Tryon 1985) or the genetic advantage of the purportedly increased outcrossing favored by closely associated prothalli (Walker 1985).

All seeds of epiphytes can be considered small by general standards; reduction has apparently not accompanied every movement into canopy

Figure 5.2. Epiphyte reproductive structures: (A) Fruit and seed of *Hydnophytum formicarium;* (B) seed of *Chiloschista lunifera;* (C) embryo of a *Laeliocattleya* hybrid infected by fungi; (D) seed of *Catasetum russellianum;* (E) seed of *Vanilla* sp.; (F) fruit and seed of *Anthurium scandens;* (G) fruit and seed of *Aechmea bracteata;* (H) winged fruit of *Encyclia cochleata;* (I) seed and fleshy funicle of *Paradrymonia gibbosa* with an unappendaged seed of

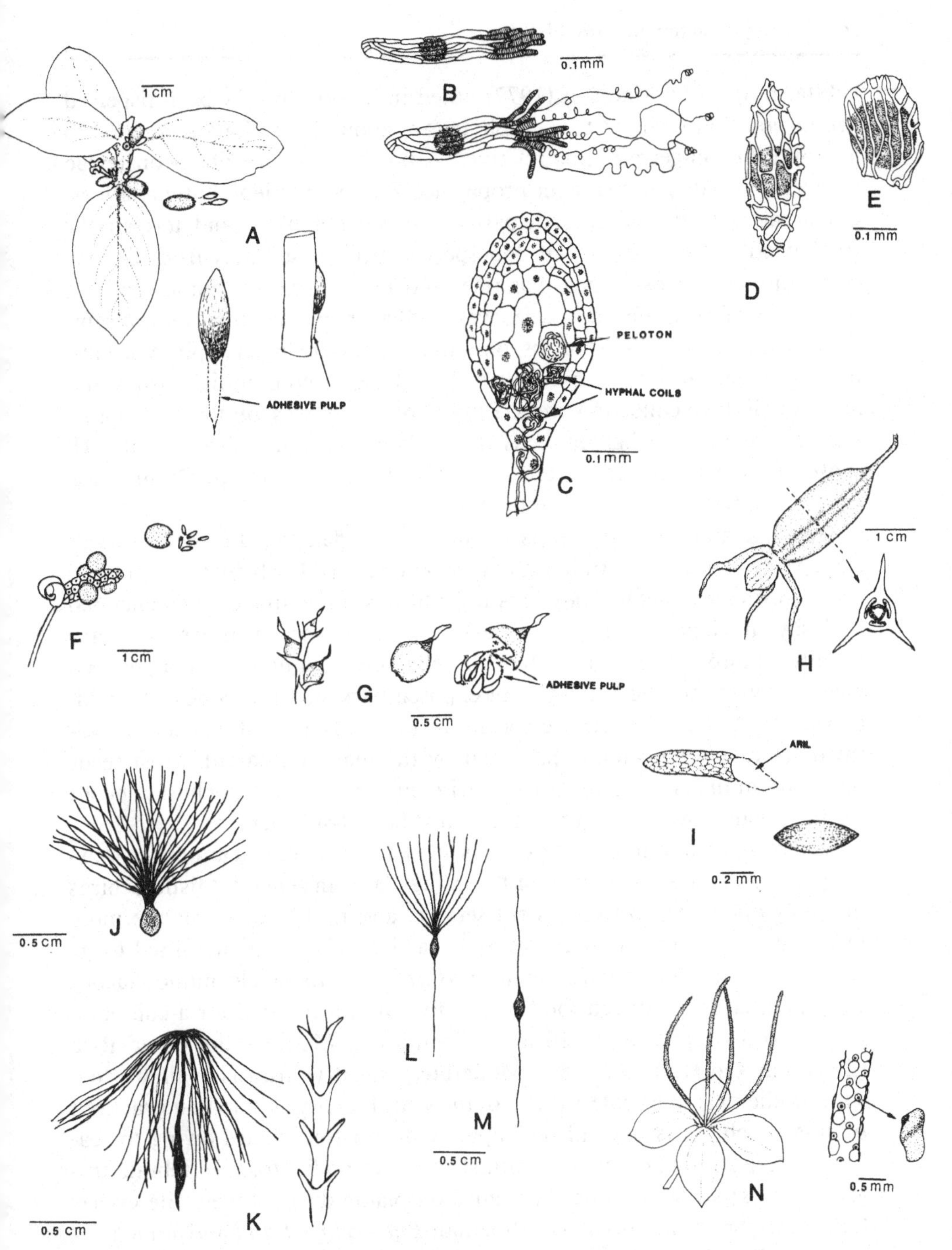

P. aurea for comparison; (J) seed of *Dischidia gaudichaudii;* (K) seed of *Tillandsia balbisiana* and part of a single coma hair; (L) seed of *Aeschynanthus* sp.; (M) seed of *Rhododendron;* (N) inflorescence and fruit of *Peperomia camptotricha.* (C after Burgeff 1936; I after Wiehler 1978.)

habitats, however. Madison (1977) noted in predominantly bird-dispersed *Anthurium* that seeds of true epiphytes are about 2 mm long compared to 4–8 mm for soil-germinating relatives. Arboreal forms in tribe Monstereae (Araceae) also disperse smaller propagules than do hemiepiphytic relatives. Cactaceae exhibit a similar disparity between epiphytes and terrestrials. Rockwood's (1985) survey of 365 species in eight families from diverse communities in Costa Rica, Panama, and Peru revealed another pattern. Seed mass of 59 epiphytic species was greater on average than that of sampled terrestrial herbs and shrubs; only tree seeds were larger. Epiphytic Gesneriaceae featured seeds that outweighed those of confamilial shrubs and herbs in all three collection areas. Rubiaceous epiphytes, on the other hand, were found to have lighter seeds than related trees and shrubs, although those of soil-rooted herbs were smaller still. Microspores of Orchidaceae were noted but not included in the tally.

Seed size among epiphytes is distinctly bimodal. Wind-dispersed seeds can be tiny – among the smallest known – and rivaled only by those of some holoparasites and achlorophyllous mycotrophs. In contrast, seeds enclosed in fleshy envelopes (normally berries) are much larger. Anemochorous taxa usually feature high fecundity but are relatively ill-equipped on a per seed basis to overcome the harshness of exposed bark surfaces. Seeds of Orchidaceae, as described later, are special in that fungal substrates are substituted for reserves normally provided by the maternal parent. Most seeds borne by animals contain more nutriment than do anemochores, hence seedlings can grow larger before they must be self-sufficient.

Seed compared to pollen dispersal is less narrowly defined among canopy-adapted flora; the same is true for flowering plants in general. Mistletoebirds probably offer the most consistent service, and nest-garden ants are more fastidious seed gatherers than are most carriers. Contrasts in animal usage can be striking. Neotropical *Ficus pertusa,* with its single monophagous wasp pollinator, produced food over a five-month interval for a coterie of 26 frugivores representing 10 avian families in Monteverde, Costa Rica (Bronstein and Hoffman 1987). Moderately nutritive fruit is available to a shifting clientele of fig eaters much of the year. Fidelity akin to that practiced by mistletoebirds is beyond the capacity of most epiphytes owing to seasonal, modest, fruit crops. In contrast to insect pollinators, vertebrate frugivores must switch among different food plants during longer life cycles. Other vertebrates are involved: Brazilian *Billbergia zebrina* and some relatives regularly occupy knotholes where their seeds are deposited following consumption by bats attracted to the sizable pungent fruit.

Mistletoes certainly demonstrate the efficacy of bird use and how reproductive success can be ensured by crops of relatively few well-provisioned diaspores (see Chap. 6). Certain baccate bromeliads, as well as *Anthurium* and *Rhipsalis* (Cactaceae) species in Trinidad, approach the mistletoe condition by fruit type (Snow and Snow 1971) and are, in fact, dispersed by some of the same avians. The importance of bird dispersal to epiphytism in some clades is illustrated by Bignoniaceae. Gentry and Dodson (1987b) reported that, except for two possibly chiropterophilous taxa, all epiphytes but only one nonepiphyte in this family were serviced by birds. Berry fruit was produced by 87% of melastomataceous epiphytes compared to 67% by soil-rooted taxa. These authors also noted that plume-seeded outnumbered wing-seeded epiphytes, whereas the opposite holds true for tropical forest trees and lianas.

Numerous mechanisms and devices promote appropriate carriage and secure anchorage. The buoyancy of orchidaceous "dust" seeds is due not only to small size but also to a large airspace between embryo and testa (Fig. 5.2D); wall sculpturing and overall shape (usually fusiform) also help to keep them aloft and may encourage attachment to rough bark. Tillandsioid bromeliad seeds feature hooked coma hairs for better attachment (Fig. 5.2K); similar devices on a much smaller scale adorn microsperms of shootless *Chiloschista* (Orchidaceae; Fig. 5.2B). Holdfast devices are probably present in other groups with appendaged diaspores; of special interest would be the clinging fruit of some *Peperomia* (Fig. 5.2N). Numerous berry fruits (e.g., *Hydnophytum, Aechmea;* Fig. 5.2A,G) contain seeds in pulp that sticks effectively to bark.

At least three predominantly epiphytic, wide-ranging *Tillandsia* species include populations whose members exhibit vivipary – that is, they produce robust inflorescence offshoots that grow much faster than seedlings whether or not pollination leads to simultaneous capsule development (Fig. 5.1). Patterns of occurrence help cast this unexpected condition in perspective. *Tillandsia flexuosa, T. paucifolia,* and *T. utriculata* in Florida and parts of Mexico and Central America produce exclusively sexual inflorescences and root solely on bark (pers. obser.). Dwarf cypress, mangroves, and other relatively open-canopy phorophytes support the densest populations. *Tillandsia flexuosa,* also a native of coastal Venezuela, differs in both habit and habitat; here, vivipary ensures especially rapid proliferation of earth-rooted specimens. Bushy shrub crowns overhead support abundant sprawling hemiepiphytic cacti and also probably offer better opportunity than most other canopies for securement of suspended *T. flexuosa* inflorescence offshoots. Typical basal ramets are also present but remain rooted to the original anchorage site. Parallel distributions of the other reproductively similar congenerics would support the hypothesis that inflorescence offshoots evolved to enhance colony expansion on the ground and perhaps where dense meshworks of fine twigs provide aerial support for multiple generations of loosely attached offshoots.

Influence of microhabitat

Because dispersal mode and seed size influence seed mobility and seedling establishment, they should be prominent factors in plant distribution within epiphyte communities. Two studies dealt with these phenomena and illustrate how additional parameters may alter expected patterns. Kelly (1985) found a marked correlation between vertical zoning and dispersal type in Jamaican wet montane forest. Nearly all the epiphytes anchored above 16 m (mostly orchids and tillandsioid bromeliads) bore small to minute wind-dispersed diaspores (Fig. 5.3). The incidence of species with fleshy fruit (mostly aroids) was greater in lower tiers, especially below 8 m. In all, 42.6% of the epiphytes had succulent fruit. Exposure and air turbulence were deemed critical for the observed segregation. Small seeds were considered the optimal type for upper-canopy life because there they can be wafted a long way by stronger wind currents. Moreover, intense irradiance reduces the need for extensive energy stores. Below, where air moves slower, animals supposedly promote more reliable dispersal, and greater reserves are required to create the leaf surface necessary to achieve autotrophy in shade light.

The analysis of Bromeliaceae in Trinidad carried out by Pittendrigh (1948), and referred to previously in connection with his categorization of these plants by location in tree crowns, also included mention of the family's two distinct dispersal modes. This parameter, however, appeared to have less influence on occurrence than did light and moisture. Taxa restricted to upper-canopy branches (Table 2.3) were exclusively high-exposure members of Tillandsioideae, the subfamily with small plumed seeds (Fig. 5.2K) – a finding consistent with Kelly's precedent. Also predictable was a split into two groups of the more shade-tolerant bromeliads in midcanopy: 14 tillandsioids and 8 baccate bromelioids. In the understory, all six species were again tillandsioids but of the most shade-tolerant category, a pattern predicted by proposed evolutionary history. As pointed out earlier, Tillandsioideae and Bromelioideae appear to have acquired their carbon and water balance mechanisms under different climatic regimens – humid and arid conditions, respectively (Benzing et al. 1985). The fundamental syndrome affecting water use and photosynthesis in Bromelioideae – namely CAM –

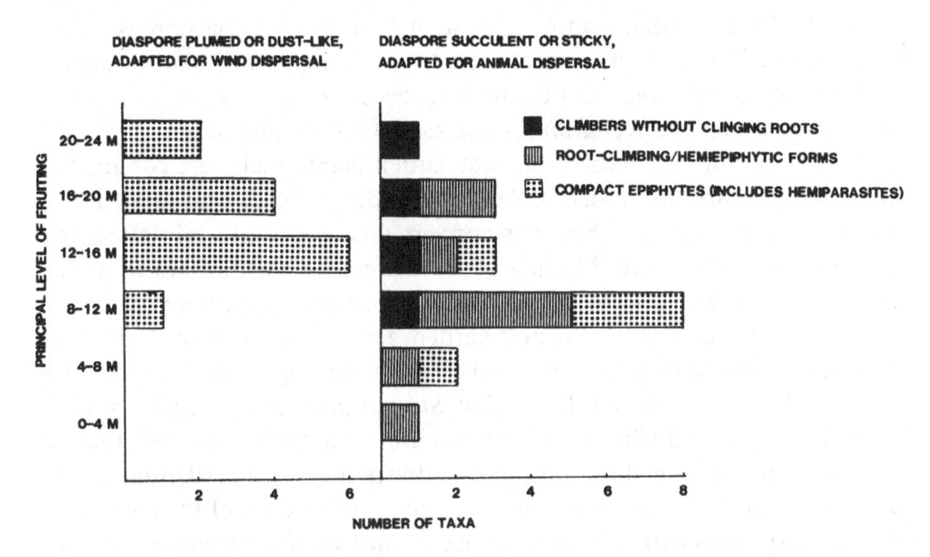

Figure 5.3. Seed dispersal modes of epiphytes in a humid Jamaican montane forest relative to height in the canopy. (After Kelly 1985.)

is usually accompanied by a high exposure requirement, although not necessarily full sun. Tillandsioideae, a basically C_3 group, is more shade-adaptable. Nevertheless, upper-canopy tillandsioids (*Tillandsia* mostly) have adopted CAM secondarily, and with it a specialized reflective and absorptive indumentum that further obliges high irradiance and, in addition, imparts sensitivity to excess humidity.

Ant nest-gardens

Epiphyte seed dispersal has been best documented for Amazonian ant nest-gardens. Here, anemochory appears to be insupportable. For instance, tillandsioid bromeliads are absent from these substrata despite the presence of numerous bromelioids in at least three genera (Fig. 5.11). Excluded are hundreds of tank-forming *Tillandsia, Vriesea,* and *Guzmania* species that, except for seed type, would seem to be equally suitable candidates for ant nest-garden habitation. Similarly, despite large numbers of epiphytic relatives in the same forests, few orchids root in ant nest-gardens. Those that do (e.g., *Coryanthes* spp., *Vanilla planifolia,* and *Epidendrum imatophyllum*) tend to be restricted to cartons and produce seeds atypical for this family. *Coryanthes* microsperms are uncharacteristically large and contain a yellowish substance between embryo and testa that may attract ants (C. H. Dodson pers. comm.). Those of *Vanilla* are even more outsized and embedded in the sticky, sweet, fragrant fluid mentioned above. Deposits that stain fat-positive with osmic acid occur in seeds of ant-associated *Acriopsis javanica* and *Dendrobium pallideflavens* in Australasia.

The question of why certain ant nest-garden plant species are confined to carton is now partially resolved: Ants responding to chemical lures, nutritive seeds, and possibly other cues convert their nests into miniature epiphyte nurseries (Fig. 5.10; Madison 1977; Davidson and Epstein in press). Fleshy arils are the reward offered by *Codonanthe,* apparently the most widely distributed neotropical nest-garden genus. The fruit of *Peperomia macrostachya* has a large, sticky, basal oil gland serving as an eliaosome. All bromelioids (e.g., *Aechmea,* Fig. 5.2G; *Streptocalyx*) and *Anthurium* (Fig. 5.2F) yield fleshy fruit from which attending ants mine seeds. All told, at least 11 species from eight families ripen diaspores which gardening insects carry into nest interiors. Amazonian epiphytes from each of the eight families lure dispersers with similar volatiles. Lemon and/or vanilla odors characterize stored seeds of *Peperomia macrostachya, Codonanthe uleana, Philodendron uleanum, Streptocalyx longifolius,* an unidentified *Epiphyllum,* and *Vanilla planifolia* (Davidson and Epstein in press). Seeds produced by nine of 10 common, taxonomically diverse ant-garden epiphytes in Amazonian Peru contained the attractant methyl-6-methylsalicylate, an essential oil closely related to the alarm/defense exocrines of some ants. This compound is the sex pheromone secreted by males of *Camponotus* subg. *Tanaemyrmex,* another group of neotropical arboreal nest-garden inhabitants. Benzothiazole occurs in fruits and/or seeds of eight members of the same nest-garden guild (Seidel 1988).

Field experiments confirm the importance of olfaction in ant nest-garden establishment (Davidson and Epstein in press). Porcelain "seeds" impregnated with *ortho*-vanillyl alcohol attracted *Crematogaster linata parabiotica* and *Camponotus femoratus;* limonene had a similar effect on an *Azteca* species. In more extensive bioassays of about 60 substituted phenyl derivatives, ants were most responsive to such 6-substituted phenyl derivatives as methyl-6-methylsalicylate. Ant nest-gardens opened in Peru contained larvae of *Crematogaster linata parabiotica* feeding on material clinging to the hard testa of harvested seeds. Quite a few ant nest-garden species feature seeds that lack the distinct oily arils of the more conspicuously myrmecochoric taxa, and some even remain attractive after passage through a vertebrate gut. So far there is no information on the quality of seed-associated tissue nor has the hypothesis that pupal mimicry promotes ingress to ant nests been adequately tested. Seeds of *Codonanthe* differ from all others produced by neotropical gesneriads and are reportedly striking mimics of ant pupae, as are seeds of *Anthurium gracile* (Madison 1979a). Seeds of *Aechmea mertensii* and *A. brevicollis,* in addition to fruit of some *Peperomia,* share the same size and shape but not the color of appropriate ant broods.

Seeds did not entice *Camponotus femoratus* at the Cocha Cashu Research Station, Peru, in the numbers expected based on local epiphyte occurrence. In fact, *Peperomia macrostachya,* the most abundant of the nest flora studied, had the least popular seeds in cafeteria-style baiting tests. Nor was ant interest predicted by immediate returns. For instance, arillate seeds of *P. macrostachya* ranked below those of *Ficus paraensis, Markea ulei,* and an undescribed *Anthurium* that bore no apparent edible tissue beyond occasional adhering fruit fragments. And there was no correlation with longer-term food supplements; important resource species like *Peperomia macrostachya* and *Ficus paraensis* were chosen less often in baiting trials than were taxa that offered no extrafloral nectar or pearl bodies and regularly mature after nests are abandoned (the undescribed *Anthurium* and *Streptocalyx longifolius*). Ability to discriminate *Ficus paraensis* seeds from those of another nongarden fig after passage through bat guts further supports the existence of important nontrophic cues to ant nest-garden epiphyte dispersal.

Longino (1986) questioned the notion of coevolution between ants and nest-garden plants. The presence of seed lures is not convincing because all myrmecochoric species produce them. Moreover, nest-garden plants may actually reduce ant fitness by clogging brood chambers and slowing temperature-dependent larval development through shading. Extensive root development eventually drives large ant species from nests dominated by certain robust acaulescent (stemless) *Anthurium* and bromeliad specimens (Fig. 5.11). Habitual poor housekeeping constitutes the best case for specific adaptation among plant-feeding ants (Janzen 1974, but see Davidson and Epstein in press). *Iridomyrmex cordatus* supposedly packs nutrient-rich refuse in absorptive chambers of *Hydnophytum* and *Myrmecodia* tubers in contrast to the habit of ejecting debris purportedly practiced by ants that simply protect their botanical partners.

Ant nest-gardens are remarkable for floristic uniformity throughout much of Amazonia, suggesting either prolonged stasis of the phytobionts or considerable mobility on their part. Fruit type and the longevity of plants and attending insect colonies may provide the answer: As noted above, nest-garden flora usually ripen seeds enclosed in edible tissues, many of which go unharvested by ants. Birds, on the other hand, consume some of these berries, and even occasional visits may promote enough dispersal (gene flow) to impede speciation. Davidson and Epstein estimated hundreds of years for the duration of entire archipelagoes containing numerous subcolonies. Although individual nests may endure only a few seasons before the vacated carton disintegrates, the plant mass sometimes lives on unattended, providing additional opportunity to disperse genes over great distances. Wind and mutualists combine to disperse some paleotropical ant epiphytes. *Dischidia rafflesiana, D. nummularia,* and several *Hoya* species, although equipped like other asclepiads with a large plumose coma, also bear an oily elaiosome. Docters van Leeuwen (1929) observed ants removing comas prior to conveying seeds to food depots. At least two *Aeschynanthus* species appear to be dispersed in the same fashion. There is also tandem animal transport. Recalling that ants mine strangler-fig seeds from bird and bat feces in order to consume adhering undigested pulp, an observer could demonstrate whether or not this second transfer leads to deposition in more propitious seed beds, as occurs with myrmecochores of some terrestrials.

Chances are good that myrmecochory has more importance in canopy habitats than is currently recognized. Quite a few New and Old World epiphytes not normally considered nest associates aggregate through the ants' scavenging activities (Ule 1906; Docters van Leeuwen 1929; Janzen 1974; Madison 1979a). The ubiquity of these arboreal insects in all but cool montane tropical forest may mean there are many additional cases. Ant biomass actually outweighed that of all other invertebrates combined in those neotropical rain forest canopies sampled by Erwin (1983).

Dispersal and establishment pattern

Madison's (1979b) study in North Borneo provides interesting commentary on wind versus animal dissemination in a heterogeneous group of 25 epiphytes. Although his survey was static in the sense that history was inferred from a single census, characteristics of the site eliminated much complicating postdispersal influence on plant distribution. Subjects were arrayed through an even-aged rubber tree plantation so that spacing, tree size, and type were relatively uniform (Fig. 5.5). Initial recruitment occurred from nearby older plantations, and also (though probably to a lesser degree) from native forest no farther away than 0.5 km. Anemochores were produced by nine orchids, six ferns, and six asclepiads; ornithochorous species were two Rubiaceae, one mistletoe, and a melastome. The rubber trees had been growing long enough to allow some, if not all, sampled epiphytes to reproduce within plantation boundaries.

Immobilization of wind-carried seeds by a homogenous target (e.g., a plantation canopy of relatively uniform height and porosity) should dimin-

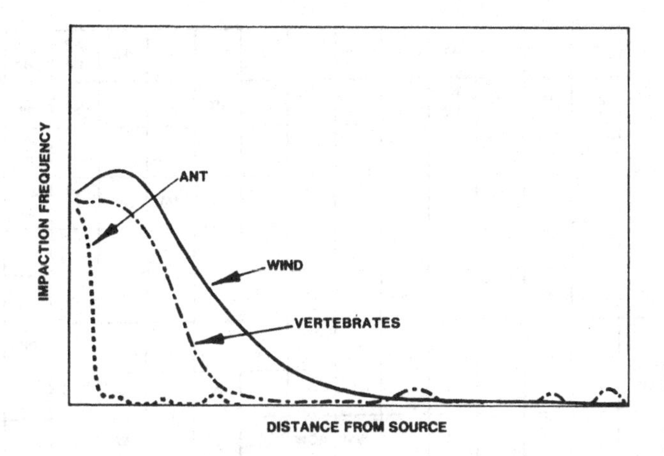

Figure 5.4. Schematic trajectories of seeds dispersed by plants utilizing ants, wind, and vertebrates as vectors. Note the patchy sedimentation of diaspores associated with transport by animals as opposed to wind.

ish geometrically with distance beyond a discrete release point; that is, sedimentation is leptokurtic (Fig. 5.4). If targets are very dense, the inverted hyperbolic curve describing seed deposition will be steep. A relatively pervious canopy will lengthen that trajectory, as would more airworthy missiles, greater wind speed, or release from a higher elevation. If, as was the case at Madison's North Borneo site, numerous seed sources surround the target area but none is too close, seed rain within the plantation will approach randomness. Once a wind-conveyed epiphyte matures inside the survey site, however, its progeny will colonize adjacent phorophytes more thickly than hosts farther away. A clumped pattern will eventually result. The exceptional storm can of course account for rare long-range dispersal whether the vector is wind or animal. Much oceanic island biota (e.g., Hawaiian) originated from single founders. Resulting populations sometimes display reduced dispersibility, a phenomenon not yet reported in an epiphyte unless ant nest-garden flora qualifies. Carton is an island to these species, and routine dispersal by wind or vertebrates would reduce seed success.

Different patterns of establishment between the anemochorous species and the four Rubiaceae examined by Madison might seem logical, considering the two vectors (Fig. 5.4). Baccate-fruited species should be arrayed according to the flight habits, perch sites, and digestive systems of the responsible avians. Particularly important was how long it took for seeds to pass through the disperser's gut. Also influential were fruit quality and abun-

J	J		JNM	X	R			JR				J			AB	AB CJ	ABC JRX	ABC HJ	AC
	K	A	ASP	DF	BL	AL	AC	AH LQ	A				RT		J	JCX	AJ	A	
JKN RU	O		AJO										A				J		
														AM	A	AH IV	AHI JT	AHI LY	AHJ
																		A	AC HQ
J				J		JL			J							J		Q	QRT
J	J	J				J				J					J	J	RH		J
			JT	JT									JT		K			JT	
					R		J	X		JVW	AGW			J	W				
	JR		J	J								JK	DE	X					

Figure 5.5. Epiphyte occurrence in a rubber plantation comprised of equal-aged trees in Sarawak. Each square represents a tree; letters in squares indicate the presence of designated epiphytes there: (A) *Dischidia collyris;* (B) *D. rafflesiana;* (C) *D. gaudichaudii;* (D) *Dischidia* sp.; (E) *D. astephana;* (F) *Hoya* sp.; (G) *Hydnophytum formicarium*; (H) *Myrmecodia tuberosa;* (I) *Medinilla* sp.; (J) *Dendrobium crumenatum;* (K) *Dendrobium secundum;* (L) *Dendrobium* sp.; (M) *Calanthe* sp.; (N) *Coelogyne* sp.; (O) *Bulbophyllum* sp.; (P) *Bulbophyllum* sp.; (Q) *Cymbidium* sp.; (R) *Drymoglossum* sp.; (S) *Lecanopteris sinuosa;* (T) *Drynaria* sp.; (U) *Davallia* sp.; (V) *Platycerium coronarium;* (W) *Lecanopteris carnosa;* (X) Loranthaceae sp.; (Y) *Hippeophyllum celebicum.* (After Madison 1979b.)

dance, and population structure. Aggregated epiphytes and the isolated individual offering high quality and/or plentiful food will discourage mobility compared to scattered or less bountiful individuals in a denser array. Ants obviously confine dissemination more than do birds.

Figure 5.5 shows that each of the fern and orchid species examined by Madison (except one *Dendrobium*) were randomly arrayed throughout the plantation as predicted. Three fleshy-fruited species followed suit whereas a fourth *(Myrmecodia tuberosa)* and all asclepiads exhibited clumping, perhaps because their elaborate comas truncated the impaction curve or because ants rather than birds were the predominant seed carriers. Or, it is conceivable that most conspecifics in dense colonies represented progeny once to several times removed from more scattered primary colonists. Nor was the reason for the overall gregariousness of these rubber plantation epiphytes entirely clear. Although individuals of most species were randomly arrayed among potential phorophytes, certain trees tended to harbor more than a single kind of epiphyte. Perhaps collections of debris around pioneers attracted ants or in some other way encouraged multiple colonizations. Although Madison's efforts were only preliminary, they do demonstrate how dispersal helps promote the spotty distribution so often exhibited by epiphytes. Imagine how many other factors also must influence spatial occurrence, among them high seedling mortality or wider scattering of acceptable phorophytes than was present at this site.

Germination

Except for the orchids and some bromeliads and mistletoes, epiphyte seed physiology has never been intensively examined. Light often stimulates germination in canopy habitats, a process mediated via phytochrome in bromeliads (Benzing 1980); photosynthetic pigments appear to serve the same purpose for mistletoes. Only a tiny fraction of the canopy flora has been assayed, but seed longevity has so far proved to be modest – generally less than a year. Viability can be quite transitory; some chlorophyllous *Utricularia* seeds perish within days under normal room conditions. Seed banks would seem to offer no particular advantage to epiphytes, and there is no evidence that they exist. Seed vivipary may be overrepresented in epiphytes (Araceae, Cactaceae, Gesneriaceae, and other fleshy-fruited taxa; Madison, 1977).

Aridity constitutes the major physical threat to early juveniles. Seedlings of most plants desiccate faster than adults simply because of their less favorable S/V ratio and small size. Especially vulnerable are slow-growing types because their chances are poorest of developing a more advantageous morphology before drought. Young epiphytes store water in diverse structures: cotyledons (*Rhipsalis,* mistletoes); hypocotyls *(Hydnophytum, Souroubea, Epiphyllum, Tillandsia);* protocorms (Orchidaceae); and first roots *(Anthurium, Hedychium).* Additional aspects of drought preparedness may be exaggerated during early ontogeny. Seedlings of Tillandsioideae bear denser layers of insulating and absorbing trichomes than do later stages, but some young *Peperomia* and Gesneriaceae appear more mesic (Madison 1977). Functional aspects have been little studied. Young of *Tillandsia deppeana* just a few millimeters long and seemingly quite xeromorphic exhibited the same C_3 photosynthesis as did the water-impounding, thin-leaf adult (Fig. 3.17; Adams and Martin 1986a). Values for $\delta^{13}C$ would reveal whether heterophylly is paralleled by altered water–carbon relations elsewhere; more anatomical details would also be informative.

Epiphyte establishment has been monitored in *Tillandsia paucifolia* (Benzing 1978b, 1981a). Germination was followed on both heavily and

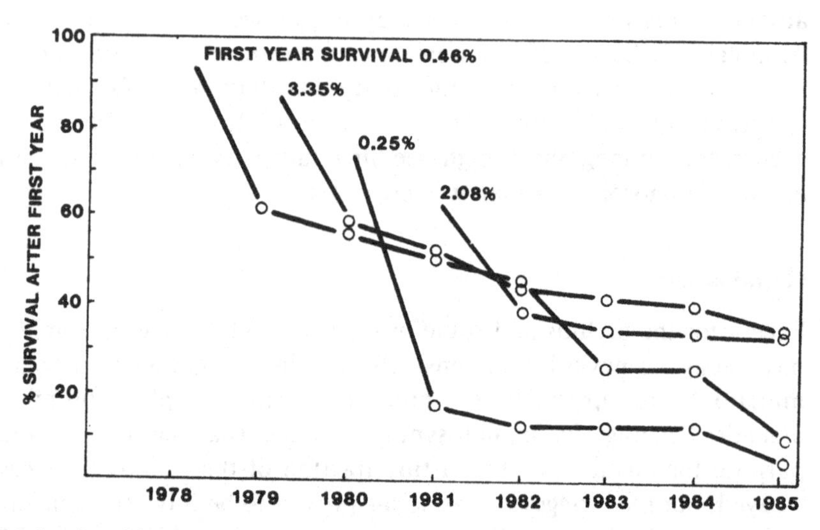

Figure 5.6. Survivorship curves after year 1 of four cohorts of *Tillandsia paucifolia* artificially affixed to *Taxodium distichum* bark in South Florida. First-year survival is indicated by the accompanying percent values.

lightly populated phorophytes, including an axenic (routinely free of attached plants) species in South Florida. Seeds collected from local populations were affixed to substrata in test patterns of 48 seeds each by applying a nontoxic permanent glue to coma hairs (Fig. 5.8), after which they were plastered snugly against bark by the first rainshower. Survivors (Figs. 5.6, 5.7) were counted one year later; success was spotty. Overall percentages were routinely low; less than 4% of seeds germinated out of thousands deployed each spring between 1978 and 1981. Complete failure was the norm on trees with few native epiphytes. Establishment in successive years tended to occur on the same trees (Table 5.1). Local phenomena, most likely exposure to light and moisture, rendered tested phorophytes (all cohorts on a tree were contiguous) either accommodating to seeds or largely axenic. Cypress *(Taxodium distichum)* proved an excellent host, as it is for most epiphytic flora in South Florida. Largely axenic *Ficus aurea* may be allelopathic; each year, ungerminated seeds were still attached to its smooth, stable surface. *Bursera simaruba,* at best an occasional host for bromeliads, regularly shed small bark fragments bearing test subjects before they could be censused. Seeds attached to lath strips under daily greenhouse misting consistently germinated at about 90%.

Tillandsia paucifolia also fared well in culture on 6–9 cm × 0.5 m limbs with seeds attached according to the practice followed outside (Fig. 5.8). Four tree species were examined: *Taxodium;* two occasional hosts (*Rhizo-*

Figure 5.7–5.12. (5.7) One-year-old seedlings of *Tillandsia paucifolia* following artificial attachment of seeds to the bark of *Taxodium distichum* in South Florida. (5.8) Seeds of *Tillandsia paucifolia* just after artificial attachment to the bark of *Taxodium distichum.* (5.9) Three-month-old seedling of *Tillandsia balbisiana* illustrating the early appearance of absorbing trichomes and retarded root development. (5.10) Small ant nest with seedlings in Venezuelan Amazonia. (5.11) Ant nest-garden largely composed of roots of bromeliads and other epiphytes in Ecuadoran wet forest. (5.12) Dwarf cypress habitat in South Florida where life table studies of *Tillandsia paucifolia* were performed.

Table 5.1. **Survival of *Tillandsia paucifolia* seeds on nine tree species in South Florida over the four years 1978–81**

	Number of trees with one or more year-old survivors (cohort = 48 seeds)				
Support	None in any year	One year in four	Two years in four	Three years in four	Every year
Quercus virginiana	5	1	0	2	5
Myrcine guianensis	1	1	0	0	0
Bursera simaruba	2	2	1	0	0
Conocarpus erecta	8	4	0	0	0
Avicennia germinans	4	3	1	0	1
Rhizophora mangle	7	2	1	0	0
Taxodium distichum	0	2	8	13	3
Ficus aurea	2	0	0	0	0
Pinus elliotii	0	2	0	5	4

Source: Unpublished data.

phora and *Conocarpus*); and *Bursera simaruba.* Subjects were maintained on a misting bench under separate timers set to irrigate for 30 minutes once every 1, 2, 4, and 6 days. Bark surfaces dried out within 3–4h following each misting, even sooner on sunny days. Fourteen weeks later, 6–35% of the 100 seeds set out for each treatment had germinated (Table 5.2). Except for subjects on *Rhizophora*, where success was fairly consistent throughout, performance was best under the three wettest conditions.

Although these results underscore the remarkable capacity of *Tillandsia paucifolia* to colonize arid substrata, seedling counts alone are somewhat misleading. Vigor was much depressed by severe drought. On all four phorophytes, subjects watered just once every sixth day were only 10–20% the size of those grown under the two wettest regimens. Individuals misted every fourth day were somewhat larger, but still only about half as long as those moistened each or every other day. Assessment of seedling vigor was hampered by persistent seed coats and intertwining coma hairs (Fig. 5.9).

More definitive data on stress tolerance could be obtained by using polyethylene glycol solutions tailored for varying water potentials. Care must be taken during such experiments, however, to ensure adequate aeration, particularly for epiphytes known to be sensitive to excess moisture. It is not clear whether atmospheric bromeliads, intolerant as they are to continuous wetting, die from anoxia or disease, or some combination of these two.

Table 5.2. **Percent germination of *Tillandsia paucifolia* seeds after 14 weeks under various misting regimens while attached to cut limbs of four supports**

	Misting regimen			
Support	One per day	One in 2 days	One in 4 days	One in 6 days
Bursera simaruba	32	35	25	7
Conocarpus erecta	33	17	21	6
Rhizophora mangle	22	24	26	20
Taxodium distichum	23	21	18	9

Note: Each sample group = 100 seeds.
Source: Benzing 1978b.

Seedlings thrive on hard, sterile agar media, suggesting that suffocation and pathogens can be eliminated through appropriate experimental design. Young epiphytes may prove to be more specialized in their water balance mechanisms than most other xerophytes. In contrast to many arid-zone terrestrials whose most drought-labile stages are completed while soil reservoirs are seasonally charged, tree crown media in all but the wettest forests repeatedly dry out long before resident seedlings reach substantial size.

Orchids and mycotrophy

Orchids deserve special note for the unique way they reap some of the advantages of large seeds without bearing the associated cost. Recall that juveniles succeed without reserves because germination and early growth are symbiotic. Establishment begins with invasion of the rudimentary endosperm-free embryo by a septate fungus, usually not a specific species (Fig. 5.2C; Hadley 1982). Development takes place for a time, nurtured through fungal saprophytism or perhaps by attack on adjacent living bark tissue. Infection is contained; intrusive hyphae are inhibited by phytoalexins and ultimately lysed to form pelotons that nearly fill hosting parenchyma cells (Fig. 4.15). Important physiological details remain obscure, but the orchid essentially acts as a necrotrophic parasite on what is normally a free-living or pathogenic fungus. Mycotrophic stages are briefer in epiphytic than in many terrestrial orchids; greening and emergence of first leaves and roots are usually completed within a few weeks, although some humus species (e.g., *Catasetum*) require more time to achieve nutritional independence. Seeds of these deep-rooted taxa, like those of numerous earth-bound relatives, may share a need for extended fungal assistance in reaching light after percolating deep into a porous substratum.

Mycotrophy has also emerged in several groups of lower vascular taxa with epiphytic membership, but even less is known about these systems. As in Orchidaceae, fungal involvement is endophytic (Mesler 1975). Likewise, infection is restricted to certain tissues, and aseptic culture is possible with sucrose or simple sugars as the sole carbon source. Achlorophyllous stages are probably longer than those of the orchids, however, perhaps extending more than a year (e.g., *Ophioglossum palmatum*). Gametophytes of *Psilotum* and *Tmesipteris* develop in humus-filled crevices or mats over bark in humid tropical forests or on the ground. Those of terrestrial or arboreal ferns alike may be free-living or mycotrophic. Green or colorless gametophytes are produced by terrestrial *Lycopodium,* but prothalli of the epiphytes are consistently fed by fungi.

Survivorship

Two methods were used to study long-term survival of *Tillandsia paucifolia* in Florida: experimental plantings and static life-table analyses (Benzing 1978b, 1981a). Germination plots have been monitored up to the present except where fires and urban development ended observations. Results from one continuing site in the Big Cypress Preserve are typical of surviving populations. Each spring from 1978 through 1981, 480 seeds were glued to the same 10 *Taxodium* hosts. Every cohort contained survivors after the first year (Fig. 5.6), but 1979 was most favorable. Failure was most common in the first few years of life; survivorship increased after that. Cause of death was difficult to document because dried-out specimens could have succumbed to drought, frost, or pathogens. Other subjects had simply vanished. No specimens had flowered by 1986, although several members of the eight-year-old cohort were near reproductive size. Growth had been quite uneven.

Life tables were assembled by felling random 85- to 200-year-old dwarf cypress trees in mixed *Pinus/Taxodium* forest (Fig. 5.12; Benzing 1981a). Ten to 15 crowns were censused in adjacent areas during three consecutive winters. The average number of specimens within crowns varied from year to year according to local differences in epiphyte density. Proportional representation among age/size classes (Figs. 5.13, 5.14) was similar, however. Phorophyte age failed to predict crown colony statistics; neither total epiphytes nor numbers of adults correlated with ring number.

Survivorship paralleled expected results (Fig. 5.14; Table 5.3). Fewer indi-

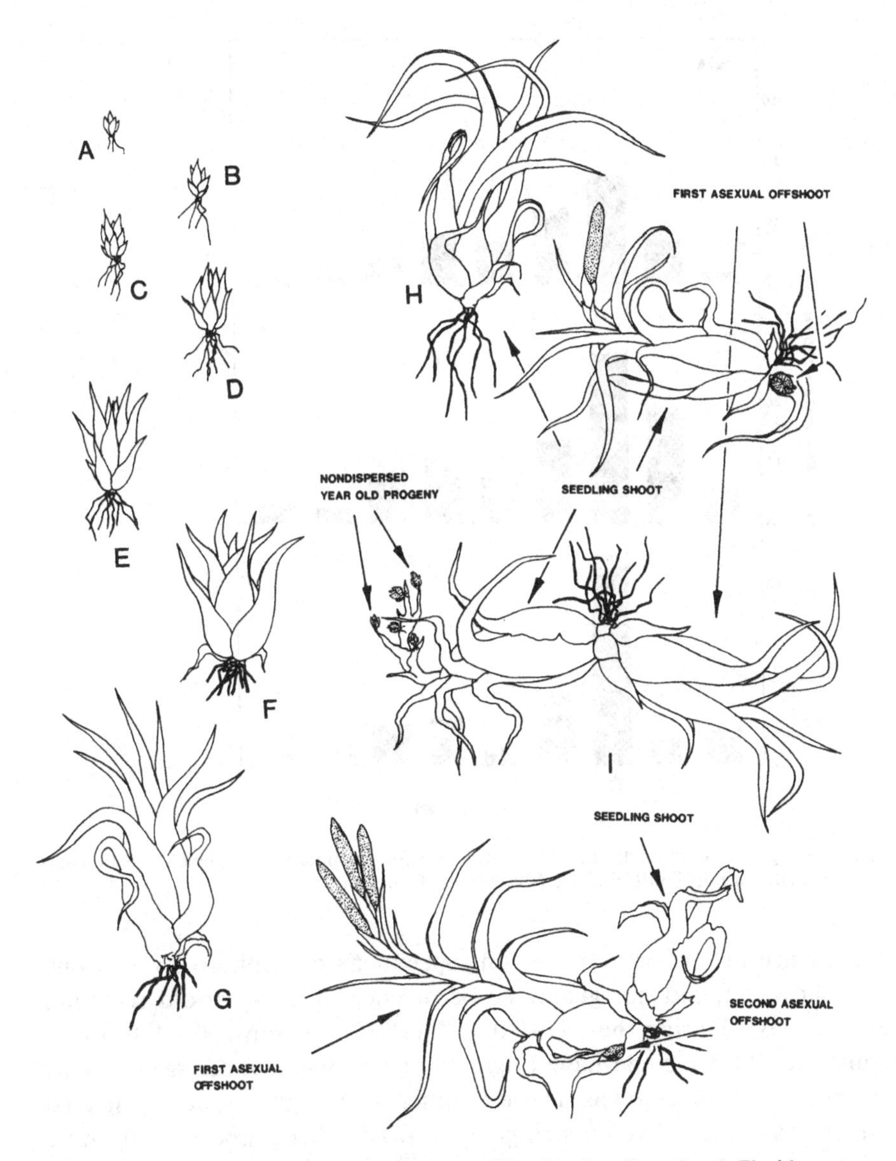

Figure 5.13. Age/size categories used to generate life table data from South Florida populations of *Tillandsia paucifolia* on felled *Taxodium distichum.* (Letters depicting age/size categories correspond here to those in Figure 5.14.)

viduals tended to fall into categories A (definite first-year seedlings only) and H (subadults, a more age-diverse group) than into intermediate growth stages. When numerous (at least 10 per host; Table 5.3, host #6), most one-year-olds were clustered within about a meter of the probable maternal par-

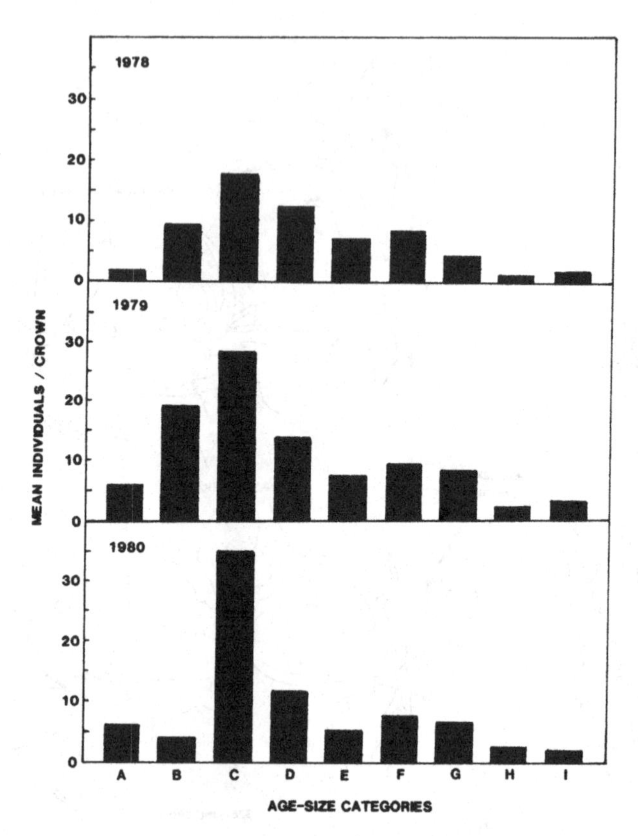

Figure 5.14. Age/size composition of *Tillandsia paucifolia* populations in the crowns of dwarf *Taxodium distichum* in South Florida. See text for details.

ent. Presence of one or more fruiting specimens on a phorophyte in one winter, however, did not guarantee occurrence of one-year-old seedlings there the next. During the 1979/80 and 1980/81 seasons, 99 of 116 new recruits on 20 trees could have originated from one or more seed parents anchored in the same crowns. Those 17 others were fugitives, having arrived on their hosts (up to five on a single tree) from other supports; there were no fallen adults nearby. Two crowns harboring fruiting material either in 1978 or 1979 received no progeny from those parents, whereas other supports with the same history bore from one to 36 year-old seedlings the following season. Seed-source size no doubt varied among fruitful crowns but could not be calculated because capsule remnants on spent inflorescences quickly fall away. Fecundity was recorded in 1981 when 11 fruiting specimens were ripening 15 capsules containing about 1500 seeds.

Table 5.3. **Colonies of *Tillandsia paucifolia* on *Taxodium distichum* sampled in winter 1979/80**

Age–size category (mm)	Host number 1	2	3	4	5	6	7	8	9	10	Mean ± SE
A 0–3	3	4[b]	2	0[b]	0	36[b]	1	5	5	0	5.6 ± 3.4
B 4 – 5	16	32	5	18	2	3	18	7	8	72	18.1 ± 6.7
C 6 – 10	29	25	10	43	21	36	15	35	21	43	27.8 ± 3.6
D 11–15	10	14	4	12	8	3	12	21	34	13	13.1 ± 2.8
E 16 – 20	7	3	5	4	3	2	5	7	22	7	6.5 ± 1.8
F 21 - 30	7	3	5	7	9	6	9	10	18	8	8.2 ± 1.3
G 31 – 50	9	3	7	7	5	7	4	13	15	3	7.3 ± 1.3
H 50+	1	3	1	4	0	4	4	2	0	0	2.1 ± 0.4
Nonfruiting adults	1	3	0	1	0	4	4	2	0	0	1.5 ± 0.5
Fruiting adults[a]	2	2	2	1	1	0	0	4	2	1	1.5 ± 0.4
Total juveniles	82	87	39	92	48	94	66	102	125	149	83.4 ± 12.6
Total adults	3	5	2	2	1	4	4	6	2	1	3.0 ± 0.5
Total epiphytes	85	92	41	97	49	98	70	108	127	150	91.7 ± 10.5
Host age (years)	—	75	49	117	—	67	131	65	140	160	

[a]Includes both seedling adults and asexual adults in fruit.
[b]Tree bore one or more fruiting epiphytes in the year prior to census.
Source: Benzing 1981a.

Category B juveniles (mostly second-season young, but perhaps some slow-growing third-year seedlings, too; Fig. 5.13) also tended to be few and aggregated, but maternity was less clear. There was little clustering by the time specimens qualified for category C. Here, larger numbers and greater dispersion among hosting crowns reflected additional coalescence of cohorts – at least four or five – caused by unequal seedling growth. Reduced sample numbers and additional dispersion through the rest of the categories reflected increased growth rate (log phase) and steady attrition. Category C's consistent numerical superiority could, of course, mean unusually heavy recruitment(s) some time between 1970 and 1975.

Each of the 20 supports examined in 1979/80 and 1980/81 harbored an average of 2.8 adult bromeliads, including 1.3 in fruit. Fourteen of the 26 specimens bearing capsules were reproducing for the first time. The remaining 12 supported fruit on one or more ramets, indicating that their present crop was not the first. Apparently, mortality rates of *Tillandsia paucifolia,* although highest early on, remain substantial throughout the entire life cycle. Because one out of approximately every four adults was new to the group each year, its life expectancy was just three more seasons. Should this census be representative, several conclusions can be drawn. Few of this epiphyte's propagules on Florida cypress survive even a first season; drought is probably the major killer of very young seedlings. Disturbance also takes a high toll because *Tillandsia* seeds lodge on cypress bark of all ages. Maturity of the epiphyte will not determine when a supporting twig or bark fragment falls to the ground unless detachment is promoted by the bromeliad's weight. These events alone would render mortality density-independent.

Several additional deductions about the life history of *Tillandsia paucifolia* can be gleaned from these observations. If the 1978–81 seasons were typical, its reproduction in the sampled forest is relatively modest but even; annual recruitment is low, compared to the size of parent colonies but not to numbers of adults. Seed success is regular, compared to the spotty performance of many terrestrial xerophytes; for example, some long-lived, soil-rooted desert succulents produce seedlings less than one year in ten (e.g., Turner, Alcorn, and Olin 1969). Despite the minimum 8 to 10-year juvenile stage of the individual *T. paucifolia* plant, a mature population may develop on dwarf cypress early enough to seed other "islands" over many decades. Lack of correlation between age of support and number of juveniles, adults, and total epiphytes further suggests that colonies begin and expand irregularly and that much of a colony's growth is autogenic.

Whether heavily infested crowns at this sample site are saturated with epiphytes is unclear. Competition among conspecifics or between *Tillandsia paucifolia* and the few other epiphytes present seems unlikely because individuals are scattered and so little of the precipitation or irradiance passing through hosting crowns contacts them. Conceivably, most of what appears to be suitable bark surface in a dwarf cypress crown is axenic, like that discovered in the germination tests. Or, perhaps disturbance (Benzing 1981b) plus modest reproductive power by the epiphyte hold colony size well below a crown's carrying capacity.

Life-history characteristics

In this final section, a topic only narrowly discussed earlier is reconsidered comprehensively: how and why the various characters responsible for plant life history are expressed and combined for survival in tree crowns. High-risk habitats such as a forest canopy favor reproductive power as stated earlier. More precisely, they favor an elevated Malthusian coefficient (r_m), a function that describes the rate at which an unconstrained population can grow. In real life, of course, propagation is limited by many factors, and actual expansion, when it occurs at all, falls well short of full potential. Small seeds and self-compatibility need be considered no further for their impact on reproductive power; several other features remain that can offer further insight into epiphyte success.

Simple algebraic functions demonstrate how reproductive schedule influences r_m. For instance, fecundity improves more by truncating juvenility than by lengthening it to achieve larger seed crops later (Cole 1954; Stearns 1976). In addition, populations grow faster under conditions to be described shortly by pursuing the single exhaustive reproductive efforts of monocarpy rather than the more resource-conserving multiseason effort of polycarpy. Juvenility is, of course, indispensable: Plants need time to acquire resources and convert them into vegetative and then reproductive organs. The interval required for this transformation varies with the organism and its environment; the time to maturity depends on the subject's habit, availability of resources, and mechanisms for their conversion to biomass. Previous discussion of photosynthesis stated that reproductive power is reduced in much epiphytic vegetation in order to gain stress tolerance. Effect on fecundity can be mitigated, however, through architectural change.

Plants produce organs serially, owing to a nodular design and indeterminate growth. Each nodular unit – in effect, each node with its associated foliage – expands in turn to nourish those that follow. Units are assembled in order by type. For each species, a characteristic number, n, of vegetative segments must be fabricated before reproduction is possible. Numerous units, each with an ephemeral leaf, constitute the shoot of a tree; pronounced axial extension is necessary in order to gain competitive height. Tall trees postpone flowering for years, presumably to some competitive advantage. More beneficial in high-mortality but less crowded habitats is less elaborate architecture (Figs. 4.25A–D, 4.27A–D). Heterochrony has a place in this discussion because it has provided the mechanisms by which growth is condensed, juvenility shortened, and stress tolerance increased.

Now, with more background in reproductive and life-history theory, it is possible to return to a subject first broached in Chapter 4: vegetative reduction. Reductive – in effect, heterochronic – evolution in Tillandsioideae (Fig. 4.25A–D) was then hypothesized as a mechanism for promoting fitness in stressful habitats by reducing investment in vegetative tissue. It was implied that tank species, by virtue of abundant impounded moisture and nutritive solids, could assemble large bodies within a few seasons. But life in more arid canopies favored reduced leaf number and overall size, in part to allow larger reproductive indices. At the extreme, roots were nearly eliminated; CAM-type foliage equipped with absorbing hairs carried out all vegetative functions. Arguably, derived atmospherics would have retained less reproductive power had they continued to produce the leafier, heavily rooted habits of less stress-adapted ancestors.

An equally advanced condition involving roots rather than shoots was cited among orchids of Sarcanthinae (Fig. 4.24A–D). Some 300 species with telescoped, monopodial stems in about a dozen genera are shootless or nearly so (Benzing and Ott 1981). Less reduced, but far more numerous and diverse, are sympodial relatives bearing several to a single expanded leaf on each of many attached determinate shoots (Fig. 1.10). As one would expect, the most diminutive species in numerous genera (e.g., *Bulbophyllum, Dendrobium, Encyclia*) tend to be xeromorphic compared to leafier congenerics. Even better designed for speedy maturation are the ephemeral equitant oncidioids (Fig. 1.16) whose simple shoots, supplied by exposed green roots, are mostly leaf tissue, a habit that favors compounding of energy capital and ultimately a high reproductive index. The only reproductive indices known for epiphytes apply to *Encyclia tampensis* and *Tillandsia paucifolia* (Benzing and Davidson 1979; Benzing and Ott 1981); these were fairly sizable.

Vegetative reduction favoring fecundity may pervade several large epiphyte clades, but the reproductive schedule necessary to maximize r_m is far less common. Except for a few species, epiphytes are iteroparous. Fitness is apparently promoted if some resources that could be committed to sexual effort are retained for continued growth. Serially produced, determinate branches originating from lower axillary or adventitious buds describe many large nonorchid epiphytes as well (e.g., most Bromeliaceae, many gesneriads, *Lycopodium;* Fig. 1.18). Indeterminate shoots are less common, although they characterize several prominent groups such as scandent aroids and asclepiads. If flowering occurs from closely placed nodes, monopodial types – for example, Sarcanthinae orchids – may differ little in resource use pattern from the short-stemmed ramet formers.

The rarity of monocarpy among epiphytes, despite its association with high fecundity elsewhere, is best appreciated by applying some theoretical constructs provided by plant demographers to compare closely related bromeliads featuring both types of reproduction. *Tillandsia* is predominantly polycarpic, although members of a few populations fruit only once [e.g., *T. utriculata* (Figs. 2.11, 2.12), *T. makoyana, T. dasyliriifolia*]. Predictably, flowers of monocarpic species can self-pollinate, sometimes without vectors. Equally significant is their resource base. Large impoundments of moisture and debris, and a biota to help process nutritive solids, appear obligatory. After about 15 years, rosettes of *T. utriculata* in South Florida are substantial enough (and incidentally, bigger than those of atmospheric relatives like co-occurring *T. paucifolia;* Fig. 1.11) to generate crops of many thousands of seeds. Monocarpy would be distinctly disadvantageous for a nonimpounding *Tillandsia;* even though first-flowering seed production could exceed that of a polycarpic counterpart with the same resource base, total reproductive effort would still be modest. Without the inflated leaf axils, much more time would be needed to marshal the same quantity of resources a mature tank bromeliad commits to its seeds. More importantly, the interval required for an atmospheric relative to reach comparable reproductive size would exact a high – probably unsustainable – toll; after all, attrition continues throughout the life cycle. Just a few percent of those *T. paucifolia* plants censused in Florida cypress were between reproductive episodes or bearing fruit for the second or subsequent times. Chances of dislodgment and other lethal events would probably reach unacceptable levels for any crown-adapted population whose members routinely needed several decades to reproduce.

The mathematics presented by Schaffer and Gadgil (1975) help clarify the relative merits of single versus serial reproductive efforts by otherwise similar epiphytes. Equation 5.1 expresses the rate (λ) at which a population of monocarpic (m) individuals – here exemplified by *Tillandsia utriculata* – multiplies (C is the probability that a seed will germinate successfully and survive to reproduce; B is the average number of seeds ripened annually by each population).

$$\lambda_m = C\,B_m \qquad 5.1$$

For polycarpic (p) *Tillandsia paucifolia,* the corresponding λ is determined by equation 5.2.

$$\lambda_p = CB_p + P \qquad 5.2$$

Here, P represents the probability that a reproducing adult will survive from one year to the next with potential for multiple seed crops. Again, B is the average size of the population's annual seed crop. For the sake of simplicity, seed success is assumed to be identical for both cases; hence, the respective C's are equal. Equating rate expressions 5.1 and 5.2 and dividing through by C produces equation 5.3:

$$B_m = B_p + \frac{P}{C} \qquad 5.3$$

It is clear that, in order for the monocarp population to grow as fast as that of the polycarp – that is, for the rates to be equal as in equation 5.3 – B_m must be greater than B_p. How much greater is obvious from the performance of Florida materials. Juvenile survival there is very low (C is <0.001 for *T. paucifolia,* whereas P is probably >0.7). A mature *T. paucifolia* specimen ripens about 300 seeds each fruiting season, whereas an average-size *T. utriculata* produces approximately 10,000 propagules in its single exhaustive reproductive effort. Should circumstances change such that life for adults becomes relatively more hazardous – that is, if multiple fruiting becomes less likely relative to juvenile attrition (P becomes smaller) – the benefits of monocarpy and the likelihood of its evolution increase. Single bouts of seed production eventually become sustainable only if P/C becomes sufficiently small – apparently a rare occurrence for plants anchored in tree crowns.

Disturbed, resource-rich (e.g., ruderal) habitats, in particular, select for monocarpy because of the regular decimation of adults (but not buried seeds) and because juvenile phases can be short, owing to abundant moisture, light, and key nutrients. In stressful but stable locations – namely deserts – longer-lived monocarps (e.g., *Agave* spp.) are more common. Epiphytic habitats, with their arid, fragmented characteristics and substantial rates of patch turnover, impose a particularly powerful set of constraints. Here, problems of achieving enough reproductive capacity to counter high, density-independent mortality are further magnified by slow growth. Except for some mistletoes, most seeds fall short of the canopy or soon fail even if they land safely. Shading, dislodgment, and tree fall, which increase as anchorage sites age, further ensure that an adult epiphyte is uncommon.

Actually, these two *Tillandsia* species differ apart from reproductive schedule – a fact that, along with the less than complete life table data for *T. utriculata,* complicates comparisons of life histories. Seeds of *T. utriculata* are heavier than those of *T. paucifolia* by about 25%; hence its C may be higher if the longer interval to adulthood does not cancel the presumably increased juvenile survivorship imparted by better-provisioned embryos. Evidence at this juncture points to the sizable resource base and corresponding capacity for large seed crops as the primary reason why *T. utriculata* can be monocarpic: A superior ability to counter the forces that depress C (principally aridity) probably does not affect reproductive options that much. However, bromeliad monocarpy could be further promoted by unusually mobile adhesive seeds and broadly adaptable seedlings. It may be no coincidence that *T. utriculata* is sometimes viewed as a weed through much of its broad Mesoamerican range.

Ramet production with associated polycarpy is the near-rule among epiphytes, but expansive clonal growth, such as that typical of a *Typha, Spartina,* or other spreading form, is rare or absent here. Resource-limited as they are, and faced with high mortality rates, the drier-growing epiphytes seem to do best where there is a balance between long-term retention of existing anchorage sites and attempts to recruit new ones. Should emphasis on either effort be too great, the appropriate ratio of mature to prereproductive individuals would deteriorate, thereby threatening the population's existence.

Although scarce resources and density-independent mortality may complicate epiphytic reproduction enough to oblige novel economy measures, tolerances can be quite relaxed elsewhere. Consider pollination in some of the epiphytic orchids. Cymbidioids serviced by male euglossines exhibit floral syndromes that often yield very poor fruit set. Highly coevolved and intricate as these insect-mediated systems may be, they are not as reliable on a per flower basis as are less restrictive arrangements (Benzing and Atwood 1984; Benzing in press). Flowering individuals that fail to set a single fruit dominate in many native populations. Nevertheless, hyperovulate ovaries, multiple-pollen delivery via pollinaria, and symbiotic germination provide enough margin to ensure adequate fecundity. Orchids more than any other group demonstrate how many disparate factors may coincide to influence life history in forest canopy habitats.

6 Mistletoes

Mistletoes are unique enough among canopy flora to merit separate treatment. Certain relic terrestrial forms parasitize roots of other plants, but they will be mentioned only in passing; the principal focus will be on aerial mistletoes which are here defined as shrubby hemiparasites growing attached to branches. These unusual plants deviate from "true" epiphytes in form, diversity, physiology, and impact on hosts. Most mistletoes belong to Santalales, a sizable, predominantly tropical, order. Xylem rather than phloem supply is reputed to be the usual consequence of santalalean parasitism, but, as noted later, advanced forms as well (e.g., *Arceuthobium*) take host substrates. Mistletoes have long occupied a place in European folklore and continue to figure prominently in certain holiday rituals of the Western world. Their destructive qualities are widely recognized. Fortunately, enough scientific curiosity has been aroused by these remarkable organisms to encourage a hard look at their biology. In fact, vegetative and reproductive activity is better known for these plants than for any other like-size assemblage of forest-canopy residents. In this chapter, that information is summarized and aerial mistletoes are contrasted with the true, fully autotrophic, epiphytes.

Systematics and biogeography

The mistletoe habit is polyphyletic, having arisen at least three or four times in Santalales and again in Laurales. The largest mistletoe family is santalalean Loranthaceae with some 900 species distributed unevenly among about 65 genera. Second in size and much more uniform in floral structure is Viscaceae, a group of perhaps 400 species in just seven genera. Eremolepidaceae and Myzodendraceae are small satellite taxa; the larger Santalaceae contain mostly root parasites. Elevation of Loranthaceae and Viscaceae from subfamily to family status has been argued convincingly on cytological, morphological, and phytogeographic grounds (Barlow 1983). Restriction of numerous primitive loranthaceous genera – including terrestrial *Atkinsonia* and *Nuytsia,* and arboreal *Cecarria* and *Muellerina,* to Australia/Papuasia; *Alepis, Peraxilla, Trilepidea,* and *Tupeia* to New Zealand; and, among others, *Desmaria, Gaiadendron,* and *Tripodanthus* to South America – points to a Gondwanan origin much like that attributed to several other predominantly Southern Hemisphere families, including Casuarinaceae, Myrtaceae, and Proteaceae. If Barlow is correct in his proposed mid-Cretaceous ancestry, Loranthaceae must be considered one of the oldest extant angiosperm families. Current distribution indicates that substantial secondary radiations occurred in Africa, Asia, South America, and Australia/New Zealand after elements of an already diverse stock were isolated by shifting continental plates. It is not clear when parasitism evolved.

Viscaceae are, by contrast, Laurasian and more recently divergent and specialized, having originated in eastern Asia (Barlow 1983) and subsequently achieved wide distribution in the Paleocene. The family is predominantly tropical today but maintains a notable temperate presence in North America, Asia, and Australia/New Zealand. Some dwarf mistletoes extend well into Northwest Canada, making this group by far the most frost-hardy of the vascular epiphytes. Viscaceae, like Loranthaceae, are parasites of dicots, but with a few exceptions. Radiation of highly reduced and specialized *Arceuthobium* as a conifer parasite seems to be rather recent and was probably favored by territorial expansion of Pinaceae during the late Tertiary global cooling. Viscaceae contrast sharply with ornithophyllous Loranthaceae (Fig. 6.1F,G) in that the former produce small, inconspicuous flowers (Fig. 6.1D,E.J) and what is usually a more reduced mycetoid endophyte (Fig. 6.2E). The two families are more alike in fruit and seed structure, dispersal, and germination. Both taxa achieve greatest diversity in humid tropical forests.

Vegetative morphology

Exposed structures

All mistletoes are autotrophic to some degree; photosynthesis occurs primarily in simple, usually opposite, leaves (Fig. 6.1A,B) or aphyllous green stems. The endophytic part of the mistletoe ranges from a discrete but often much expanded haustorium with a ball-in-socket configuration (Fig. 6.2F) to a largely internal thalloid apparatus composed of ramifying parenchymatous strands; examples are *Arceuthobium* (Fig. 6.2D,E), some other Viscaceae, and a few Loranthaceae. Exposed growth among the loranthacean mistletoes is usually extensive and persistent from germination onward. There are intervals, notably juvenile stages, when advanced viscaceans are completely embedded in host tissue. Recorded lifespans based on observation of shoots has exceeded 100 years.

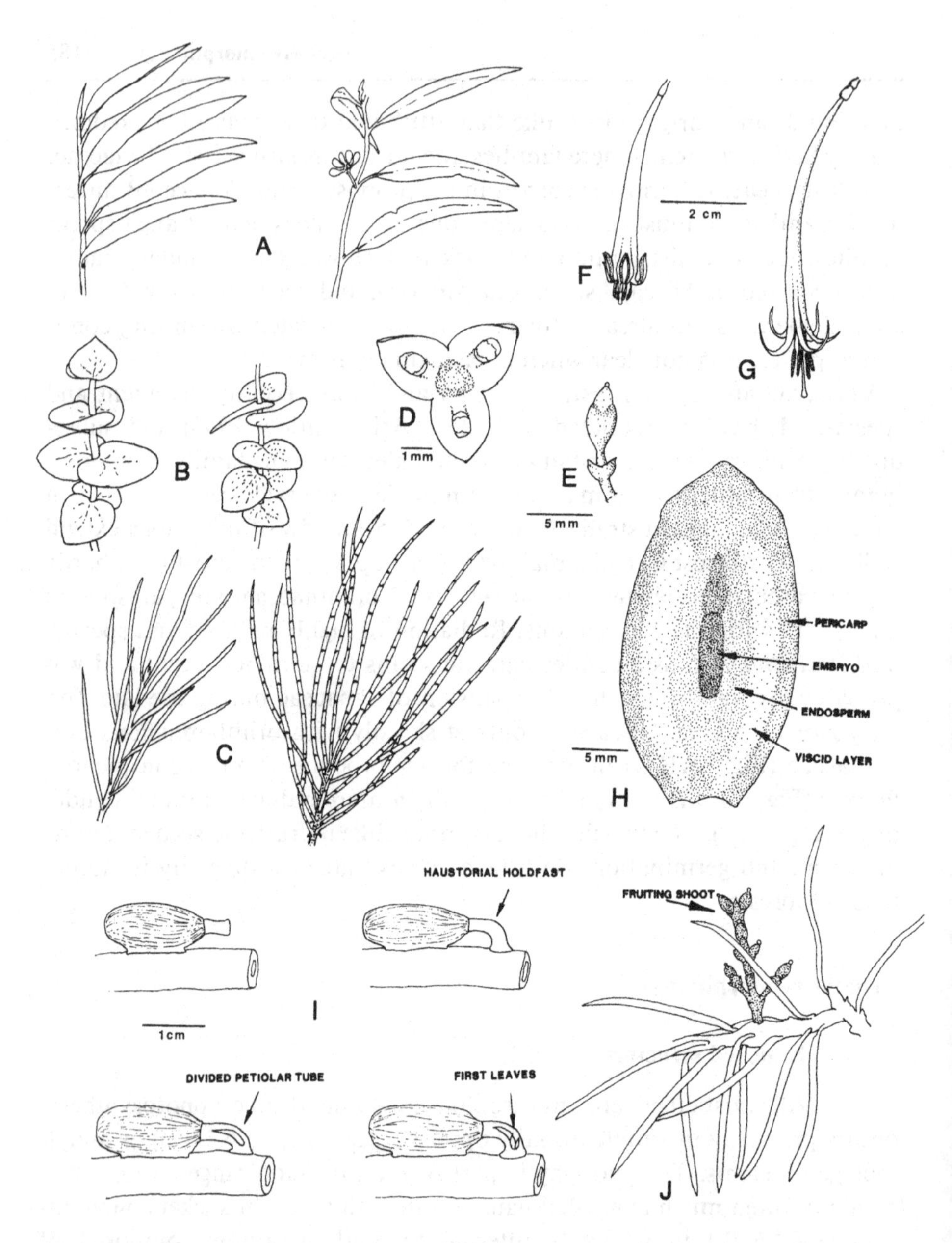

Figure 6.1. Aspects of mistletoes. The mimetic foliage of (A) *Amyema pendulum* (left) and *Eucalyptus paucifolia;* (B) *Dendrophthoe homoplastica* and *Eucalyptus shirleyi;* (C) *Amyema cambagei* (left) and *Casuarina torulosa.* Staminate and pistillate flowers of (D) *Arceuthobium campylocentrum* and (E) *A. minutissimum,* respectively. Ornithophilous flowers of (F) *Macrosolen platyphyllus* and (G) *Actanthus macranthus.* (H) Longitudinal section of pseudoberry of *Psittacanthus cuneifolius.* (I) Germination and initiation of parasitism by *Tristerix tetrandrus.* (J) Small fruiting shoot of *Arceuthobium* sp. on *Picea* sp. twig. (C after Barlow and Wiens 1977; D–G after Kuijt 1969; H after Bhatnagar and Johri 1983; I after Hoffmann et al. 1986.)

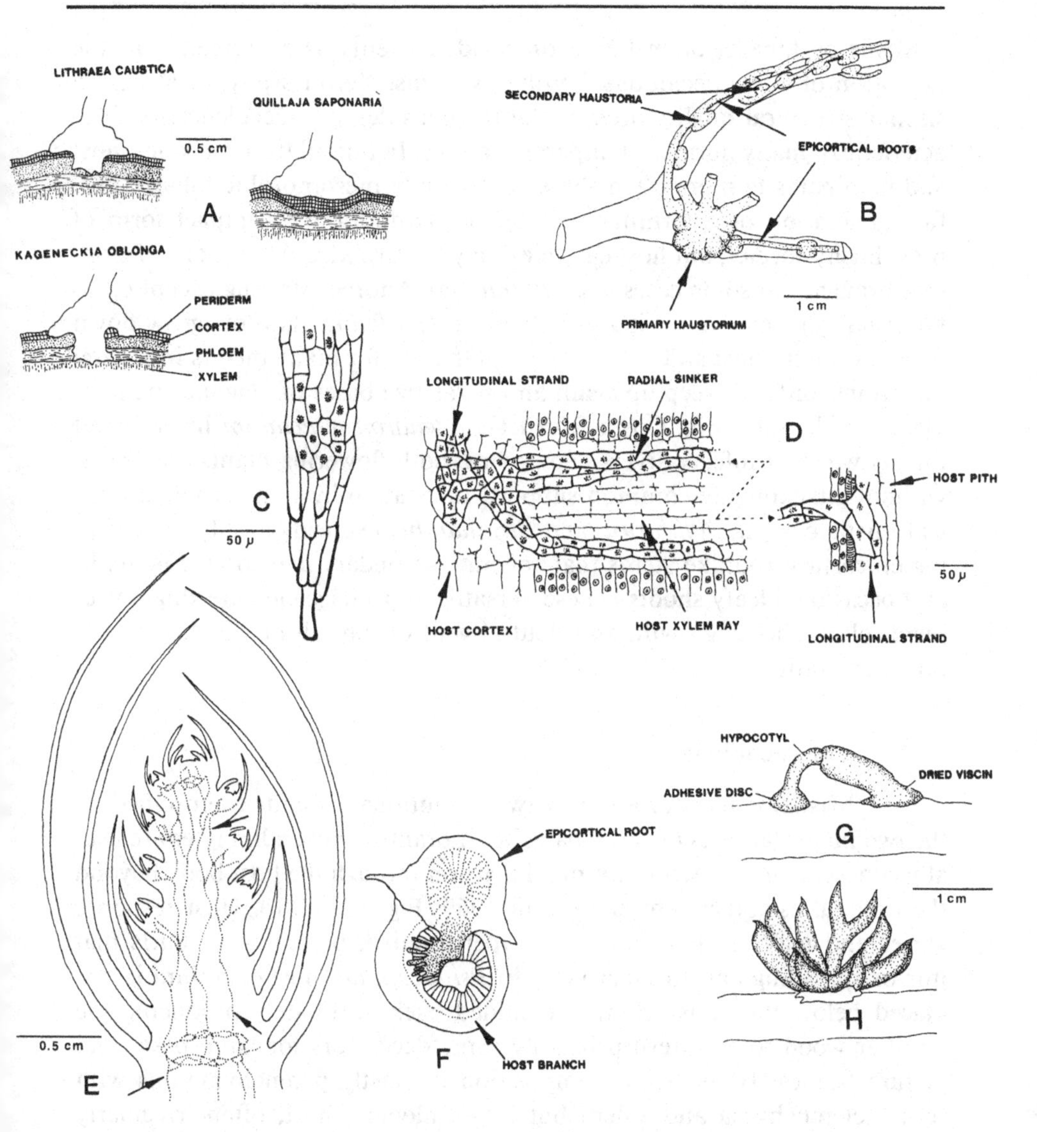

Figure 6.2. Aspects of mistletoes: (A) Haustorial penetration of bark of two incompatible and one host species by *Tristerix tetrandrus* – only the infection of *Kageneckia oblonga* succeeds; (B) *Eremolepis* sp. growing on branches of *Weinmannia* sp.; (C) apex of cortical strand of *Viscum album;* (D) Endophytic system of *Arceuthobium americanum;* (E) distribution of endophyte of *A. americanum* in dormant bud of broom on *Pinus contorta;* (F) cross section of haustorium of *Macrosolen cochinchinensis* on *Achras sapota;* (G) germinating mistletoe illustrating adhesive disc. (H) *Psittacanthus schiedeanus* seedling. (A after Hoffmann et al. 1986; B–H after Kuijt 1969.)

Mistletoe foliage, often borne on pendent stems, is evergreen with the exception of a few deciduous *Loranthus* forms. Xeromorphy, or at least a similar structural quality often including considerable succulence, is characteristic of many northern temperate natives. In humid tropical zones, host and hemiparasite more often share comparably mesomorphic foliage. Profuse generation of determinate shoots accounts for the compact form of most bushy species, but host canopies may be shrouded by festoons of elongate branches in some cases (e.g., *Phthirusa*). Another striking exception to the usual mistletoe habit is *Phrygilanthus acutifolius:* Its roots grow down from a primary host and extend through the soil for many meters in several directions, only to creep up again and parasitize bases of neighboring trees via multiple haustoria. Costa Rican *Gaiadendron punctatum* lives in wet forests where it infects diverse epiphytes, both flowering plants and ferns, without attacking the common support! Infestation by some arboreal tropical mistletoes (e.g., *Phthirusa* and *Struthanthus*) spreads locally via creeping epicortical rootlike stems that generate secondary haustoria (Fig. 6.2B) and occasional leafy shoots. These versatile appendages feature thigmotropism, photophobic growth, and neutral geotropism, all of which promote host penetration.

The endophyte

Mistletoe haustoria follow two evolutionary trends exemplified by the two major families of aerial parasites. Loranthacean endophytes are usually coherent entities with minimal intrusion through the host body beyond the original establishment site (Kuijt 1969; Fig. 6.2A,B,F). Host response varies and may result in massive, hypertrophied, placenta-like burls that mirror spreading haustorial growth. In *Psittacanthus,* host cambium is displaced below the haustorium and folded back, ultimately producing the familiar wood-rose pattern prized by some decorators for lamp bases and picture frames. Haustorium composition is mostly parenchymatous with some sclerenchyma and xylem but little phloem. Short, often irregularly shaped, tracheary cells create limited continuity with host xylem, sometimes via adjacent perforation plates. The enigmatic graniferous tracheary inclusions which characterize so many terrestrial parasite haustoria are generally absent in aerial mistletoes.

Viscacean endophytes consist of numerous, often vascularized, cortical strands radiating longitudinally from the intruding shoot base (Fig. 6.2D,E). Some are thickened by asymmetric secondary growth. Advancing apices resemble root tips, complete with a meristematic zone preceded by caplike cells (Fig. 6.2C). Radial strands (sinkers) arise from the inward sides of longitudinal strands and may eventually become enveloped by expanding host wood. In other cases, sinkers penetrate wood rays and axial parenchyma (Fig. 6.2D). Vascular continuity is all apoplastic; sieve tubes are totally absent in some of the dwarf mistletoes *(Arceuthobium)*. Viscacean endophytes can be aggressively invasive, even penetrating dormant buds well beyond infection sites (e.g., *Arceuthobium americanum* on *Pinus contorta;* Fig. 6.2E). Shoots may occasionally emerge from cortical strands well removed from germination sites. Extremely reduced forms generate aerial branches primarily to reproduce; the bulk of the organism is endophytic. *Phoradendron libocedri* and some other dwarf species sometimes remain buried in host tissue for years, clearly disproving claims that all mistletoes tap only xylem. Greater evolutionary emphasis on endophyte than on shoot development, and utilization of host photosynthate, may have been promoted by repeated dieback of exposed, frost- or drought-sensitive growth.

Pollination

Loranths are known for producing spectacular floral displays featuring ornithophilous (Fig. 6.1F,G) flowers in bright shades of red, orange, and yellow (e.g., *Macrosolen, Actanthus, Gaiadendron, Psittacanthus, Tapinostemma*). Elaborate springlike responses to probing beaks sometimes shower visitors with pollen. Explosive corolla opening is convergent in Loranthaceae with different mechanisms serving diverse genera (e.g., *Englerina, Erianthemum, Globimetula, Plicosepalus, Tapinanthus,* and *Vanwykia;* Feehan 1985). A minority, including members of *Struthanthus, Tupeia,* and *Barathranthus,* feature much less colorful flowers and appear to be pollinated by insects. Tropical American forms are routinely patronized by hummingbirds. The same service is provided in Australasia by mistletoebirds (Dicaeidae) and honeybirds (Meliphagidae), and in Africa, by sunbirds (Nectariniidae). Flowers are usually hermaphroditic but dichogamous to promote allogamy. Cross-pollination favored by protandry and effected by birds is characteristic of *Amyema* spp. in southeastern Australia (Bernhardt 1983) and may reflect broader patterns. Most of the observed species showed some self-compatibility, however, especially where host range was narrow or supports were short-lived or widely scattered. For some Loranthaceae in Indonesia, abundant fruit set occurs in plants shielded from pollinators, and occasional cleistogamous populations further illustrate the point that consistent fruiting can be exceedingly important. Diverse mistletoes should be surveyed for reproductive as well as other life history correlates of substrata. Particularly interesting would be the incidence of autogamy and homozygosity across the entire breadth of host specificity and community types.

Viscacean flowers are invariably unisexual, participating in either monoecious or dioecious breeding systems (Fig. 6.1D,E). Occasionally, corollas are bright red *(Arceuthobium)*, but most are yellow to green and quite small. Inflorescences are also much condensed. Insects – mostly Hymenoptera – are their pollinators. Anemophily, a mode of pollen conveyance not yet documented in any true epiphyte, may operate in *Viscum album* and some *Arceuthobium*. Alternatives to birds are mandatory when plants are too small to generate satisfactory trophic rewards for vertebrate pollinators. Similar floral syndromes among larger Viscaceae, however, point to a long history of entomophily throughout the entire family.

Dispersal

Loranthacean and viscacean fruits are best described as pseudoberries rather than true berries or drupes. Seed coats are much reduced, perhaps for reasons associated with germination and host penetration. Explanation for the odd embryo sac development and fertilization sequence will not be attempted here. Suffice it to say that complete seeds never form; instead, the mature pseudoberry consists of one or, less often, two embryos surrounded by a three-layered pericarp (Fig. 6.1H). The outer layer (epicarp) is leathery, heavily cutinized, and astomatous. The middle layer (mesocarp) contains viscin, a relatively undigestible cellulolytic material in a pectic matrix. When wet, it acts as a "water sponge"; when dry, as an adhesive. (In bygone days, small avians were captured by spreading "birdlime," prepared mainly from sticky *Viscum album* fruits, on tree branches.) The innermost, very thin endocarp encloses green endosperm tissue. Dispersal is effected by one of two mechanisms – most often by ornithophily but sometimes by forceful ejection as mounting hydrostatic pressure finally ruptures a weakened pericarp (dwarf mistletoes and a few other Viscaceae). Liberated sticky propagules may travel several meters; as expected, species with ballistic seed release colonize forests much more slowly and evenly than do those whose seed is dispersed by frugivores.

Successful seed dispersal does not always require regurgitation or passage of the fruit through an avian gut. But *Tristerix tetrandrus* rarely established new plants unless seeds were first voided by birds (Hoffman et al. 1986); this is also the primary, but not exclusive, dispersal mode in Australia and southern Asia (Liddy 1983), and in Indonesia (Docters van Leeuwen 1954).

Poorly protected as they are, mistletoe seeds would surely fail to pass intact were it not for the accommodating guts of specialized vectors. Dicaeidae, in particular, possess a digestive system that voids mistletoe embryos complete with intact viscin in just 20–30 minutes. Passage into a reduced gizzard is prevented by a sphincter that allows insects to enter and be ground, but contracts as mistletoe seeds pass by. Birds with less selective alimentary tracts (in effect, potential seed predators) may be discouraged by the intoxicating qualities of crushed psuedoberries.

Mistletoebird behavior further aids plant survivorship. Dicaeidae have been observed, in southern Asia and Australia, defecating seeds upon rather than off branches. Droppings were sometimes wiped directly onto bark surfaces. Adhesion was effected by residual viscin and considerable fluid and sometimes by extrusion of seeds in attached strings. Mistletoebirds (e.g., *Dicaeum hirundinaceum*) feed almost exclusively on mistletoe fruit and may well have acquired appropriate behavior and food-processing mechanisms through coevolution. Being broadly polyphagous, avian dispersers are less dependent on mistletoes at higher latitudes (e.g., *Turdus falklandii* and *Curaeus curaeus* in temperate Chile; Hoffman et al. 1986).

McKey (1975) places the aerial mistletoes in model 1 of his three bird-dispersed categories – the high investment type. All species belonging to this group have relatively accessible germination sites and produce a few seeded fruit with highly nutritive (protein- and lipid-rich) flesh. Fruit supposedly ripens continuously over long seasons; indeed, many pseudoberries do so. Also consistent with model 1, aerial mistletoes are often dispersed by a few small specialized birds (Dicaeidae, tanagers, and silky flycatchers) that depend heavily upon the edible fruit wall for sustenance. These birds are dwarfed by generalist frugivores such as the larger parrots. Mistletoe fruit is not large in the absolute sense as called for by McKey's construct, but it is sizable relative to the needs of its major dispersers (Godschalk 1983) and thus does belong to model 1. The high seed success of at least some mistletoes also satisfies a model 1 criterion.

Germination and establishment

Recent research has dispelled some misconceptions about mistletoe germination and further supported McKey's assignments with respect to the group's dispersal and establishment characteristics. Except for the diaspores of temperate species which must be stratified, and those of several lower-latitude taxa that require after-ripening, germination is usually possible immediately upon separation of seed from fruit. Under suitable environmental conditions, dispersal often triggers renewed embryo growth, a behavior unknown elsewhere. Predispersal dormancy in *Amyema preissii* is enforced by the largely gastight pericarp rather than by chemical inhibitors present in viscin or the seed proper (Lamont 1983a). It has been recorded that C_i in intact pseudoberries reaches 27%, or 760 times ambient! An important role for CO_2 in germination is likely. Cleaned *Amyema preissii* seeds maintained in CO_2-enriched air failed to germinate regardless of accompanying oxygen levels. Embryo activation was never observed at a C_i as high as 20%. There is no adequate explanation so far for the well-established stimulatory effects of H_2O_2 on seeds of many mistletoes.

Light and moisture requirements for germination vary. Most tested species yielded higher germination percentages upon irradiation, and some would not respond at all in total darkness. The only mistletoe totally free of a light response was terrestrial *Nuytsia floribunda* whose seeds are achlorophyllous. Lamont (1983a) recorded 22–40% transmittance of PAR through pericarps of four species; these values are presumably routine since all epiphytic mistletoe fruit is translucent. He also noted that light stimulation appears to be mediated by Chl rather than by phytochrome in *Amyema preissii.* Direct products of photosynthesis, perhaps simple sugars, seemed to be responsible; starch reserves may not be immediately available, at least not very rapidly. Elongation of an already protruding hypocotyl (Fig. 6.2G), the first sign of germination, occurred in CO_2-free air faster in the light than in the dark by embryos fixing their own respired CO_2. Ripe *Arceuthobium* fruit with green embryo and endosperm tissues reassimilated as much as 43% of respired carbon (Knutson 1983). Red rather than blue light was found to be most effective in germinating *Viscum album* (Lamont 1983a), and red was superior to far-red irradiation for *Arceuthobium campylopodum* (Beckman 1964). Further experimentation is needed to clarify the process.

Arid-land species sometimes tolerate remarkably dry conditions during establishment; their seeds may even be damaged or destroyed by excess moisture. An exception is *Tristerix tetrandrus,* whose seeds routinely failed on hosts occupying the driest microsites in Chilean matorral (Hoffman et al. 1986). Polyethylene glycol solutions were used to demonstrate that *Amyema preissii* seeds germinated at -1.5 MPa, the permanent wilting point for many soil-rooted species (Lamont and Perry 1977). In contrast, *Arceuthobium abietinum* seeds, whose moisture content varied greatly, germinated under humidities ranging from zero to 90% (Lamont 1983a). *Viscum album* establishment was inhibited by a saturated atmosphere as was germination in a number of similarly xerophytic mistletoes. A second, more mesophytic group required humid air, whereas seeds of still other taxa from wet forests remained quiescent unless periodically soaked.

Very young aerial mistletoes face impediments that are unusual even for epiphytes. Haustoria must penetrate bark without dislodging the germinating seed, a process often requiring several weeks. In the meantime, only external moisture sources are available, seemingly a poor supply for utilization by emerging mistletoe seedlings. Both difficulties are mitigated, however, by the behavior of the green, geotropically neutral, negatively phototrophic hypocotyl which is sometimes capable of circumrotation. Once this organ contacts a suitable medium, a flattened, terminal, holdfast, haustorial disc quickly develops as a prelude to parasitism (Figs. 6.1I, 6.2G). (A true radicle giving rise to a primary haustorium is missing, at least in advanced taxa.) Growth into surface irregularities and the presence of copious hardened secretions soon create a tight junction between hypocotyl and host. Parasitism begins by penetration of bark via a structure consisting of a central core of provascular tissue surrounded by glandular cells. Progress to the xylem is probably aided by enzymatic digestion and hydraulic expansion, but recorded details are scarce. If seed reserves are depleted before the epicotyl develops, invasive growth can proceed through activity of the two or more green, fleshy cotyledons (Fig. 6.2H).

Host specificity is not immediately expressed, at least not in those systems studied to date. *Phoradendron tomentosum* seeds planted on three different hosts all germinated at about 90%, but the haustorial disc formation needed for establishment was less consistent (Clay, Dement, and Rejmanek 1985). Successful establishment of properly matched mistletoe seed can be quite high, well above that for some, if not most, free-living epiphytes (e.g., *Tillandsia paucifolia;* Table 5.1). For example, after 37% of dispersed *Arceuthobium tsugense* seeds reached suitable substrata (Lamont 1983a), 62% germinated and more than half (57%) of those germinated went on to establish parasitism. Much lower, but still substantial, establishment values were recorded for bird-dispersed *Amyema miquelii* (6%), *A. cambagei* (7–18%), and ballistic-seeded *Arceuthobium pusillum* (4.6%). Champion of the tested group was *Viscum album,* establishing 69% of germinated seeds in apple orchards (Lamont 1983a). Success was much lower for *Phoradendron tomentosum* seeds in some Texas forests, but the year observations were made was marked by severe drought and an abnormally cold winter (Clay et al. 1985). Size of seed crop suggests that the mistletoes, on average, need less fecundity compared to many other plants in the same forests, including their hosts – a rather surprising fact given the seeming vulnerability of the parasitic existence.

Hemiparasitism: its variable nature

Mistletoes are obligate hemiparasites but differ in the benefit they extract from hosts. Given that neither moisture nor nutrient ions can be procured in sufficient quantities elsewhere, host xylem flow must be captured via haustoria. Tapping of host photosynthate varies among species from negligible to substantial. Hellmuth (1971) demonstrated an example of the former when he noted that *Amyema nestor* was vigorously autotrophic and even outproduced its Australian *Acacia* supports on a leaf area basis. Another interesting study examined endophytic tissues of a *Phoradendron* that received transported labeled metabolites from its own foliage rather than from the much closer host phloem (Hull and Leonard 1964). Also found was a requirement for substantial host input in some *Arceuthobium* species that contained as little as ⅕ to ⅒ of their partners' chlorophyll concentration [Chl] and met only about ⅓ of their own carbon needs. Until the atrophied primary axis of these dwarf mistletoes is replaced by aerial shoots from the expanding endophyte, energy dependence is of course absolute. Complete heterotrophism may reoccur periodically later in life.

Trends toward reduced emphasis on external versus endophytic development in both Loranthaceae and Viscaceae support the view that mistletoe parasitism was originally xylem-based and remains so in all but a few advanced genera. Less obvious is whether the need for moisture or key nutrients has been the more powerful selective force in mistletoe evolution. Preparasitic mistletoes were undoubtedly terrestrial and quite possibly native to arid soils. But such media are usually also infertile, particularly in N, and just as likely to foster adaptations for oligotrophy as for xerophytism among plants growing there.

Water balance

True epiphytes, mistletoes, and their supports not only obtain but also utilize water in different ways. Cited earlier were the relatively less negative values for Ψ_{crit} and Ψ_S as well as the seemingly limited osmotic adjustment found in true epiphytes. Phorophytes, on the other hand, exhibit more negative values, and the values for mistletoes are lower still (Table 3.3). The host–mistletoe water potential gradient must be steep in order to divert an adequate transpiration stream from host to mistletoe because short, distorted tracheary elements in haustoria, and limited xylem–xylem continuity, impose considerable hydraulic resistance. These choke points probably impede flow no matter whether the parasite interdicts part or, as in the case of *Amyema miquelii* on *Eucalyptus,* all of the xylem supply, a deprivation that eventually kills distal branch segments of the infected host. Succulence – that is, the elevated water capacitance displayed by many arid-land and temperate mistletoes – may exist to provide reserves for continued transpiration while evaporative demand temporarily exceeds supply from the host (Whittington and Sinclair 1988). A more conductive haustorium would reduce the parasite's isolation from its water source, but the host's vulnerability to drought would increase accordingly.

Solute potentials are ordered even among serial parasites. For instance, when *Phthirusa pyrifolia* was found attached to *Dendrophthora gracilis* parasitizing a tree in Jamaica, Ψ_S became progressively more negative from host to primary to secondary parasite (Harris and Lawrence 1916). Mistletoes also prevail over hosts through unusual stomatal behavior. They out-transpire supports (Fisher 1983), particularly during dry weather, because their guard cells retain turgor despite considerable stress. Gas exchange rate remains substantial as xylem tensions approach -3.0 to -4.5 MPa, values usually well below those required to maintain high foliar conductance in hosts. A water potential gradient of -0.1 to -0.3 MPa favoring the parasite may persist through the night (Whittington and Sinclair 1988). A number of Australian mistletoes lost moisture up to 8.9 times faster than did their supports; nevertheless, stomatal conductance shifted synchronously in host–parasite pairs (Ullmann et al. 1985). Their regular nocturnal closure is further evidence that mistletoe stomata are fully functional.

Unabated photosynthesis in mistletoes compared to droughted hosts may have a hormonal basis. Abscisic acid could be scarcer or cytokinins more abundant in these than in other types of vegetation under similar stress. El-Sharkawy, Cock, and Hernandez (1986) attributed the relative insensitivity of mistletoe stomata to physical rather than biochemical phenomena. Working with *Phthirusa pyrifolia* on *Citrus reticulata,* they recorded the usual lower photosynthetic rate, less favorable WUE, and poor correlation between foliar conductance and WVPD in the parasitic partner (Fig. 6.3). In addition, leaf mesophyll in *Phthirusa pyrifolia* leaves was found to be sufficiently compacted to minimize localized water deficits resulting from peristomatal transpiration. Moisture lost from guard cells reputedly permits sensitive plants to alter stomatal aperture according to changing local WVPD. If El-Sharkawy et al. are correct, leaf anatomy in such mistletoes as *P. pyrifolia* ensures tighter coupling between bulk-leaf and guard-cell moisture status: As long as moisture content of shoots is adequate, leaf conductance remains high irrespective of evaporative demand. An additional feature of *P. pyrifolia* which supposedly contributed to its guard cell

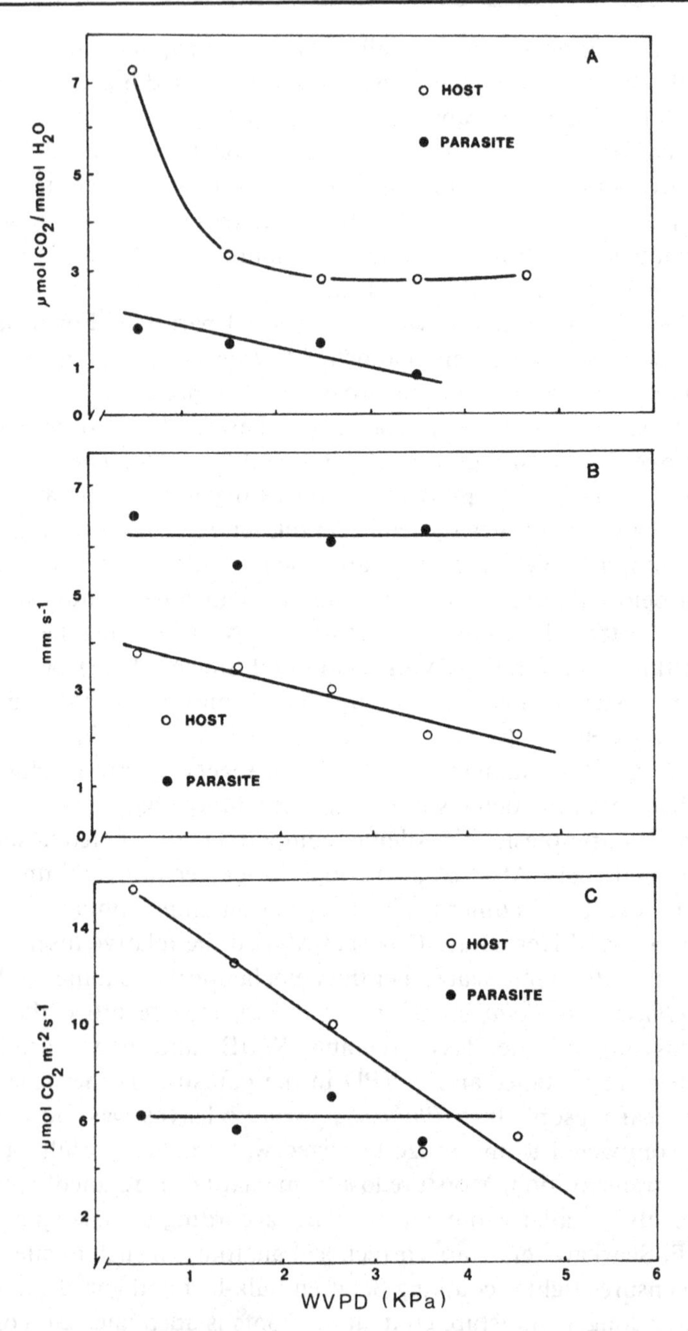

Figure 6.3. Effects of air humidity on (A) water use efficiency (WUE), (B) leaf conductance, and (C) photosynthetic rate in *Phthirusa pyrifolia* growing on *Citrus reticulata.* (After El-Sharkawy et al. 1986.)

insensitivity is the low density of stomata in the mistletoe compared to its host *Citrus reticulata* (adaxial: 98 vs. 130 mm^{-2}; abaxial: 136 vs. 540 mm^{-2}).

Mineral nutrition

As unusual as the mistletoe's water use pattern is its element composition. Were mistletoes simple direct xylem feeders, relative mineral concentrations would be similar to those in supporting tree branches. In fact, they are not. Concentrations of K, P, Mg, Mn, Na, N, and Fe in mistletoe foliage generally exceed those in infected trees, often by wide margins; [K] can be 13 times greater and [P] not far below (Table 6.1; Lamont 1983b). Closer to parity is [N], usually not exceeding 1.5 times that in parasitized branches. *Amyema preissii* contained 0.90–5.54 times more of eight elements compared to its Australian host *Acacia acuminata* (Lamont and Southall 1982). Sulfur and most trace nutrients have been little studied in host–mistletoe liaisons. Pronounced affinity for Cu, a toxic element at high concentrations, was demonstrated during an attempt to control *Loranthus pulverulentus* by injecting host wood with copper sulfate solution. Seasonal variations in element content may also be considerable: [P] and [K] in *Viscum album* were usually greatest (3–16 times host level) in summer or autumn and lowest (1.2–2.1 times) in winter. There is no information on differential expenditure of one element relative to others during reproduction or on mobilization and loss when foliage is abscised.

As Ca is the least phloem-mobile mineral element, [Ca] can be used as a standard in analyses of exchange dynamics and identification of the responsible vascular tissues. If phloem transport of elements other than Ca plays a major role in aerial parasitism, the ratio [element]/[Ca] should be higher in mistletoe foliage than in that of the infected support. El Sharkawy et al. (1986) noted a [K]/[Ca] ratio of 1.61 in *Phthirusa pyrifolia* foliage compared to just 0.56 for hosting *Citrus reticulata*. Phosphorus and N always showed the same trend, but Mg and Na did not: Their ratios were sometimes lower in mistletoe than in support foliage (Lamont 1983b). Whether xylem or phloem is the more important delivery system in determining mineral content in mistletoes, intervening parenchyma tissue probably influences nutrient flux into the parasite.

Maintaining N sufficiency may be the greatest nutritional problem for xylem-tapping mistletoes. There are several organic and inorganic nitrogenous compounds in a host's transpiration stream, but quantity and proportion differ with the tree and season. The predominant organic species in xylem sap tends to be asparagine, with lesser amounts of glutamine, gluta-

Table 6.1. **Ratio of element concentration in mistletoe versus that in infected branch of host on a dry weight basis**

Mistletoe/host pair	P	N	K	Ca	Mg	Mn	Fe
	2.92	1.52	1.84	—	—	1.27	1.12
Dendrophthoe falcata on	3.00	1.50	4.00	—	—	2.00	1.60
Mangifera indica	—	1.07	—	—	—	—	1.60
Loranthus europaeus on							
Quercus petraea	—	—	6.64	3.39	1.84	—	—
Viscum album on *Pyrus malus*	6.61	5.65	7.80	0.29	—	—	—
Viscum album on *Juglans*	2.52	—	—	—	—	—	—
nigra	9.68	—	3.29	—	—	—	—
Amyema preissii on *Acacia*	—	1.09	—	0.95	0.98	—	0.39
acuminata	—	1.14	—	0.92	—	—	0.90
	—	—	—	—	1.72	—	—
Viscum laxum on *Albies alba*	—	—	12.99	6.28	—	—	—

Source: After Lamont 1983b.

mate, and aspartate (Knutson 1983). The most plentiful inorganic ion seems to be NH_4^+, although Stewart and Orebamjo (1980) found that *Tapinanthus bangwensis* and its host had similar nitrate reductase activities: Both could convert NO_3^- to glutamine or glutamate in vitro. Activity was inversely related to the parasite's N content, however; these authors agree that their African subject usually utilizes forms of N other than NO_3^-.

Insight into the mechanism of ion movement between mistletoe and host is hampered by too little investigation and too much variation among species pairs. Poor xylem continuity at the host–haustorium interface may explain some of the oddities of hemiparasitism. Absence of as much, or any, direct contact with phloem complicates the picture even more. Labeling experiments indicate that a particular mistletoe's sink strength varies depending on the nutrient and point of entry into the system. Radiophosphorus and ^{35}S injected into host branches beneath attachment points of *Arceuthobium campylopodum* and two *Phoradendron* species moved horizontally along the branch as readily as they did up into the mistletoe shoots (Lamont 1983b). But when ^{32}P was placed on host leaves distal to attachment points, it entered endophytes at twice the proximal application rate; almost certainly, phloem was involved, although not as a direct conduit across the host–mistletoe junction. Radiophosphorus provided to aerial stems and leaves of *Arceuthobium campylopodum* failed to reach the endophyte; however, carbon from host foliage did in the case of *A. tsugense*

(Miller and Tocher 1975). In stark contrast are scattered reports of mistletoes feeding hosts; In one instance, *Amyema pendulum* remained active for three years on a leafless *Eucalyptus* specimen (Kerr 1925).

Data on physiology and morphology, in addition to that on element composition, provide evidence for active transport in haustorial function. Abundant endoplasmic reticulum (Kuijt and Toth 1976), as well as high phosphatase activity in endophyte parenchyma, and accumulation of Fe and DNA in opposed host tissues indicate substantial metabolic activity on both sides of the critical host–mistletoe interface. But no plasmotic unity has been reported, except for minute *Viscum minimum,* although pits in host tracheary elements and those on endophytic parenchyma are sometimes juxtaposed. "Half-plasmodesmata" between host parenchyma and *Arceuthobium* sinkers have been recorded, a condition that prompted Alosi and Calvin (1985) to suggest leakage into, and subsequent uptake from, a common, shallow, apoplastic junction. Once in the haustorial symplast, transport of photosynthate to parasite shoots is purportedly effected by a phloem-like mass flow aided by differential starch formation. Ion transport probably also has an apoplastic step mediated either via direct xylem continuity or through an intervening host–parasite ground tissue continuum.

Hormonal involvement in mistletoe parasitism remains to be proven, but there is some tantalizing evidence pointing that way. An indication of possible cytokinin action exists in the apparent capacity of the mistletoe to mobilize nearby nutrients. For instance, Lamont (1983b) found that [P] was highest in mistletoes, next highest in uninfected branches, and lowest in infected branches. This order does not always obtain for K, Mg, and N: Although highest in mistletoes, concentrations of these ions may be greater or less in infected branches vis-à-vis uninfected ones. As for carbon gain, dark respiration and photosynthesis in spruce branches infected by *Arceuthobium pusillum* are increased over those in uninfected axes (Clark and Bonga 1970), perhaps providing additional benefit to this specialized parasite. Quite possibly, hosts of dwarf mistletoes suffer less deprivation from their infections than carbon losses would indicate. Increased photosynthesis in remaining leaves on plants subjected to partial defoliation is a widely available compensatory response. A fuller explanation of phloem tapping must await better understanding of source–sink coordination. Clearly, there need not be sieve-tube or any other type of symplastic continuity across the host–parasite junction for the parasite to have access to host photosynthate. Appropriate mechanisms are fundamental to vascular plant function. Apoplastic junctions are regularly traversed during leaf phloem loading everywhere, and probably also again when sieve tube contents move out to growing tissues and storage organs. Similarly, host–mistletoe concentration gradients may be effected by mechanisms like those operating in roots where the endodermis and stelar parenchyma control the chemistry of xylem sap and hence tissue composition upstream.

Water versus nitrogen as the growth-limiting resource

A recent survey of xylem-tapping mistletoes was carried out to determine whether foliar conductance in these plants is more closely correlated with mineral nutrition or with, as in more conventional vegetation, water–carbon balance (Ehleringer et al. 1985). Most terrestrial species regulate guard cell turgor primarily to conserve moisture – short of unnecessarily impeding photosynthesis, of course. In this study, the authors' hypothesis states that the high TR maintained by mistletoes practicing xylem parasitism evolved principally to promote not water but nutrient procurement, particularly of N. Species parasitizing numerous, some leguminous, hosts in semiarid regions in South Africa, Central Australia, and North America served as subjects.

Several facts influenced experimental design, including choice of species pairs. First, woody legumes produce more N-rich xylem sap than do most other trees. Second, mistletoes and virtually all of their hosts except succulent cacti and euphorbs are C_3 species and thus show more negative values for carbon isotope ratios than do C_4 and CAM types. Moreover, $\delta^{13}C$ values among C_3 plants vary according to C_i at the time of photosynthesis. Illuminated, nonstressed foliage maintains a C_i just high enough to saturate carboxylation capacity and expends only enough water to do so; transpiration offers no additional advantage beyond creating the xylem flow necessary to supply nutrients to aerial organs. The more xeric the C_3 plant, the steeper the C_a/C_i gradient, the higher the WUE, and the less negative the $\delta^{13}C$. So it is that bulk leaf $\delta^{13}C$ values can provide a measure of diurnal C_i, an integrated index of WUE, and, under appropriate circumstances, information on the nutrient procurement mechanisms of a xylem parasite.

In these studies (Fig. 6.4) as in others (Fig. 6.3B), mistletoes exhibited higher leaf conductance than did hosts. Consequently, instantaneous WUE at the recorded intervals was always lower and C_i higher in the parasite. Moreover, continuous performance discrepancies were documented by carbon isotope data: $\delta^{13}C$ values were always more negative among mistletoes compared to their hosts (Table 6.2). Most telling support for the N-procurement hypothesis of Ehleringer et al. (1985) were comparisons of carbon

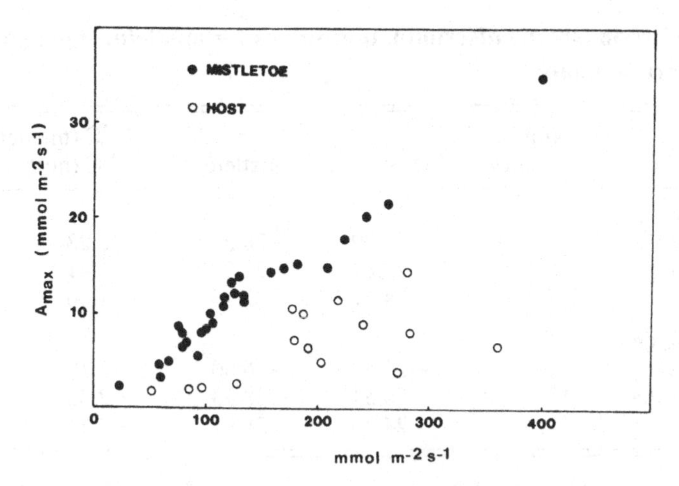

Figure 6.4. Relation between daily maximum rate of photosynthesis (A_{max}) and concomitant leaf conductance to water vapor under field conditions for mistletoes and host plants in Central Australia. (After Ehleringer et al. 1985.)

composition between the two types of hosts; differences in $\delta^{13}C$ values between parasite and tree were smaller if the host was a legume, regardless of absolute values. Thus on all three continents, mistletoe and host WUEs were closer if [N] was higher in the host's transpiration stream. Xylem sap chemistry was not actually measured, but foliar [N] was higher in mistletoes tapping an N_2 fixer. Schulze and Ehleringer (1984) also discovered that *Phoradendron californicum,* a leafy North American parasite, grew more luxuriantly on nodulated hosts than on adjacent nonfixing species. Even greater support for the nutritional hypothesis came from the higher WUE recorded by this hemiparasite on the first as opposed to the second set of supports.

Evolution of aerial parasitism

Restricted taxonomic involvement in aerial parasitism is made all the more intriguing by the widespread abundance of mistletoes and the much greater diversity of true epiphytes. Why many more thousands of plants from so many higher taxa root nowhere but on bark yet never invade host vasculature is puzzling until aspects of water balance are considered. The presence in Santalales of haustorial terrestrials but no epiphytes suggest that aerial parasitism arose from root parasitism; more direct aerial transitions were

Table 6.2. **Carbon isotope discrimination values for mistletoe–host pairs from three arid regions**

Region	Number of pairs	Host	Mistletoe	δ ^{13}C (mistletoe) – δ ^{13}C (host) (‰)
N-fixing host				
United States	7	−26.29	−26.51	−0.23
Central Australia	28	−26.87	−28.28	−1.41
South Africa	4	−24.67	−25.73	−1.06
Non–N-fixing host				
United States	8	−23.43	−26.60	−3.18
Central Australia	19	−26.54	−28.83	−2.30
South Africa	11	−24.70	−26.91	−2.21

Source: After Ehleringer et al. 1985.

probably precluded by aridity. The mistletoes' profuse transpiration even during drought in order to procure sufficient N from host xylem may be central to the problem. This behavior represents the antithesis of an epiphyte's usual conservative water use pattern. Failure of so many epiphytic lineages to achieve parasite status simply because invasive organs are difficult to acquire seems unlikely in light of the diverse earth-based species that tap roots. During an historical crossover, the transition between the epiphyte's high solute potential (Table 3.3), low maximum turgor, and sensitive stomata on the one hand, and the parasite's three to five times more concentrated osmotica (Harris and Lawrence 1916; Harris 1918) and corresponding higher foliar conductance at low Ψ_L on the other, would require buffering which is less available in canopy than in terrestrial habitats. Most of the root hemiparasite's absorptive apparatus continues to function in the conventional manner, drawing upon ground moisture and thus reducing the vulnerability of intermediate forms. Consistent with the proposed terrestrial origin of branch parasitism is the discovery that eight herbaceous Scrophulariaceae also transpire more profusely than do hosts (Press, Graves, and Stewart 1988).

Host–mistletoe specificity

The presence of a particular mistletoe on a certain host or hosts extends from monospecific to broadly polyphagous. European *Viscum album* subsp.

album attacks 230 kinds of trees, including at least 186 introduced species. Immensely versatile *Dendrophthoe falcata,* the most adaptable mistletoe documented so far, had 343 recorded Indian hosts and at least 500 over its extensive Australasian range. Representative of an intermediate group is *Phrygilanthus aphyllus,* a parasite known to attack only eight genera of cacti. Restricted to a still greater degree is *Amyema lucasii* whose anchorage is provided solely by *Flindersia.* Likewise, *Muellerina bidwillii* is adapted for *Callitris,* as is *Diplatia grandibractea* for *Eucalyptus* (Knutson 1983). One *Arceuthobium* attacks a single conifer (e.g., *A. minutissimum* on *Pinus griffithii*), and no angiosperms are infected by any of its 27 congeners.

Closely related parasites may differ greatly in substratum specificity. *Viscum album* is polyphagous whereas *V. cruciatum* attacks only the olive *Olea europea.* Considerable differentiation is possible within individual species. There are three recognized races of *Viscum album* (Sallé 1983): *V. album mali,* which parasitizes numerous angiosperms; *V. album pini,* which attacks mainly *Pinus sylvestris* and *P. nigra;* and *V. album abietis,* which infects only *Abies.* Clay et al. (1985) conducted experiments to determine whether host races exist in eastern American *Phoradendron tomentosum.* Previous surveys had revealed suggestive geographic differences in affinity for particular trees. Three hosts were used; an *Ulmus,* a *Prosopis,* and a *Celtis.* Seeds placed on experimental supports matching seed–source identity succeeded much more often than those placed on other-species hosts. Acceptable trees, whether few or many, need not be closely related. *Phoradendron bolleanum* subsp. *bolleanum* grows equally well on its two accommodating genera, *Arbutus* and *Juniperus* (Kuijt 1969). No major vascular taxon, including the pteridophytes, escapes entirely. Even the perpetrators themselves are vulnerable: *Dendrophthora epiviscum, Phacellaria* of Santalaceae, and *Ixocactus* all parasitize other mistletoes exclusively. Self-parasitism is also fairly common.

Host–mistletoe specificity is based on disparate determinants, some peculiar to these parasites and others evidenced by free-living epiphytes as well. Everwet tropical forests reputedly promote polyphagy (Barlow and Wiens 1977), owing to high canopy diversity and hyperdispersion of conspecific trees. Limited host ranges are more common in savannas and temperate zones, reputedly in part because woody floras there are less speciose. Restriction of dwarf mistletoes to conifers, a group that tends toward monoculture, could be offered as proof for this ecological hypothesis except that another factor complicates the issue. *Arceuthobium,* being significantly heterotrophic, extracts considerable organic material from its supports, a condition that could impose tighter criteria for compatibility than does strictly xylem-based parasitism. Utilization of the transpiration stream alone may require some biochemical specialization. Two host-specific races of *Phoradendron californicum* varied at five genetic loci that population biologists use to measure polymorphism (Glazner, Devlin, and Ellstrand 1988), but there is no evidence that the coded allozymes affect host–parasite relationships.

Hypersensitivity leading to abnormal growth and/or death of infected bark further suggests a physiological component in host–mistletoe compatibility. Kuijt (1969) has stated that extensive hypertrophy indicates a high degree of incompatibility and that here mistletoe growth is often slower and fruit production lower. Recognition of a host is thought to be based in part on sensitivity to phenolics (Knutson 1983). Good subjects for comparative studies are available to interested investigators: For example, in France there are *Arceuthobium*-resistant individuals of *Pinus ponderosa* and marked variability in susceptibility of *Populus* clones to *Viscum album*.

Parasite rejection is sometimes effected or accompanied by a histological response. Haustorial penetration of Chilean *Quillaja saponaria* by seedlings of *Tristerix tetrandrus* elicited production of corky tissue (Hoffmann et al. 1986; Fig. 6.2A). No such barrier developed in *Lithraea caustica* or *Kageneckia oblonga,* although xylem connection was not established in the former. An infection in susceptible *K. oblonga* prevented establishment of additional mistletoe seedlings, evidence that a "host exclusion" mechanism operates in this shrub (Hawksworth and Wiens 1972). Some potential supports may remain mistletoe-free for the same reasons that noninvasive epiphytes are also absent – namely, unstable or toxic bark or overly dense canopies. Smooth surfaces or torrential rains may be insurmountable impediments at some locations. Although many mistletoes seem to require high exposure, this may be illusory and related instead to the perching and foraging behavior of frugivorous birds. There is little information on light response in mistletoe photosynthesis.

It appears that polyphagous root parasitism is the primitive condition. Haustoria of extant soil-based taxa *(Atkinsonia, Gaiadendron, Nuytsia)* are unusually indiscriminate; virtually every annual and perennial within range, as well as telephone cables and clods of dirt, are attacked (Fineran and Hocking 1983). Multiple parasitic liaisons by a single plant are standard in relic Loranthaceae, as in many other root hemiparasites. Possible advantages of simultaneous polyphagy in a primitive santalalean stock disappeared as mistletoes evolved the arboreal habit. With it came greater host specificity that presumably obliged extensive morphological, physiological, and ecological changes.

Host–mistletoe crypsis

Australian loranths are unique among mistletoes for their frequent resemblance to primary hosts (Fig. 6.1A–C). Leaves in particular are so similar in form and presentation to a human eye several meters away that tree and parasite are indistinguishable, although closer examination often reveals less faithful coloration and texture. *Amyema, Dendrophthoe, Diplatia, Lysiana,* and *Muellerina* are mimetic mistletoes; their hosts are most often *Acacia, Casuarina,* and *Eucalyptus.* Parasitic mimics with compressed or terete foliage attack *Casuarina* (e.g., *Amyema cambagei* on *Casuarina torulosa*). Those with flat leaves (e.g., *Amyema sanguineum* and *Dendrophthoe glabrescens*) infect *Eucalyptus,* other Myrtaceae, and phyllodinous *Acacia.* Several mangroves harbor mistletoes with comparably thickened and shaped foliage. Perhaps the most striking example of convergence involves *Dendrophthoe homoplastica* on *Eucalyptus shirleyi* (Fig. 6.1.B).

Resemblance is greatest among species pairs occupying arid rather than humid parts of Australia (Barlow and Wiens 1977). Excluding the mistletoes from everwet northeastern forests (which show relatively broad host specificity) and the two endemic terrestrials, 78% of Australian loranths bear foliage resembling that of their hosts too closely for coincidence. Barlow and Wiens and others before them have postulated that vegetative similarity constitutes genetically based mimicry which has arisen independently many times in response to selective pressures and confers fitness by hiding mistletoe foliage within matrices of less palatable forage. Leaf form is supposedly especially convergent in Australia because the two dominant tree genera – *Acacia* and *Eucalyptus* – have become so well defended (i.e., relatively unpalatable because of sclerophyllous leaves containing the essential oils of *Eucalyptus* and the phenolics of *Acacia*) that the parasites have become attractive food items. Mistletoes are therefore all the more liable to seek, as it were, safety in crypsis in order to escape their putative selective agent, the arboreal possum *Trichosurus* and other possible foliage feeders.

Support for the predator avoidance hypothesis was obtained from analyses of leaf Kjeldahl N, hence protein content and nutritive quality (Ehleringer, Cook, and Tieszen 1986). If mistletoe foliage is more nutritive than host foliage, cryptic mimicry could be especially advantageous. Of 22 Australian mistletoes that exhibited mimicry, 17 bore foliage richer in N than that of hosts. (None of the five mimics of *Eucalyptus* had higher [N], but presumably "hid" because these hosts are so well defended.) Fifteen of 26 nonmimetic mistletoes produced foliage with [N] averaging about a third below that of their supports. Similar findings have been reported for a smaller group of New Zealand mistletoes (Banister 1989). Nitrogen-poor parasites might be expected to benefit less from cryptic mimicry than by standing out against the more nutritive food around them; vertebrates are quite adept at sorting out the most desirable herbiage.

Atsatt's (1983) morphogen hypothesis states that evolution of host specificity was sometimes accompanied by "genetic selection for hormonal compatibility." Concordant morphogens were especially important in hosts with "highly reduced or otherwise specialized foliage," and presumably an equally extraordinary set of growth factors that excluded incompatible mistletoes. But this proposal ignores the basic tenets of plant morphogenesis. The shape and size of determinate organs are established by the behavior of localized, temporary meristems in expanding primordia. Intercalary, plate, marginal, and other transitory mitotic loci remain active for precise intervals and in programed order as the appendage takes form. Uneven versus uniform activity along expanding embryonic margins creates a lobed rather than an entire lamina. Should a basal intercalary meristem shut down quickly, a sessile instead of a petiolate appendage results. Regulator and modifier genes are probably largely responsible for leaf shape, and ubiquitous mitogenic and growth-influencing hormones, not form-specific morphogens, serve as their mediators. Existence of as many specific translocatable agents as there are different or unusual laminar shapes stretches credulity and draws no support from the literature. Hall et al. (1987) used radioimmunoassays to identify cytokinin bases, ribosides, *O*-glucosides, and nucleotides in xylem sap of two *Amyema* species and three *Eucalyptus* populations that support them. Closer relationships were found between the type and concentration of cytokinins in one mimetic host–parasite pair than in two form-distinct combinations of tree and mistletoe, but assignment of cause to occurrence of these common growth factors would be premature.

Calder (1983) makes a plausible case for dispersal as the primary advantage of host simulation. If a mistletoe resembles its support well enough to be imperceptible at a distance, birds must learn to key on the more conspicuous (larger) host–crown signal to locate pseudoberries. Such a mechanism ensures regular recruitment of uncolonized supports because appropriate frugivores, some with guts already containing seeds, must explore all potential food-bearing trees to discover whether or not fruiting mistletoes are located there. Of course, advantage diminishes as host range widens, but so it would if protective concealment were the purpose. The fact that mimetic mistletoes, especially those native to relatively humid forests, are not all host-specific is attributed to relaxed selection pressure associated with late Tertiary and Quaternary extermination of herbivorous marsupials (Barlow and Wiens 1977). Changes in phytophagous insect fauna, if any, would be inconsequential; these predators usually cue on food source by olfaction rather than by sight.

It might seem surprising that no one has apparently entertained the notion that host–mistletoe mimicry is not really mimicry at all but a manifestation of functional convergence. After all, in warm arid zones especially, leaf proportions normally track regional meteorological conditions because inappropriate foliar morphology could have especially dire consequences for water economy and heat dissipation. Predominant leaf shapes in fossil floras have long been known to be useful indicators of paleoclimate. But such a proposal is not found in the literature, and for good reason. Variation among co-occurring species – those in a single temperate forest, for instance – attest to the absence of fine-tuned selection for uniform shapes at the local level. In addition, gas exchange patterns, by influencing energy budgets and placing constraints on the evolution of leaf form, differ substantially between trees and their aerial parasites, as noted above. Finally, homoplasy in response to similar climatic stresses fails to explain the variety of leaf forms shared by sympatric, mimetic host–parasite combinations involving models with widely disparate leaf architecture (e.g., *Acacia, Casuarina,* and *Eucalyptus*) and their mistletoes. Microclimates surely do not vary much among crowns of these trees. More study is needed to test the first and third – and possibly additional – hypotheses concerning vegetative form in Australian mistletoes. Scattered reports (Atsatt 1983) of comparable mimicry in other parts of the world should also be verified.

Mistletoes as pathogens

What few data are available on the deleterious impact of mistletoes in natural forests come from western North America, where each year *Arceuthobium* reportedly reduces wood yield by 500 million cubic feet in the United States alone (Hawksworth 1983). Mistletoes are second only to fire as damaging agents in certain Mexican forests where 10–24% of trees are infected. Significant economic hardship is experienced in many parts of the world by

Table 6.3. **Some important disease-inducing mistletoe genera, their hosts, and regions of serious crop losses**

Family/genus	Hosts	Region
Loranthaceae		
Amyema	several, including eucalypts and acacias	Australia
Dendropemon	several trees, including citrus	Puerto Rico
Elytranthe	rubber, cashew	Malaya
Oryctanthus	cocoa	Costa Rica
Phthirusa	rubber, cocoa, erythrinas, citrus, mango, coffee, avocado	Costa Rica, Brazil, El Salvador
Tapinanthus	cocoa	Ghana, Nigeria
Tolypanthus	citrus	India
Viscaceae		
Arceuthobium	conifers	North America, Mexico, China, India, Pakistan, Kenya, Mediterranean area, Middle East
Dendrophthora	rubber, mango, avocado, cocoa	South America
Korthalsella	acacias, eucalypts	Hawaii, Australia
Phoradendron	coffee, avocado, teak, erythrinas, citrus, cocoa, various forest trees	Bolivia, Trinidad, Central America, Mexico, United States, West Indies
Viscum	rubber, fruit trees, deciduous trees, walnut, persimmon, fir, pine	Europe, Australia, China, Asia, Africa

Source: After Knutson 1983.

plantation owners growing such crops as *Citrus,* fig, mango, rubber, coffee, and cocoa. Table 6.3 lists important disease-causing mistletoe genera in both Loranthaceae and Viscaceae and the regions of greatest distress. Numerous xylem-tapping species are involved, but others forming diffuse endophytes are particularly destructive.

Symptoms of disease vary with the host–parasite combination. The literature contains frequent statements about visible damage, and there are occasional references to trees being killed by heavy infestations. Injury caused by *Loranthus langsome* can be so severe that heavily infected *Lansium domesticum* (mahogany) lives only seven to eight years. Fruit yield of English apple trees has been depressed 40% by *Viscum album* (Hawksworth 1983). Common reactions to mistletoe attack are localized hypertrophy at the attachment site, dieback of infected branches, deformation caused by death of the leader, reduced overall vigor, and formation of witches'-brooms. (Brooming is the profuse production of abnormal host twigs and foliage.)

Mistletoes are potentially more destructive than free-living epiphytes, no doubt owing to their direct use of host resources and possibly because of physiological consequences of the tap-in. Moisture stress is heightened during drought, damaging infected more than uninfected trees. Sometimes much host foliage is shaded by heavy infestations. There is also evidence of considerable hormonal involvement: Cytokinin levels in *Arceuthobium douglasii* have been found to exceed those in adjacent host tissue as much as tenfold. Brooming similar to that induced by *Phoradendron* and dwarf mistletoes is caused by growth factors where certain microbes are the infectious agents. Likewise, hypertrophy adjacent to the endophyte is undoubtedly under hormonal control, but whether the agents involved originate in tree or parasite is not known. Hormonal imbalance can be systemic and contribute to the general decline of diseased hosts; stressed foliage has been shown to contain reduced cytokinin and elevated abscisic acid activity, but these conditions may be more directly related to water deficits caused by the parasite's transpiration than to direct metabolic intervention. Like other obligate parasites, mistletoes lose fitness if too virulent, and indeed many infections seem to create no serious problems for supports. Further inquiry might reveal that damage is greatest on occasional or unnatural (exotic) hosts or those trees already weakened by other factors. At this point, pathogenicity seems unrelated to host specificity.

Several means have been tried to control mistletoes in orchard crops and shade trees; In Europe, 2,4-dichlorophenoxyacetic acid (2,4-D), 2,4,5-trichlorophenoxyacetic acid (2,4,5-T), 4-(2-methyl-4-chlorophenoxy)butyric acid (2,4-MCPB), and dichloroethane killed *Viscum album* shoots on *Abies* with little host damage (Hawksworth 1983). Such less expensive agents as copper sulfate and diesel oil have been used against *Dendrophthoe* and other Indian mistletoes. Unfortunately, chemical control of *Arceuthobium* has not been successful. Increasingly, statutory restrictions will probably deter widespread application of broad-spectrum herbicides. Attempts to eliminate avian dispersers would be unwise. Acceptable alternatives might be based on insect predators or fungal pathogens, of which several are known. In the meantime, the ancient practice of pruning infected branches will probably have to suffice in most instances.

7 Ecology

Epiphyte ecology has been a recurrent topic throughout the preceding chapters but little or nothing has been said about community organization, succession, associated phytotelmata, phorophyte specificity, or influence of epiphytes on supports and other biota. Although documentation of cause and effect is scarce, it is clear that canopy-dwelling flora can help shape forest structure and economy; processes as fundamental as community-wide mineral cycling and productivity are affected. Influence on animals is no less pervasive. Without plant resources beyond those provided by earth-rooted vegetation, much of the immense and diverse fauna characteristic of humid tropical woodlands (Erwin 1983) would not exist. This chapter will emphasize the role of epiphytes as members of communities and substrata for other organisms.

Host specificity

Only the exceptional epiphyte has but one acceptable phorophyte (Table 7.1). Far more commonly, anchorage occurs on several kinds of supports, although usually not with equal frequency. Valdivia (1977) studied the distribution of 153 vascular epiphyte species on 45 different trees in east-central Mexico. Only *Acacia cornigera,* a myrmecophyte that is aggressively defended against other insects and encroaching vegetation by its ant colonies, hosted no epiphytes. The remaining 44 each supported more than one species; the record was 107, demonstrating that certain trees offer especially suitable crowns. Few phorophytes provide anchorage to every potential colonist, however, nor does occurrence always follow expected patterns. *Aechmea bracteata* is abundant on several trees in semievergreen forest in the Sian Ka'an Reserve, Mexican Yucatan (Olmsted and Dejean 1987). *Tillandsia balbisiana* was recorded at even higher frequency there, but only in crowns of *Bucida pinosa,* a dominant taxon in the area that is invariably free of *Aechmea bracteata* specimens. Closely related *Tillandsia dasyliriifolia* shared several types of supports with *Aechmea bracteata* and also flourished in local *Bucida* crowns. *Encyclia krugii,* a bark epiphyte endemic to dry forest in southwestern Puerto Rico, also tended to occur on the more

Table 7.1. **Specificity of common epiphytes on phorophytes in an Australian forest**

Epiphyte/phorophyte	Number of epiphytes sampled	Percentage of epiphytes on phorophyte species	Number of other phorophyte species utilized in sample plot
Ferns			
Dictymia brownii on *Pennentia cunninghamii*	25	52	33
Drynaria rigidula on *Cryptocarya* aff. *hypospodia*	22	77	3
Platycerium hillii on *Cryptocarya* aff. *hypospodia*	28	61	5
Dicot			
Myrmecodia sp. on *Cryptocarya* aff. *hypospodia*	31	55	4
Orchids			
Dendrobium malbrownii on *Cryptocarya* aff. *hypospodia*	57	60	3
Dendrobium falcorostrum on *Nothofagus moorei*	23	100	—
Sarcochilus hillii on *Backhousia sciadophora*	123	69	5

Source: After Wallace 1981.

common trees (Ackerman, Ontalvo, and Vera 1989). Samples were too small, however, to determine whether phorophyte frequency or biological compatibility was responsible for the observed plant combinations. Surface texture had no decisive effect. In plots containing 10 smooth-bark and 10 rough-bark tree species, epiphytes anchored on six of the former and four of the latter. As in the case of many uncommon epiphytic orchids, occurrence was patchy with relatively few scattered individuals comprising each colony ($\bar{x}$ = 1.4 plants per occupied phorophyte). Subjects were often absent where acceptable trees abounded, suggesting that factors other than favorable germination sites limit population growth. Unexplainable patchiness also exists on a larger scale; for instance, Richards (1952) reported uneven colonization by an extensive flora in Guyanan wet forest, ranging from 16% to 38% of the larger trees in five plots.

Germination on earth soil might relax host specificity for the secondary hemiepiphyte, but later root development apparently does not. Todzia (1986) studied 20 primary hemiepiphytes from six families on Barro Colorado Island in Panama. Infestation rate varied considerably among potential supports; 58% of *Hura crepitans* but only 1% of *Quaraibea astrolepis* specimens were hosts. Almost 21% of 160 tree species supported African *Ficus* spp. in an Ivory Coast woodland (Michaloud and Michaloud-Pelletier 1987). Trees featuring epiphytic ferns and orchids in well-illuminated crowns fostered fig success. There was no consistent relationship between a support's dominance in the forest and its use by hemiepiphytes.

Features associated with, or responsible for, epiphyte–tree pairings are usually obscure. Degree of intimacy between the two has no consistent bearing on exclusivity. Mistletoes and orchids alike often exhibit extreme fidelity, proof that parasitism is not a sine qua non for specificity. Not surprisingly, epiphytes (and mistletoes) known for their strict substratum requirements tend to inhabit low-diversity forest. Few workers have documented substratum specificity; of those who have, Went (1940) was most convinced of its ubiquity, and was able, in fact, to differentiate between unidentified *Castanopsis* species in Tjibodas, Java on the basis of epiphyte colonization. Johansson (1974) and Sanford (1974) drew a different conclusion from separate orchid surveys in West Africa. Wallace (1981) tested the hypothesis in Australia and found that a small minority – just three ferns, one dicot, and three orchids – showed 52–100% constancy (Table 7.1). Even *Dendrobium falcorostrum,* the single species registering exclusive association with one tree *(Nothofagus moorei)* on sample plots, had a few individuals on other supports elsewhere. Wallace also commented on *Dendrobium acmulum,* an Australian orchid known in several forms, each usually confined to a particular tree. One of these populations represents the only epiphyte known to colonize *Eucalyptus* bark, although some *Cymbidium* species flourish in knotholes and wherever else humus collects in these trees. A second population of *Dendrobium acmulum* grows almost exclusively on the upper portion of rough-textured bases of myrtaceous *Tristania conderta,* another otherwise axenic phorophyte.

Questions concerning host specificity may be researched from several vantage points. Wallace's (1981) phytosociological approach, like that of Ackerman et al. and Valdivia, is the obvious first step; this sort of preliminary documentation on occurrence could save needless effort where the phenomenon is more apparent than real. Surveys may show that, in fact, many of these epiphytes are not the faithful colonizers they seem; there are just fewer suitable anchorage sites available on some trees than on others. Absolute axeny could turn out to be quite rare. *Rhizophora mangle* in South Florida supports few or no epiphytes where other phorophytes carry light to moderate loads, but at the occasional location with abundant arboreal flora, this mangrove harbors at least a few bromeliads. *Rhizophora* may be colonized in the second instance simply because denser seed rains reveal the few favorable anchorage sites in what are largely uninhabitable crowns. Or perhaps the quality of a particular medium varies with the site.

Numerous and disparate factors determine where a particular epiphyte grows. Phorophytes sometimes exhibit what seem to be especially accommodating qualities – for example, the persistent thick mantle of absorbent leaf-base remains that ensheath *Vellozia* stems. This penetrable substratum is used exclusively by *Pseudolaelia vellosicola, Oncidium warmingianum, Encyclia ghillanyi,* and *Constantia cipoense* in southeastern Brazil (Dressler 1981). Unfortunately, reports never mention whether or not other local epiphytes fail to colonize these same phorophytes. Should such exclusions occur, the case would be far more interesting.

Tight association sometimes reflects specialized plant needs. A deciduous Ecuadoran *Cyrtopodium* orchid, requiring copious amounts of moisture while leafy, occurs only in humus-filled leaf bases of a single palm. *Ophioglossum palmatum* grows on trunks of the cabbage palm *Sabal palmetto* in most of its Florida habitats; much more common *Vittaria lineata* is almost as consistent (Fig. 7.1). Perhaps no other local tree provides as much well-rotted debris for nurture of delicate gametophytes. Trash-basket ferns rather than trees can also provide nursery beds, and invaders may be specific; *Cymbidiella rhodochila,* for example, grows solely in the humus accumulated by *Platycerium madagascariense.* A support can present serious impediments to some colonists but not to others. Neotropical *Psidium* has a coarse exfoliating bark on older axes that prevents securement of all but the occasional bark epiphyte; nevertheless, its twigs and even leaves host a characteristic flora of fast-maturing orchid miniatures, including species of *Ionopsis* (Fig. 1.3), *Psygmorchis* (Fig. 1.16) and *Rodagesiana,* some of which anchor nowhere else. Corky ridges may aid securement of these small heliophiles, but alone seem insufficient to explain fully the regular relationships.

Factors less closely related (or even unrelated) to plant culture may dictate occurrence. Several *Ficus* stranglers congregate in crowns of *Hura crepitans* on Barro Colorado because of the secure anchorage provided to seeds and juveniles by abundant stem spines (Todzia 1986). Foraging and perching habits of dispersers are also important determinants of host specificity among Panamanian hemiepiphytes. *Platypodium elegans* was largely free of canopy residents despite hospitable bark, at least in part because its wind-

Figures 7.1–7.6. (7.1) Persistent leaf bases of *Sabal palmetto* harboring *Vittaria lineata* in South Florida. (7.2) Dense festoons of *Tillandsia usneoides* in the crown of a declining *Quercus virginiana* in central Florida. (7.3) Dwarf *Quercus virginiana* infested with *Tillandsia recurvata* and other epiphytes in a sterile coastal strand habitat in southwestern Florida. (7.4) Vigorous *Quercus virginiana* supporting few epiphytes in central Florida. (7.5) *Aechmea bracteata.* (7.6) A leaf covered with epiphylls in humid Ecuadoran forest.

dispersed seeds do not attract frugivorous birds and bats seeking to supplement earlier meals of ripe figs. Mistletoe occurrence is strongly influenced by the behavior of avians, some of which have become highly specialized berry feeders.

Moisture, exposure, chemistry, and bark morphology determine the suitability of a tree once a seed anchors in its crown. Juveniles must be sensitive to moisture retention and drying qualities of substrata. Subjects held too far above an irregular phorophyte surface are wetted only while rain is falling. Conversely, close contact with bark may promote loss of plant moisture during drought unless embolism blocks that route (Fig. 3.20). Microclimate seems to affect host–epiphyte specificity through its influence on bark moisture and on transpiration; colonization diminishes within groups of conspecific trees arrayed along ecological gradients. Much reduced density of several bromeliads on *Taxodium distichum* within meters of wet depressions in South Florida's Big Cypress Swamp appears to parallel gradients in atmospheric humidity. Once a trend is identified, measurements using soil-testing techniques could reveal whether tree species that first became axenic possess barks with the least serviceable water relations.

Chemistry could influence substratum exclusivity at many levels. Bark pH correlates with bryophyte distribution in some north temperate forests (Barkman 1958), but causal mechanisms remain obscure. Bark constituents are critical in Java according to Went (1940), but no direct evidence was offered. Frei and Dodson (1972) demonstrated a distinct parallel between presence of phenolics in bark extracts and epiphyte use. A bioassay using *Encyclia tampensis* seeds yielded anticipated results; inhibition was greatest in those cultures prepared from substrata most repellent to orchids in the field. Experiments based on sterile filtered, rather than on autoclaved, infusions would have been more credible, however.

Still, phorophyte specificity must have a more complex chemical or biological basis in some cases than has been demonstrated thus far. Epiphytes with different reactions to the same host may be disturbingly similar in all the obvious qualities that would seem to dictate substratum compatibility. Why, for instance, do sympatric *Psygmorchis pusilis* and *P. glossomystax* rarely share hosts? The former species, despite broad host and geographic ranges, virtually never colonizes *Psidium,* yet the latter more insular South American taxon will grow on little else. The differences distinguishing these short-lived oncidioid orchids seem too slight to explain their mutual exclusivity. Perhaps the answer lies in seed or seedling biology. Epiphytic orchids, like related terrestrials, utilize fungi to germinate and nourish juveniles, and the bark of certain trees may not support an appropriate microflora. If so, there would be an answer for complete but not partial orchid repellence. Compatibility between epiphytic (although less so terrestrial) orchids and their fungi is generally nonspecific, and exclusion based on selective infection should be the exception. *Psidium* certainly provides extensive inoculum if findings in an Ecuadoran guava orchard are typical (Benzing and Bermudes in press): Every root examined from a small sample of *Ionopsis* and *Rodriguezia* was heavily infested with hyphal coils. In order to solve such problems as the *Psygmorchis* enigma, trees must be screened for bark-inhabiting fungi competent to germinate and sustain some but not all local epiphytes.

Epiphyte, particularly orchid, occurrence on certain hosts has another dimension that transcends the hospitality of a given tree species: Why are certain supports within a group of fairly even-aged conspecifics sometimes exceptionally burdened with crown flora? Do these hosts provide better access to conventional resources (light, water, or nutrients), or is this proliferation simply the result of extraordinary recruitment by early colonists? Did uneven site preparation by moisture-retaining lichens and bryophytes or use of a tree as a bird or bat roost play a part? Were beneficial ants unusually abundant? Might the occasional heavily infested tree be diseased, perhaps with an organism that promotes orchid germination or in some other way conditions its crown for greater orchid success? Alex Hirtz, a remarkably sharp-eyed Ecuadoran collector, has suggested that infrared sensing be used to detect infection in phorophytes that exhibit no clear symptoms of pathology. Earth satellite technology is routinely employed by foresters and agronomists to monitor epidemics and predict crop yields.

Ant-dispersed epiphytes, particularly nest-garden forms, represent a special situation because an animal chooses the phorophyte and, by tending the sown garden, reduces its own as well as the plant's dependence on the tree. Complex animal needs affect mutual occurrence; Davidson and Epstein (in press) report preferential ant colonization of Peruvian trees with certain attributes: extrafloral nectaries (e.g., *Inga*), suitability for Homoptera (e.g., *Calyptranthes*), spiny surfaces that prevent ascent of predatory ants (e.g., *Tococa*), and aromatic oils. The importance of tree products to the ants is underscored by the rapid desertion of dead or dying phorophytes by nest builders. Trees weakened by deep shade or other factors remain largely nest-free regardless of identity because their resources are inadequate for ants. Ant-fed ant-house epiphytes in northeastern Australia often occur on trees with loose bark that provide effective cover for Homoptera. Rubiaceous epiphytes inhabited by *Iridomyrmex cordatus* are overrepresented on *Calo-*

phyllum incrassatum (Guttiferae) and *Ploiarium alternifolium* (Theaceae) where ants tend numerous scale insects (Davidson and Epstein in press).

Numerous Amazonian epiphytes can be considered restricted to ant nest-gardens. Three of the most abundant species observed at Cocha Cashu were rarely seen rooting elsewhere (Davidson 1988) Individual plants utilizing other than the usual substratum were *Peperomia macrostachya* (5 of 675 sightings), *Ficus paraensis* (1 of 202 sightings), and *Anthurium gracile* (6 of 261 sightings). Alternative media did not appear to exist for several additional species, including *Vanilla planifolia* and a *Codonanthopsis,* but samples were too small to allow definitive judgments. Nest size, its exposure to irradiance, and ant identity influenced garden floristics. *Anthurium gracile* was statistically overrepresented on small ant nests, whereas *Codonanthe uleana* succeeded without respect to carton size. All of the relatively common ant nest-garden species except *Neoregelia* were more likely to co-occur with others than to live alone on a carton. Near obligacy was further evidenced by reduced density of most epiphytes on deteriorating substrata. Bromeliads, probably because they draw resources from materials trapped in leaf axils, appeared to be least vulnerable to carton erosion. All but two of the observed epiphytes occurred at lower frequency on *Azteca* gardens compared to those tended by *Camponotus femoratus* and its usual nest associate, *Crematogaster linata parabiotica.*

Finally, constraints affecting both carbon and ion balance interact in complex ways to help determine which trees support which epiphytes. Leachates of some hosts are more ion-rich than others (Tukey 1970; Table 4.3), but how this might affect canopy flora varies with the epiphyte. Oligotrophic taxa, being especially effective scavengers with modest nutrient needs, may be little constrained by ion-poor canopy fluids, and in fact can benefit from related stress experienced by the phorophytes. Dense populations of atmospheric *Tillandsia paucifolia* throughout much of extreme southern Florida occur on scrub cypress (Fig. 5.12) stunted in part by impoverished soil (Benzing and Renfrow 1971c). For this air plant, a sparse canopy is obligatory; light seems to be the principal limiting resource, although resident epiphytes are also relatively deprived of scarce ions. Better-nourished trees in deeper earth produce a more nutritive bark (Table 4.5) and stemflow, but their relatively opaque crowns support many fewer, albeit individually larger and better-provisioned, bromeliads (Benzing and Renfrow 1971c; Benzing and Davidson 1979; Fig. 7.9). Shortages of N, P, and K have an especially telling effect, whereas access to Mg (Figs. 7.7, 7.8) and to a number of additional elements seems to be sufficient to meet requirements fully at

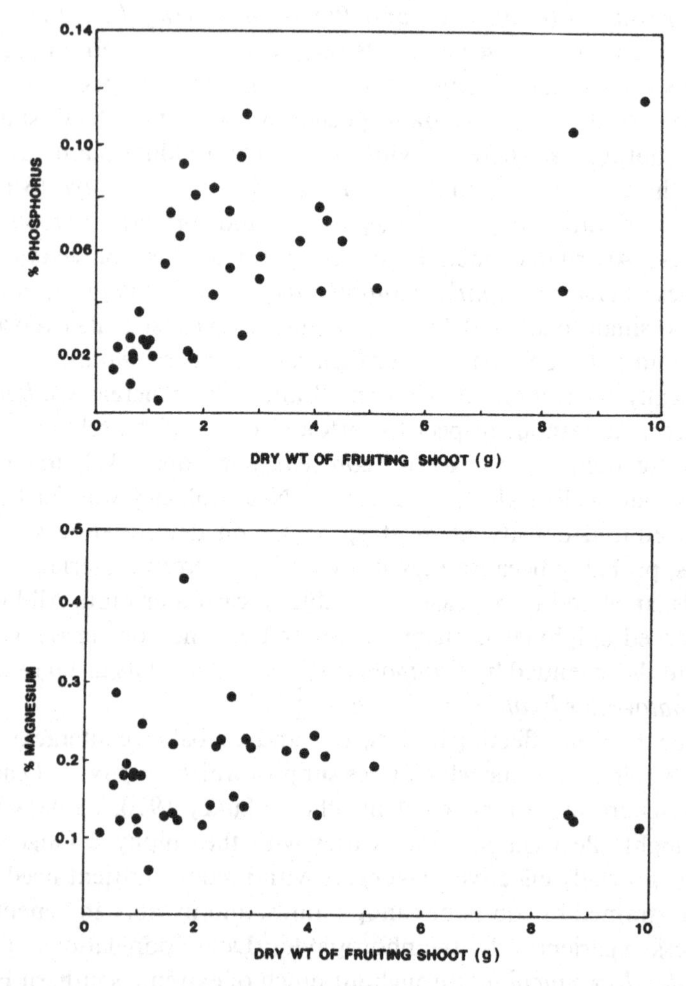

Figures 7.7, 7.8. Relationship of shoot P and Mg content to mature plant mass of *Tillandsia paucifolia* growing in diverse South Florida habitats. (After Benzing and Renfrow 1971c.)

both lower- and higher-quality sites. There could well be an inverse relationship between epiphyte success and fertility on the one hand and energy level on the other. Perhaps certain canopy flora must be able to tolerate either shade for the sake of more nutrients and/or better moisture retention, or infertility for the sake of irradiance. In the absence of supplements from ants or tanks, specialization for full exposure certainly obliges adjustment to dilute, sporadic ion sources as well as to drought.

Because epiphytes can damage heavily infected hosts through several

Figure 7.9. Nutritionally stressed (A) and better provisioned (B) specimens of *Tillandsia paucifolia.*

mechanisms discussed below, axeny has its advantages. Putz (1979) reported that some neotropical trees with long leaves and flexible axes remain free of enshrouding lianas, partly owing to these aspects of habit. Cause-and-effect was implied. Early successional trees (e.g., *Cecropia*) are usually free of epiphytes (but not of herbaceous vines), most likely because of their rapid growth, often smooth bark, and short life-span. Certain mature forest dominants may support only sparse colonization because they repel or shed these plants. Bark texture and stability, chemistry, and production of substances that attract defending ant colonies are all amenable to selection, suggesting that suitability of a particular tree as a substratum for epiphytes need not be coincidental.

Circumscribing, labeling, and documenting the epiphyte community

Concepts of how epiphytes should be organized into phytosocietal units have varied with historical period and an author's broader views on classification of vegetation. Complicating the issue have been differing emphases on dynamism. Some workers recognized ecological succession, whereas others offered no judgments about age or seral stage. Epiphyte associations were circumscribed according to composition and location without regard to history or potential for change. Every approach has taken some account of adaptive biology, however. Schimper (1888), working before plant sociology became a science, observed distinct tiers of epiphytes that shared common physical tolerances and growth requirements in humid woodlands. From a more advanced philosophical base, Oliver (1930) maintained that plants in canopies form distinct ecological units in the individual forest. Concepts of dominance and succession were applied. Sun and shade epiphytes constituted separate synusiae representing assemblages of convergent taxa distinct from, but variously dependent upon, each other.

Segregation (in terms of water–carbon balance mechanisms) of canopy-adapted floras into vertical strata has already been described (e.g., Pittendrigh 1948; Benzing and Renfrow 1971c; Griffiths and Smith 1983). Johansson (1975) was able to identify five life zones in some West African trees, three at different depths in the crown plus one on the upper and one on the lower trunk (Fig. 7.10). Each had its unique set of orchid occupants, albeit with a certain amount of overlap: Percentages of the total orchid flora at each location varied from 4% at the crown periphery to 48.5% at midcrown, where shade was moderately deep and moisture more abundant. Percentages for the crown center and upper and lower trunk zones were 27.7, 10.9, and 8.9, respectively. Presumably, survival on outermost twigs was tested most by drought, whereas light was probably the limiting resource deeper in the canopy. Gentry and Dodson (1987b; Fig. 2.20) and Catling, Brownell, and Lefkovitch (1986) noted similar epiphyte apportionment at sites in Ecuador and Belize, respectively. Aspects of tree ontogeny and architecture (described later on) further ensure that highest epiphyte densities occur within rather than at the edge of or below the crown.

Hosokawa (1955 and earlier papers) rejected reliance on then-accepted practice. He deemed a synusial classification inappropriate for epiphytes, arguing that such a system was designed to describe terrestrial vegetation according to height and stratification. (An epiphyte's physical stature, of course, has less bearing on its location within the broader community than do its shade and drought tolerances.) Hosokawa's three-level hierarchy, to date the most elaborate one specifically developed for vascular epiphytes, runs parallel in concept to the phytosynusial framework. Categories by increasing inclusivity are *-epies* ("society"), *-epilia* ("association"), and *-epido* ("alliance"). Hosokawa also developed another classification for epiphytic life forms. Johansson's (1974) decision to recognize the epiphyte community as any group of three or more species growing in close proximity (the dominant species must occur within specified distances) seems most practical. It certainly avoids cumbersome terminology and complex hierar-

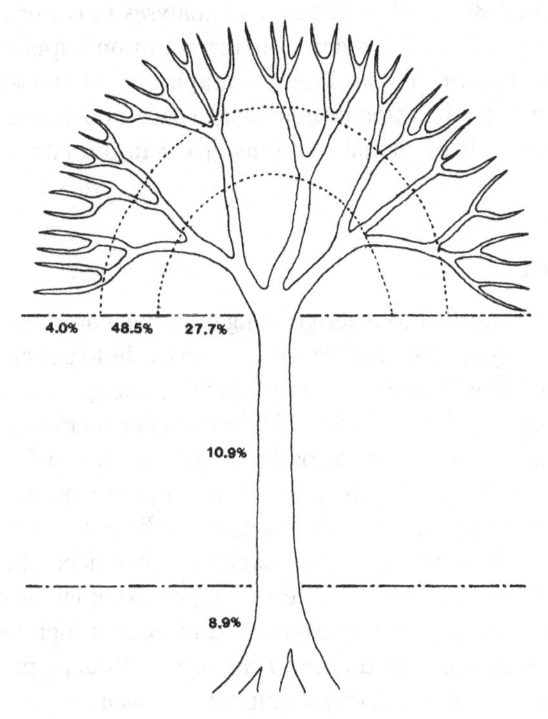

Figure 7.10. Different zones in the host tree and the percent of orchids found in each zone in a West African forest. (After Johansson 1975.)

chies of questionable meaning and utility. His vegetational units have equal status and receive names embodying those of the two most characteristic species.

Epiphytes have been identified, sampled, and recorded by various methods. Attempts to tabulate every individual or genet for spreading forms, and all life stages as well, demand close inspection of felled trees (e.g., Benzing 1981a). Nondestructive estimates have been made from a distance with field glasses (e.g., Sanford 1969). Hosokawa surveyed 30 × 30 m quadrants in Micronesia. Grubb, Lloyd, Pennington, and Whitmore (1963) used physiognomic cross sections throughout representative forest sites. Gentry and Dodson (1987b) assembled florulas from meter-wide transects; every individual was counted. Johansson (1978) discussed recording methods, including one he and subsequently Wallace (1981) employed which lists epiphytes according to phorophyte stem circumference. Vertical alignment of species occurrence on the resulting chart reveals patterns. Catling et al. (1986) and

Catling and Lefkovitch (1989) used mathematical analyses to confirm seral differentiation among groups of epiphytes utilizing common supports (see below). Nadkarni (pers. comm.) is developing a graphic computer-assisted system that places epiphytes in three-dimensional space and promises to allow more powerful analysis of spatial relationships than does any earlier method.

Community composition

The makeup of humid compared to drier-growing epiphyte communities is broad; species representing diverse families often coexist, but spatial biases are common. Ferns are inordinately common close to the ground everywhere but in the wettest sites (Table 7.2). Only occasional genera (e.g., *Pyrrosia,* some *Polypodium*) occur where exposure is extreme or rainfall at all seasonal. Aroids, gesneriads, and many other dicots follow similar distributions because they also tend to be shade-tolerant, inefficient water users. Atmospheric *Tillandsia,* in contrast, cannot survive either deep shade or abundant moisture. Orchids often dominate the upper canopies of multistratal forest, but numerous species also adjust to dim light at high humidity. Bromeliaceae as a whole are versatile, and *Peperomia,* although predominantly mesic, contains a few stress-tolerant epiphytes as well.

Wallace (1981) mapped epiphytes on several Australian phorophytes. Of eight species restricted to the shaded trunk of just one *Ficus watkinsiana* specimen in a subtropical rain forest, six were ferns, one was a dicot, and the other a hemiepiphytic aroid. An assemblage at midcanopy level was mixed: Five orchids shared space with a smaller fern contingent, three dicots, and one nonorchid monocot. Another five orchids, including four species of *Dendrobium,* were anchored in fullest exposure. Kelly (1985) counted epiphytes in a 26- to 28-m-tall Jamaican lower montane rain forest: Among compact (nonvining) epiphytes at 12 m and above were nine orchids, five bromeliads, three ferns, and a single *Anthurium.* In midcanopy (4–12 m) was a mixed assemblage of seven orchids, two bromeliads, three ferns, two dicots, and the single *Anthurium.* In the 0 to 4-m level resided one bromeliad and three ferns. Numerous vining hemiepiphytes grew throughout all three levels of the same canopy. Hosokawa (1955) recorded a strong orchid dominance of upper wet forest canopies on several Pacific islands (Table 7.2). Dodson and Gentry (1978) listed eight aroids, four dicots, eleven orchids, and seven ferns as compact shade-tolerant epiphytes in rain forest at Río Palenque, Ecuador. There were many additional secondary hemiepiphytes representing Araceae, Cyclanthaceae, and the ferns.

Table 7.2. **Number of epiphyte species representing different major taxa at upper (U), middle (M), and lower (L) levels in humid forest at five locations**

	Australia: a single tree in moist subtropical forest[a]			Jamaica: 24-tree sample plot[b]			Palau, Micronesia: sample quadrant[c]			Yap, Micronesia: sample quadrant[c]			Kusaie, Micronesia: sample quadrant[c]		
	U	M	L	U	M	L	U	M	L	U	M	L	U	M	L
Ferns and allies	0	3	6	3	3	3	0	4	6	1	4	4	2	4	9
Nonorchid monocots	0	0	1	6	3	1	0	0	0	0	0	0	0	0	0
Orchids	5	5	0	9	7	0	6	1	0	2	0	0	3	0	0
Dicots	0	3	1	1	2	0	0	0	0	1	0	0	0	1	0

Note: Definitions of height categories vary among authors.

Sources: [a]Wallace 1981; [b]Kelly 1985; [c]Hosokawa 1955.

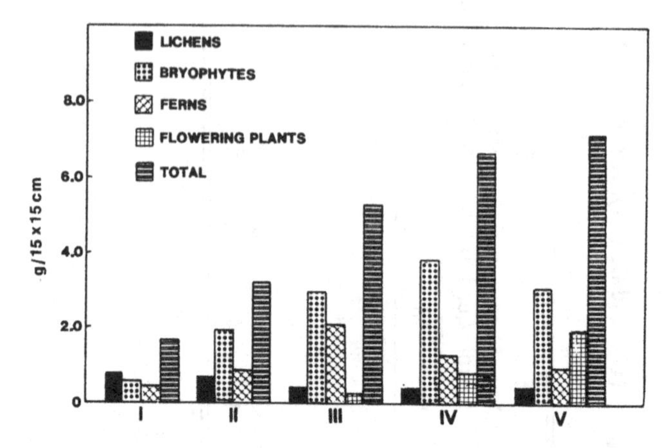

Figure 7.11. Biomass of epiphytes in five different girth classes on *Quercus floribunda* in a mixed oak–cedar forest in the Kumaun Himalayas. (After Tewari, Upreti, Pandey, and Singh 1985.)

Succession

All reports of epiphyte succession are based on simultaneous side-by-side observations rather than on records of change on a single surface. Three types of substratum are involved: initially naked bark; trash-basket debris; and neotropical ant nest-gardens. Janzen's (1974) ant-fed ant-house plants in Malaysia probably represent a fourth medium. Succession on sterile branches can begin with lichens and bryophytes (Fig. 7.11). Dudgeon (1923) divides succession in a Himalayan *Quercus* forest into six stages, beginning with the arrival of crustose lichens on three- to four-year-old stems. Foliose and fruticose lichens follow after three to four more seasons, and then pioneer mosses become established. After 16 more years, a climax moss stage exists, and ferns eventually root in the resulting mat, joined later by angiosperms. Van Oye (1924) recognizes a three-stage sequence in Java. Again, nonvascular plants are seen to condition the site for stress-resistant ferns and then for more demanding ferns and angiosperms. Oliver (1930) lists a similar progression including a terminal stage where bark exfoliation, host death, or some other accident dislodges the community.

True succession involves displacement of early seral stages by later ones, accompanied by at least a partial turnover of species. Johansson (1974) did indeed find remnants of additional vascular epiphytes in siftings from four of 13 communities in a Liberian forest. In one instance, 10 new species were viable occupants, whereas at least six more had been established during earlier stages. Several former residents were noted flourishing in nearby younger communities, apparently as colonists. Catling and Lefkovitch (1989) identified four epiphyte associations between 0.3 and 2.5 m aboveground in a 2-ha plot of Guatemalan cloud forest. Composition ranged from two to five species per association, and occurrence was influenced by age (thickness of axes) of the medium. Of two groups, one early and one late in developing, the former contained fewer but closely related ferns or orchids. Members of the more diverse associations were larger and tended to have seasonal rather than continuous flowering, but complete life histories are needed to confirm roles as pioneers and successors. Wallace (1981), studying what he called "phorophyte size and age effects," examined three widely exploited phorophytes at three different locations. Epiphyte loads were considered relative to trunk diameter at breast height (presumed to reflect age). When height was plotted against number of species added to number of individual epiphytes present, correlation proved to be positive; diversity and trunks usually expanded together. Colonization began on young trees, but pioneers were not displaced by later arrivals.

One of Wallace's phorophytes *(Backhousia sciadophora)* failed to add species as stem diameter increased, but, compared to the other two, shed bark faster. Moreover, climate at this sample site was less conducive to epiphyte establishment than at the other forest plots. Succession occurred on the first two trees in the sense that epiphytes arriving well into the sequence were humus types excluded earlier by hostile conditions. As Wallace correctly surmised, diversification depends on various agencies that alter substrata and encourage plant immigration. In addition to the role of earth builder, pioneer epiphytes attract fauna, especially ants, that help improve rooting conditions, accumulate nutrients, and bring in seeds. Long-standing supports offer weathered bark that, now purged of potential germination inhibitors and perhaps more supportive of orchid fungi, provide improved conditions for epiphyte establishment. Finally, older trees develop more surface area for interception of diaspores from additional carriers. All of these phenomena expand spatial heterogeneity and potential for habitat partitionment by colonizers. Several other observers have noted more diverse and populous flora on large phorophytes (Went 1940; Richards 1952; Johansson 1974).

Yeaton and Gladstone (1982) recorded epiphyte colonization of *Crescentia alata* in Guanacaste Province, Costa Rica, where the canopy flora is relatively impoverished. Sampled trees were growing in a mixed-age plantation. Seed sources in surrounding dry, deciduous thorn forest were no more

Table 7.3. **Percentage of small and large *Crescentia alata* supports occupied by the four most common epiphytes at a site in Costa Rica**

Epiphyte	Percentage of 94 small trees occupied by epiphytes	Percentage of 63 large trees occupied by epiphytes
Encyclia cordigera	50	13
Laelia rubescens	10	27
Brassavola nodosa	20	35
Oncidium cebolleta	20	25

Source: After Yeaton and Gladstone 1982.

than 100 m distant from any surveyed phorophyte. Consistent with the findings of Wallace, Catling, and Lefkovitch, epiphyte assemblages grew more diverse as supports aged or, in this case, developed broader canopies. Of nine epiphyte species present on censused supports, as many as seven occupied a single tree – here, the largest one. Sociality among co-occurring epiphytes was quantified by comparing relative abundance of the four commonest orchid species and aspects of their location on small versus large trees. Measured parameters were height aboveground, angle of supporting branch, position on the branch (top, side, bottom), and identity of nearest neighbors. All four orchids increased in relative abundance from small to large tress, with the exception of *Encyclia cordigera* whose dominance on young trees diminished in proportion to the others as time passed (Table 7.3). Among the four orchidaceous epiphytes, there was no segregation by height off the ground, branch angle, or position on the branch. Moreover, each species was most often its own nearest neighbor Table 7.4, and only *Brassavola nodosa* had an association with a second species, *Encyclia cordigera. Weberocereus glaber,* a humiphile, alone required older trees for colonization. All four orchids evidenced similar if not identical capacity to colonize relatively unmodified bark. Yeaton and Gladstone (1982) concluded that plant interaction affected community composition very little. Except for the single cactus, epiphytes colonized phorophytes primarily as a function of seed rain density. Hundreds of thousands of microsperms must have been released from each orchid capsule in nearby forest; cactus and *Tillandsia* fecundities pale by comparison. The heavier plumose tillandsioid seed and utilization of bird dispersal by *Weberocereus* may also account for the later appearance of these species.

Table 7.4. **Nearest neighbors of randomly chosen individuals of the four most common epiphytes on *Crescentia alata* at a Costa Rican site**

		Identity of nearest neighbors (%):			
	Number examined	*Encyclia cordigera*	*Laelia rubescens*	*Brassavola nodosa*	*Oncidium cebolleta*
Encyclia cordigera	44	66	9	14	11
Laelia rubescens	40	23	55	15	8
Brassavola nodosa	76	37	9	43	11
Oncidium cebolleta	62	19	3	15	63

Source: After Yeaton and Gladstone 1982.

Neither Wallace nor Yeaton and Gladstone noted a marked frequency of certain compass orientations as epiphyte communities developed, but Bennett (1987), working with *Catopsis nutans* and *Guzmania monostachia* (Fig. 2.15) in a South Florida swamp forest, did. Eastern exposures for *G. monostachia* and northern sides of trunks for *Catopsis nutans* were least accommodating, for no obvious reason. Both species are relatively shade-tolerant in Florida, although the *Guzmania* is more heliophilic in Costa Rica and Trinidad. Frost may play some role. Subfreezing winds accompany severe cold fronts arriving from the northwest, but small-diameter axes, like many of those sampled, would seem to offer little protection in any quadrant. *Tillandsia pruinosa* exhibited a clear orientation to the east side of supports in a second survey, whereas *T. flexuosa* (Fig. 4.25B) grew equally well whether facing either east or west (Bennett 1984).

Succession in trash-baskets has been analyzed in Australia, where it was found to be reasonably predictable and species-specific to particular ferns (Wallace 1981). *Platycerium bifurcatum* and *P. hillii* may typify the general pattern for larger forms. Development begins with establishment of a sporeling on more or less vacant bark. As the fern produces ever larger, sterile, clasping fronds (Fig. 4.19), a deepening bed of humus and ramifying roots accumulates. First invaders are often dicots germinating from fallen seeds. Other pioneers are pteridophytes with mycotrophic gametophytes and pendent sporophytes that emerge between basal fronds. Maturity and decline are characterized by a richer flora comprised of numerous higher ferns, *Lycopodium* (Fig. 1.18), a dicot or two (e.g., *Fragraea, Hoya*), and orchids (e.g., *Cymbidium*). Eventually, the nest plant dies and certain earlier arrivals, especially *Davallia, Nephrolepis,* or *Cymbidium madidum,* dominate final stages until anchorage fails and the whole aggregation falls to the ground.

Neotropical ants establish nest-garden composition; additional agencies seemingly bring about changes later. Davidson and Epstein (in press) noted displacement of pioneer *Peperomia macrostachya* by a series of more light-demanding "nest parasites" in Amazonian Peru. Heliophilic *Codonanthe uleana* is also overwhelmed by the same assemblage of slower-growing aroids, bromeliads, and woody epiphytes. These later arrivals offer fewer food rewards to ants and may drive them out of the nest by filling carton galleries with roots (e.g., *Aechmea* spp. in Ecuador; Fig. 5.11). Older gardens are often partitioned; parasites are known to dominate nest tops above a ring of pendent *Peperomia macrostachya.* Of interest is how the opportunistic forms invade vegetated nests.

Tank bromeliads provide similar colonization sites in some neotropical forests. Older specimens become totally obscured by vigorous aroids, ferns, gesneriads, and peperomias, and large rosettes occasionally provide the necessary conditions for strangling *Clusia* and *Ficus* species. Neither the successional character of these communities nor the possible roles played by ants have been investigated. Plant communities built upon orchid root masses and *Anthurium* rosettes seem to lack the complexity of those centered on the larger bromeliad tanks (e.g., *Aechmea, Streptocalyx;* Fig. 1.17). Perhaps unequal access to abundant, reliable moisture supplies in leaf axils is responsible.

Summarizing briefly: There are parallels and differences between succession on the ground and in the forest canopy. Similarities include classical stages of nudation, influx of relatively stress-tolerant colonists, substratum conditioning prior to arrival of more vulnerable humiphilous taxa, and progressive community diversification. If competition contributes to floristic change in tree crowns, it remains to be demonstrated. Less equivocal are differences in timing. Epiphyte succession must be either rapid or truncated compared to the forest as a whole – no longer than the life of the supporting tree. Even faster turnover on a finer scale is mandated by the shorter lives of individual branches, twigs, and bark fragments. Time constraints on succession are most evident in arid zones where stress slows progress toward climax (if such a stage can be said to exist). Aggregations formed by bark and twig epiphytes tend to be most loosely integrated and poorly defined of all. If interaction among members is a requisite for community status, then assemblages of dry-growing epiphytes probably do not qualify, nor should changes in their diversity and numbers on a site necessarily constitute succession.

Relationships with canopy fauna

Types of interaction

Diverse biota are drawn to epiphytes for many reasons. Visitation by animals to collect floral rewards and edible fruit has already been considered, as have occupancy of ant-fed ant-house taxa and creation of ant nest-gardens, but enemies and protective symbiosis were barely mentioned. Very little is known about herbivory or pathogens in nature. Janzen (1974) and Huxley (1978) noted insects consuming *Hydnophytum* foliage in Malaysia. Signs of predation are common among additional taxa, but extensive defoliation is rare in neotropical epiphytes (pers. obser.). Occasionally, orthopterans severely graze *Guzmania monostachia* shoots in South Florida. Inflorescences of several local *Tillandsia* species ordinarily harbor insect larvae, but some capsules escape destruction. Bromeliads are remarkably immune to herbivores beyond some Homoptera in greenhouse culture; so are the more xeromorphic orchids and epiphytic cacti. Viruses pose serious problems for cultivated orchids, but there is little information on natural occurrence. Sclerophylly and succulence may reduce palatability and nutritive quality of much epiphytic vegetation relative to adjacent, generally softer, and less durable tree foliage. There seems to be little information on toxic metabolites or other deterrents beyond some casual observations on bad-tasting foliage in Malaysian myrmecophytes (Janzen 1974). Several terrestrial orchids are notable for their alkaloids and phytoalexins; whether or not these agents provide important services to relatives in the forest canopy remains to be seen.

The prime contribution made by canopy flora to animal welfare lies in provision of safe harbor in a world of abundant predators and climatic extremes. Among the families whose members offer unusually inviting nesting and breeding sites are Araceae, Orchidaceae, Polypodiaceae (sensu lato), Rubiaceae, and especially Bromeliaceae. Epiphytic vegetation of all types promotes carrying capacity simply by humidifying the forest canopy and roughening and expanding its surface. Vascular and nonvascular epiphytes alike can render conditions so broadly equable that earthworms and numerous other normally terrestrial fauna can populate the suspended humus shrouding older branches and trunks.

The importance of epiphytes to numerous insects, reptiles, and amphibians is widely recognized; not so the benefits sought out by higher vertebrates. N. M. Nadkarni and J. Matelson (in press) documented close ties between Costa Rican montane avifauna and canopy-based vegetation. Surveys in primary forest and trees isolated in nearby pasture revealed that many birds depend on epiphytes. Overall, 15% of visits to tree crowns by 81 bird taxa over 289 hours showed some type of epiphyte use. Perching occurred in 67% of all visits to canopy flora; feeding in 48%; vocalizing in 22%; and gathering of nest materials, drinking, and bathing in 2%. Nine bird species, including hummingbirds, flycatchers, and the common bush tanager, paid some attention to epiphytes during more than half of their visits to trees. Those avians that showed greatest dependence were rarely or never seen in isolated trees, despite dense colonization by epiphytes; less dependent birds with broader foraging patterns were seen about as often in forest as in degraded areas. Implications for conservation are obvious: Unless sizable tracts of contiguous forest are left by developers, certain specialized avifauna, like many large mammals, will almost certainly vanish.

Epiphytes deemed most attractive to birds in Costa Rican montane forest belong to relatively few of the many families that populate canopies there. Appeal was influenced by numerous factors, including quality and variety of rewards and density of plants. Exceeding all others in popularity because of their abundance and diverse uses were Bromeliaceae. Ericaceae, Gesneriaceae, Loranthaceae, and Marcgraviaceae were also heavily visited but primarily for nectar or fruit. Bryophytes provided nesting material and food (presumably associated invertebrates). Without Bromeliaceae, the situation may be quite different in the Old World, but probably in degree only. About 180 avian–epiphyte liaisons, mostly in the neotropics, have been reported, but many of the visited plants are nonbromeliads with close relatives in Africa and Australasia. Birds tended to belong to Thraupidae (tanagers) and Trochilidae (hummingbirds) with lesser contributions from Furnariidae (ovenbirds), Tyrannidae (flycatchers), Fringillidae (finches), Parulidae (warblers), and Turidae (thrushes).

Vascular plants that offer resources in order to establish prolonged contact with canopy fauna fall into five not entirely exclusive categories: (1) ant nest-garden, (2) ant-fed ant-house (also termed ant-inhabited or trophic myrmecophyte), (3) ant-guarded, (4) trash-basket (also called nest-forming or debris-collecting), and (5) phytotelm (tank) forms. Epiphytes that regularly "parasitize" ant-fed ant-house plants (Janzen 1974) in Southeast Asia represent a sixth group whose needs are met indirectly through animal activity.

Ant nest-garden epiphytes

Ant nest-gardens, cultivated by carton-constructing arboreal ants, have already been described with respect to establishment, floristic composition, succession, plant nutrition, and reproduction. Warm, humid forests containing many small trees on fertile soil support highest neotropical nest-garden densities. Paleotropical counterparts exist in the sense that certain Australasian epiphytes regularly root in ant-constructed abodes (e.g., *Dischidia nummularia* planted by *Iridomyrmex cordatus*), but the resulting dwellings are generally less conspicuous than those in tropical America. Thirty-four well-developed nests, each nourishing a sizable garden, were recorded on a single 10 × 10 m plot in an Amazonian Caatinga (Madison 1979a). Citrus tree crowns in a northern Trinidad grove each contained 3–10 discrete ant nest-gardens (pers. obser.). Ant nest-garden aggregations, or what were probably the products of fragmented, polygynous colonies of *Camponotus femoratus,* occurred on 16–39% of the trees in five Peruvian forest habitats (Davidson 1988). Colonization was highest in regularly flooded, thus disturbed, open communities. Exposed microsites are utilized either because the host trees are heliophilic, a possibility explored more fully below (Kleinfeldt 1978; Davidson and Epstein in press), or because the ants do not tolerate brood chambers which would become too wet or too cool in rainy, shady areas.

Nest-garden builders include members of *Anochetus, Azteca, Camponotus, Crematogaster, Dolichoderus,* and *Solenopsis.* A shared character, reflecting the high energy cost of carton construction, is consumption of such calorie-rich food as aphid honeydew. Occupancy is often parabiotic, with up to three species partitioning common living space. In Amazonian Peru, gardens are usually shared by extremely aggressive *Camponotus femoratus* and much smaller *Crematogaster linata parabiotica* (Davidson 1988). Both species contribute to nest construction: The latter builds thin layers of carton over runways, nest sites in crevices, and long-term food sources such as extrafloral nectaries and Homoptera colonies; the former enlarges some of these shelters with decaying leaves and other detritus to form a medium for epiphyte roots.

Perhaps a better index of importance than colony frequency is the density of foraging workers. Species known to occupy ant nest-gardens dominated the arboreal fauna at Tambopata, Peru (Wilson 1987). Tree crowns fogged with insecticide in four types of lowland forest yielded enormous numbers of taxa – 43 species representing 26 genera in one tree crown alone, more than the entire native ant fauna of the British Isles! *Crematogaster linata parabiotica,* here the nest companion of *Camponotus femoratus* and *Monacid debilis,* occurred in almost half of the 513 samples. Wilson attributed the great success of ant nest-garden species at Tambopata to their ability to produce capacious nests with the aid of symbiotic epiphytes. He further speculated that close study of the microdistribution of arboreal ant colonies will provide "a good deal of valuable information on the evolution of species diversity."

Reciprocity between ant nest-garden plants and ants may be broader than previously thought. Several of the plant species produce attractive extrafloral secretions (e.g., *Codonanthe uleana, Codonanthopsis ulei,* and *Philodendron myrmecophyllum;* Madison 1979a). *Codonanthe* secretes nectar between calyx lobes and from a second type of gland scattered on abaxial leaf surfaces. Secretory tissue forms a ring around the petiole apex of *Philodendron myrmecophyllum.* Pearl bodies develop on *Ficus paraensis.* Ants sometimes tend Homoptera colonies in their nest-gardens (Kleinfeldt 1978), but not even this mode of tapping plant products seems adequate to support the population; foraging must extend well beyond the garden itself – up to 10 m according to one study of *Camponotus femoratus* in Peru (Davidson 1988). Favored foraging and farming sites are often ephemeral – for example, the young fruit and pedicels of *Ficus* and flower buds of *Coryanthes.* More durable stems and leaves are also utilized, but usually less intensely.

According to Janzen (1974), neotropical and African myrmecophytism evolved primarily for defense, whereas the preeminent impetus in Australasia was improved nutrition. In fact, ant nest-garden and ant-fed ant-house epiphytes can receive both services from attending ants (Davidson 1988). Certainly these insects protect their nest-gardens in tropical America, but they also provide nourishment, however incidentally (Table 4.2). At Río Palenque, Ecuador, small but pugnacious ants swarmed by the hundreds over the flowering inflorescence of an unusually robust *Aechmea augustifolia* specimen occupying their nest in a large citrus tree (pers. obser.). Other animals, humans included, ran considerable risk visiting this bromeliad; on several occasions, ants stung or bit investigators, sometimes actually leaping from nearby foliage to attack their tormentors. *Camponotus femoratus* tending membracids on stems of nest-garden plants in Amazonian Peru reacted to insect herbivores by spraying them with mandibular gland secretions, including formic acid (Davidson and Epstein in press). Branches of patrolled *Ficus paraensis* were substantially more fruitful than those from which parabiotic *Crematogaster linata parabiotica* and *Camponotus femoratus* had been excluded by sticky collars. Comprehensive analysis is necessary to determine net effects, however. The presence of defending *Oeco-*

phylla longinoda colonies on African *Ficus campensis* reduced seed dispersal fourfold because nocturnally active workers farming Homoptera on ripe synconia also deterred fruit bats (Thomas 1988). Fig dispersal was reduced to that conducted by a set of secondary daytime carriers.

Demonstrating that ants reinforce their nest-gardens not only by carton building but also by sowing seeds in them and protecting the resultant vegetation is but one more step toward understanding the dynamics of resulting microcosms and broader ramifications for hosting ecosystems. Turnover rate is also important. Kleinfeldt (1978) observed a high density of ant nest-gardens attributable to the single ant *Crematogaster longispina* and the plant *Codonanthe crassifolia* at La Selva, Costa Rica; gardens occupying clustered trees constituted colony fragments whose parts – the individual nests – were interconnected by covered trails. Kleinfeldt discovered that subunits can turn over rapidly through displacement of the original garden builders by more aggressive but less industrious ants (*Solenopsis picta* in this study). Within months, neglected cartons deteriorated, and previously insulated plant roots became exposed. Davidson (1988) reported 118 abandoned nests compared to 758 housing active ant colonies along a forest trail at Cocha Cashu, Peru. Should these figures be typical, mineral cycling at the ecosystem level could be significantly affected.

Ant nest-garden mutualisms differ in important but little-known ways beyond the identities of involved biota. Such associations over young alluvial soils along the Andean slopes, for instance, are quite distinct in basic biology from those farther east on well-leached, ancient white sands (Davidson and Epstein in press). Amazonian systems are often built around a single plant, frequently a tank bromeliad, without obvious ant dispersal. These poorly developed and sometimes almost unrecognizable gardens populated by single queen colonies (e.g., *Azteca* or *Hypoclinea*) must affect the forest far less than do the more complex, nest-based communities constructed by those polygynous species that so often populate richer sites with higher densities of some of the world's most aggressive ants. On impoverished lands, zoobionts appear less numerous and fearsome; further study might indicate that the requisite investment in ant resources is not possible for the plant community. Also needed are determinations of how paleotropical ant nest-gardens, by nature and impact, compare with those in tropical America.

Ant-fed ant-house epiphytes

Ant-fed ant-house epiphytes are somewhat less diverse than ant nest-garden flora in the sense that fewer families (though not necessarily fewer species) are involved. Ant associates exhibit the reverse pattern: Many more ant taxa appear to be housed by epiphytes than live in and tend nest-gardens. Australasian ant-house plants are more numerous and in some cases more specialized for ant habitation than American counterparts. Participating genera are typically small or, if larger, contain only a few species that regularly host ants (e.g., *Tillandsia, Dischidia, Hydnophytum*). *Myrmecodia* is the exception: All of its approximately 50 species are myrmecophytic. Ant-inhabited Hydnophytinae offer resources to ants over and above superior housing (Fig. 1.4); stem surfaces are often deeply ridged and sometimes equipped with stout spines (Fig. 4.24A), as if to deter large predators from preying on ants. Plant defense could also explain this armature because some insectivores are powerful enough to tear open domatia. Leaf axils harboring sessile flowers and partially embedded edible fruits may provide additional security for foraging colonists or discourage frugivorous birds in favor of seed-dispersing ants. Honeydew is available from various Homoptera farmed on adjacent host branches or on the ant plant itself. Here, as in nest-garden symbioses, ant colonies inhabiting single hypocotyls may be large (>15,000 workers) and may travel 50 m or more from the home site (Janzen 1974). Inhabited plants often simply house colony fragments, thus relegating the bulk of the insect population to more spacious quarters in the rotted interior of a phorophyte.

Like neotropical ant nest-garden plants, ant-fed ant-house epiphytes are most abundant in regenerating open forest. A 25 × 9 m quadrant sampled by Janzen contained 494 *Hydnophytum,* 31 *Myrmecodia,* 20 *Dischidia rafflesiana,* and 2 *Lecanopteris* specimens plus a variety of associated botanical nest parasites. A tally of 79 woody plants over 1 m tall showed that *Hydnophytum* was supported by 44, *Myrmecodia* by 13, *Dischidia* by 20, and *Lecanopteris* by 2. At least one member of each of the 11 local tree species had one to several epiphytes attached to it. Ant habitation approached 95% of the available domiciles. Janzen felt that the lower density of ant plants in more mature nearby forest reflected fewer edible plant products. Ants are not required for plant success; numerous *Hydnophytum* species (e.g., *H. kajewskii*) intercept moisture and/or quarter frogs, centipedes, cockroaches, and a variety of other fauna without attracting ant colonies (Jebb 1985).

There is much more to be learned about the occurrence of ant-house epiphytes and the activities of their diverse zoobionts. Jebb (1985) reported the presence of 19 ant species belonging to 14 genera in *Hydnophytum* domatia in Papua New Guinea. Seventeen species of ants representing 11 genera in four subfamilies were found in bulbs of *Tillandsia bulbosa* in a census of the forest reserve at Quintana Roo, Mexico (Olmsted and Dejean 1987). Ant occupancy of neotropical myrmecotrophs, although sometimes lower than that reported for ant-fed Hydnophytinae, is still substantial: Well over half of the *Tillandsia butzzi* and *T. caput-medusae* populations observed in Mexico and Costa Rica were occupied (Benzing 1970a) and every *Schomburgkia tibicinis* plant examined by Rico-Gray and Thien (1989) contained nesting ants. No bromeliads produce obvious myrmecochores or continuous food sources specifically for their fauna. Ants supposedly mine sugar-laden tissue in *Solenopteris* rhizome tubers (Fig. 4.24F) to improve them for occupancy. Edible appendages on sori are provided by the Australasian counterpart *Lecanopteris;* chambers are lysogenous or the ants inhabit crevices located between the bark and solid rhizome (Fig. 4.24G). Light requirement appears to be equivalent (relatively high) on both sides of the globe.

Myrmecotrophism predicated on plant modification to entice foundresses or to house fragments of growing colonies must be beneficial only under strict conditions; otherwise more taxa would qualify. Perhaps many plants may, like nest-garden species, obtain nutrients from mutualists at lower cost by simply tapping the foundation materials around their roots. Organs modified to accommodate ant colonies are probably most cost-effective where N and P supplies are meager, turnover of plant organs is slow, and demand for ant nesting sites is high (Thompson 1981). Benefits gained from the presence of aggressive ants, like access to nutritive ant products, probably vary with circumstances.

Five ant species occupied a Mexican colony of *Schomburkia tibicinis* with different results depending on the combination (Rico-Gray and Thien 1989). Hollow pseudobulbs (Fig. 4.24H), capable of absorbing solutes from radiolabeled ant carcasses, were probably able to extract nutrients from nest materials as well, but ant behavior influenced plant fitness unevenly in a second way. Despite a relatively timid nature, *Crematogaster brevispinosa* managed to dominate about two thirds of all peduncles. However, it failed to deter destruction of numerous flowers by a common curculionid beetle. Enough additional damage was inflicted by farmed mealybugs to reduce substantially plant fecundity compared to orchids patrolled by *Camponotus planatus, C. rectangularis, C. abdominalis,* and *Ectatomma tuberculatum.* Beetle damage and cultured Homoptera were absent on inflorescences supporting these larger, more aggressive ants.

Ant-guarded epiphytes

The incidents just described documenting ant protection of nest-gardens and *Schomburgkia tibicinis* in tropical America demonstrate the overlap between animal-assisted categories. Another example is that of myrmecophytic *Dischidia;* plants rooted in carton runways sometimes house ants in pouch leaves (Weir and Kiew 1986). Some other ant-fed ant-house epiphytes also receive multiple benefits from mutualists. In Malaysia, inhabitants of rubiaceous myrmecophytes (mostly *Iridomyrmex cordatus*) clearly do not deter herbivores by day but possibly venture forth at night (Janzen 1974) to forage for fungal spores and mites – a valuable service to vulnerable plants. Some Hydnophytinae and *Lecanopteris mirabilis* readily hosted greenhouse herbivores (Jebb 1985) – evidence that, like myrmecophytic *Acacia,* biochemical defense for these species has been lost in the process of eliminating redundancy. But some Central American ants definitely avoid confrontation with large enemies: Those prodded with a stick in *Tillandsia caput-medusae* bulbs fled and sometimes moved their broods to another plant within the next few hours (pers. obser.).

Although few orchids root in ant nest-gardens or are inhabited by ants, they are the most numerous of the insect-protected epiphytes. Extrafloral nectaries are common on pedicels and not uncommon on inflorescence bracts (e.g., *Cyrtopodium*). Rewards for ant guards are usually transitory. Between flowering cycles, no payment is offered that would induce longer residence. An exception is *Caularthron bilamellatum.* Of 16 orchids producing extrafloral nectar in Panama, this species alone housed ants (11 species in all; Fisher and Zimmerman 1988) and maintained food supplies on mature as well as developing and flowering shoots. Ant distribution also suggested opportunism rather than tight mutualism; most residents were about as abundant on trees with, as on those without, orchids. Nonmyrmecotrophic epiphytes may gain advantage via another attractant: Several *Begonia* species generate pearl bodies or similar edibles on abaxial leaf surfaces.

Suggestions have surfaced of yet another symbiosis involving ants and epiphytes, one result of which could be fitness imparted to phorophytes by resident canopy myrmecophytes. Trees harboring *Aechmea bracteata* in seasonally inundated and drier semievergreen woodland in the Sian Ka'an Reservation at Quintana Roo, Mexico, exhibited little evidence of attack by numerous local leaf-cutting ants compared to uncolonized conspecifics (Olmsted and Dejean 1987). Despite its phytotelm status, this large, vase-shaped bromeliad (Fig. 7.5) regularly housed divers aggressive ants in upper portions of central green-leaf chambers, 95% of which sheltered colonies. Workers sometimes numbered in the thousands. *Pachycondyla villosa* and *Hypoclinea bispinosa* were the most common defenders, but six other species in four additional genera occasionally nested there as well. An even richer ant fauna inhabited outer dead leaves of 97% of dissected plants. Benefit gained at low (if any) cost in food for ants fending off such destructive herbivores as leaf cutters would certainly compensate for almost any imaginable disadvantage imposed by the presence of nest-providing epiphytes.

Another mutualism featuring coinvolvement of ants, trees, and epiphytes occurs in Malaysia (Weir and Kiew 1986). Phorophytes usually belong to the genus *Leptospermum*; the epiphytes are *Dischidia astephana* and *D. parvifolia*; the ants are of the genus *Crematogaster*. Individual trees host single large ant colonies in extensive cavities mined from rotting xylem throughout the trunk, major branches, and older roots. Compared to uninfested supports, crowns of specimens harboring ants remain largely epiphyte-free except for the two asclepiads – which incidentally do not produce the ascidiate leaves (Fig. 4.24C) characteristic of more specialized myrmecotrophic *D. rafflesiana* and *D. complex*. Somehow, only these two species among the extensive local epiphyte flora avoid cropping by ants and survive to tap insect-provided resources via extensive root systems that grow down insect burrows to proliferate where the ants reproduce. Some debris was packed under dome-shaped *D. astephana* foliage, but no brood was found there. *Leptospermum* grows without *Crematogaster*, although Weir and Kiew postulated diminished nutrient supplies in cases where insects had not yet purged the bark of epiphytic nonasclepiads. Ants, in contrast, were always associated with trees at the study site. An attempt to document myrmecochory in the two *Dischidia* taxa was inconclusive.

Trash-basket epiphytes

Unlike ant nest-garden species, trash-basket epiphytes create their rooting medium by intercepting falling litter; ants may enter the picture later as conditions permit. Frequency of insect colonization has not been reported. Production by some *Drynaria* and *Platycerium* of nectar, rich in amino acids, indicates that significant plant benefit accrues from the symbiosis. Birds'-nest *Anthurium* (Fig. 1.5), *Asplenium*, the other impoundment ferns (Fig. 4.19), and the trash-basket orchids collect debris but little or no free moisture among leaves or upward-projecting roots. A terrestrial microcosm results. Ants reportedly (Paterson 1982) promote drainage from catchments of *Platycerium coronarium* by chewing holes in the impounding fronds. Presumably resident fauna are air breathers and distinct from those attracted to the water-filled centers of tank bromeliads. There is no clear distinction between debris collectors and many other epiphytes. Large specimens of ramet-producing orchids and rhizomatous ferns, for instance, must furnish to animals substrata similar to those offered by a mature *Platycerium* or *Cyrtopodium* specimen. Older leaf axils of tank bromeliads also contain relatively dry humus, but there have been no comparisons between the fauna of each substratum.

Parasites of myrmecotrophs

Parasites on the rubiaceous myrmecotrophs include several orchids (e.g., *Dendrobium crumenatum, Sarcanthus machadonia*), as well as *Dischidia gaudichaudii* and *Pachycentria tuberosa.* Either growth is close enough to allow rooting into ant-inhabited hypocotyls, or the seeds themselves are carried inside (Janzen 1974). *Pachycentria tuberosa* appears to be totally serviced by *Hydnophytum formicarium* at a site in Sarawak. Seeds germinate within nest chambers, and stems grow out through natural ports to produce thick leaves and pink berries. *Dischidia gaudichaudii* is planted by ants just as is myrmecophytic *D. rafflesiana* (Fig. 4.24C). Should pouch leaves be produced by the latter, roots of nearby *D. gaudichaudii* usually invade them. Edible products may represent worthwhile ant resources because *Iridomyrmex cordatus* makes no attempt to eliminate nest parasites. Behavior similar to the pruning of encroaching vines and possibly epiphytes by ants inhabiting neotropical bullhorn *Acacia* might eliminate equally undesirable invaders, but has not been reported. Alternatively, ants may be unable to distinguish seeds or other organs of nest parasites from those of the true mutualist. Needed are tests for olfactory attractants like those reported by Davidson and Epstein (in press) in seeds of neotropical ant nest-garden species. Those neotropical ant-cultivated epiphytes of unknown dispersal mode described above that displace earlier arrivals proffering known ant rewards also belong under the parasite rubric.

Evolution of ant–epiphyte associations

Relationships between ants and epiphytes have not been studied enough to describe origins with confidence (Huxley 1986), but no single explanation would suffice anyway. Ant and plant benefits vary in degree and kind with the environment and identities of participants, indicating several selective agents and more than one evolutionary pathway. One point is, however, already clear: Ubiquity in canopy habitats, diet, sociality, year-round activity, and the division of labor between reproductive and sterile (worker) castes, ensured that ants, above all other insects, would evolve the most specialized associations with epiphytes (Davidson and Epstein in press). Especially critical for mutualism is the workers' discharge of relatively dangerous extranidal tasks. As a result, the parent colony and its insular queen(s) have considerable immunity from predators and a potential for long life – often as long as, or longer than, that of the plants they utilize and serve.

Several bits of evidence support coevolution between animal benefactors and ant plants; plant morphology, ant behavior, and the obligatory nature of some of the alliances. *Camponotus femoratus,* one of the commonest ant nest-garden builders, seldom if ever occurs without its epiphytes (Davidson 1988). Certain nest flora in turn occur nowhere but on cartons, and in some cases (e.g., *Coryanthes* spp.), cultivation on horticultural media is difficult. Facultative joinings are suggestive of beginnings – for example, those between trash-basket types and primitive ant taxa (e.g., *Pachycondyla* and *Odontomachus*) that also inhabit rotting wood and other plant chambers, and between the occasional bulbous *Tillandsia* and its nonaggressive, primarily noncarnivorous *Crematogaster* populations. Weedy ants exhibit what may be a general proclivity to utilize plant cavities; they occupy 10–15% of the *Tillandsia paucifolia* in South Florida (Fig. 1.11; Benzing and Renfrow 1971c). Correlation between ant service and plant habit was illustrated in *Tillandsia* native to Quintana Roo, Mexico (Olmsted and Dejean 1987). Four of five co-occurring epiphytes listed by these authors in order of watertightness and volume of bulb chambers were inhabited by ant colonies (numbers in parentheses denote frequency of ant occupancy): *Tillandsia flexuosa* (0%); *T. baileyii* (30%); *T. balbisiana* (42%); *T. bulbosa* (41%); and *T. streptophylla* (53%).

Such phenomena as the sprouting of *Ficus paraensis* seeds in carton laid down over fruit to shield ant-tended Homoptera, and use of seed for purposes other than planting, could have helped foster regular nest-garden status (Davidson and Epstein in press). Several arboreal ant species exhibit appropriate behavior but fail to establish nest-gardens. Seeds sown in thin brittle cartons constructed by some Peruvian *Hypodinea bidens* colonies consistently failed to germinate (Davidson 1988); those cultivated by often parabiotic species farming alone (e.g., an *Azteca* sp. and *Crematogaster linata parabiotica*) produced seedlings that never matured.

An important question is whether certain ancestral ants (by behavior, life history, and diet) and particular plants (because of habit and seed qualities) were destined for myrmecophytism of a specific type. The answer is a qualified yes. Arboreal ants that perhaps had a need for seed oils to maintain fungus-free nests, or that manufactured large cartons incorporating debris, soil, and feces, were more apt to become plant cultivators than those that used less nutritive and absorptive substances to reinforce smaller domiciles.

The nest-garden plants that today produce seeds laced with methyl-6-methylsalicylate, benzothiazole, and other bioactive substituted phenyl derivatives could be convergent for these chemicals as a result of previous, less specialized associations with ants. More likely, nest-garden life arose after ancestors began to produce compounds that attracted terrestrial carriers (Davidson and Epstein in press). *Peperomia macrostachya* and *Ficus paraensis* both have congeneric myrmecochoric but uncultivated relatives. Some gesneriads on the forest floor, like epiphytic *Codonanthe* species, are dispersed by ants. Stepped-up synthesis of chemical attractants may be an outcome of the nest–garden partnership, however. Whatever the mode of origin, results can be quite precise: Congeners of some nest-garden taxa (e.g., *Camponotus sericeiventris*) are strongly repelled by the same seeds that their agrarian relatives avidly collect.

An evolutionary pathway to advanced ant-fed ant-house status is best documented by Hydnophytinae, where nest cavity structure points to at least four independent but parallel emergences of myrmecophytism (Jebb 1985). This taxon also provides the strongest case for trait magnification. Dimorphism of chambers within the swollen hypocotyls of these extraordinary myrmecophytes must be ant-related. There are two purportedly coevolved refinements here: secretion of moisture on the rough-walled absorptive compartments, which encourages ants to deposit nutritive debris rather than raise young; and foraging to fill these chambers beyond the trophic needs of the ant occupants (but see Davidson and Epstein in press).

Ant nest-garden mutualisms and those of Hydnophytinae (and probably additional ant-house myrmecotrophs as well) share ecological qualities, suggesting that evolution involved accelerated growth of plant and ant colonies and greater competitiveness in the zoobionts. Ants choose their botanical partners more on the basis of habitat quality than of epiphyte identity. The most refined and obligate ant–plant symbioses involving both types of myrmecotroph are largely restricted to exposed sites, likely in part because strong irradiance allows the plants to produce living space and food rapidly. Resident ants are typically those with multiple queens and vigorous colony expansion aided sometimes by fragmentation. The same plant growing in shade normally supports what are often more docile insects that produce slower-growing single-queen colonies with less rigid housing requirements (e.g., *Iridomyrmex scrutator* in *Myrmecodia* spp.). Occasionally, one of these less competitive ant taxa occupies an ant plant in a well-illuminated site, perhaps because the resident queen, aided by her first workers, managed to fend off late-arriving foundresses of the normally dominant species (Davidson and Epstein in press).

In order to sustain a vigorous ant colony and in turn receive high nutrient input and/or strong defense, the myrmecotroph must be able to grow fast enough to produce housing quickly; that is, a feedback loop must be operating. Hydnophytinae illustrate the initial condition in the progression toward greater plant vigor. Nonmyrmecophytic taxa tend to be shade-tolerant and slow-growing, and to maintain relatively low ratios of leaf to total biomass (Jebb 1985). Related myrmecotrophic species are heliophilic, mature faster, and produce proportionally larger amounts of green tissue. Plant lineages that were to receive improved services by promoting ant aggressiveness have not had equal opportunity (Davidson 1988). Even the most vigorous ant-fed ant-house plants must allocate resources to nest cavity walls – resources that could have been invested in additional green tissue. Freed of a similar requirement, ant nest-garden forms can and do grow faster and often produce food for more mutualists with higher metabolic demands. Ant colonies may provide a parallel because, after production of workers and sexual brood, nest construction, particularly that of carton, requires the largest amounts of energy, time, and material (Sudd and Franks 1987).

Study of the evolution of ant–epiphyte systems should capitalize on the variety of interactions between partners, ranging from opportunism to obligation and from commensalism and parasitism to unequivocal mutualism. Four subjects should receive special attention: (1) ant behavior at the level of the individual worker and that resulting from the particular colony hierarchy; (2) use of plant products as ant metabolites or fungistatic agents to protect their nestlings; (3) increased plant fitness due to better photosynthetic output and protection by ants; and (4) ant colony energetics. The suggestion has been made by Davidson et al. (in press) that species-sorting mechanisms involving both insect and plant qualities account for the composition of particular ant–epiphyte associations. If this hypothesis proves true, valuable insight into myrmecophytism in arboreal and terrestrial habitats will result.

Phytotelm epiphytes

Fauna

Phytotelm epiphytes create catchments which trap both litter and enough precipitation to support aquatic microcosms (phytotelmata) that can be indispensable to much forest fauna – and costly to humans now and then. Plants with water-impounding shoots pose a nuisance and even public hazard where they harbor larvae of bloodsucking midges, horseflies, and disease-carrying mosquitoes. About half of all phytotelm plants are bromeliads (Fish 1983), which routinely create veritable aquatic hanging gardens throughout tropical America that accommodate diverse animals seeking shelter, breeding space, or sustenance. Fish estimated that the several tillandsioids occurring at the densities reported by Sugden and Robins (1979) in a Colombian cloud forest impound over 50,000 liters of water per hectare of ground area. Even the solitary specimen can create considerable high-quality habitat; *Glomeropitcairnia erectiflora* provides both a reservoir containing several liters of water and, for less aquatic fauna, a drier impoundment of debris collected among older leaves.

Surveys of bromeliad phytotelmata (e.g., Picado 1913; Laessle 1961; Maguire 1971; Fish 1976; Frank 1983; Table 7.5) have documented extensive fauna and even some endemic vascular flora – namely, carnivorous *Utricularia humboldtii* in *Brocchinia tatei* (Fig. 1.14) and one *Vriesea* on Cerro Neblina, Venezuela. Laessle recovered 60 different invertebrates from rosettes in Jamaica; Picado encountered 130 in Costa Rican specimens. Frank reported that about half of 470 identified arthropod species reported in bromeliad tanks were mosquitoes. Invertebrates present in bromeliad shoots and suspended humus in cloud forest and nearby drier woodland greatly exceeded that on the ground except for "hot spots" in rotten wood (M. G. Paoletti, B. R. Stinner, and D. H. Stinner pers. comm.; Fig. 7.12; Table 7.5). There were more decomposers than predators in most substrata. Soil-type mega- and meso-invertebrates were 10 to 100 times more densely packed per unit volume of canopy substratum than were those on the ground. The quantity of terrestrial media (thus numbers of ground fauna) exceeded that of the bromeliad tank substratum by three orders of magnitude, nevertheless the variety of residents at the two locations was not significantly different, according to several diversity indices. Utilization of phytotelmata by invertebrates varies. Laessle found three ostracods (*Metacypris*) to be strictly dependent on bromeliads, but others were amenable to alternatives. Of special interest are the movements of macroinvertebrates to and from tanks; the possibilities are numerous and the consequences potentially substantive. Gastropods regularly hide among bromeliad leaf bases by day and feed on nearby greenhouse vegetation at night (pers. obser.). If similar migrations are commonplace and the variety of participating animal groups is great, a phytotelm epiphyte must influence many kinds of events in the canopy well beyond its own perimeter.

Little is known about the dynamics of bromeliad microcosms beyond some preliminary inquiries, and most of these concern mosquitoes, water

Table 7.5. **Invertebrate taxa present in collections taken from bromeliad tanks and forest floor at Rancho Grande, Venezuela**

	Bromeliads	Forest floor
Gastropoda	2	1
Enchytraeidae	0	2
Glossoscolecidae	12	0
Glossoscolecidae, cocoons	14	2
Hirudinea	4	0
Oniscoidea	57	32
Diplopoda	68	46
Chilopoda	13	24
Symphyla	0	11
Diplura, Japygidae	0	6
Diplura, Campodeidae	0	3
Collembola	2	2
Gryllidae	1	2
Blattodea	15	4
Dermaptera	0	3
Psocoptera	0	1
Hemiptera	29	8
Carabidae	4	1
Carabidae, larvae	0	1
Dytiscidae, larvae	1	0
Histeridae	0	2
Staphylinidae	5	26
Pselaphidae	2	23
Scydmaenidae	0	15
Nitidulidae	0	1
Coccinellidae	2	0
Lucanidae	0	1
Catopidae	0	5
Liodidae	0	4
Ptilidae	3	1
Dascyllidae, adults	0	1
Dascyllidae, larvae	66	0
Curculionidae	0	1
Coleoptera, phytophagous larvae	3	10
Diptera, Nematocera larvae	37	1
Diptera, Brachycera larvae	1	4
Diptera, Drosophilidae	0	2
Formicidae	456	64
Chelonethida	0	7
Pedipalpida	0	2
Phalangida	1	3
Acari, Mesostigmata	4	7
Acari, Trombidiidae	0	1
Acari, Prostigamata	0	2
Acari, Orbatei	0	4
Araneida	12	17

Source: M. G. Paoletti, B. R. Stinner, and D. H. Stinner pers. comm.

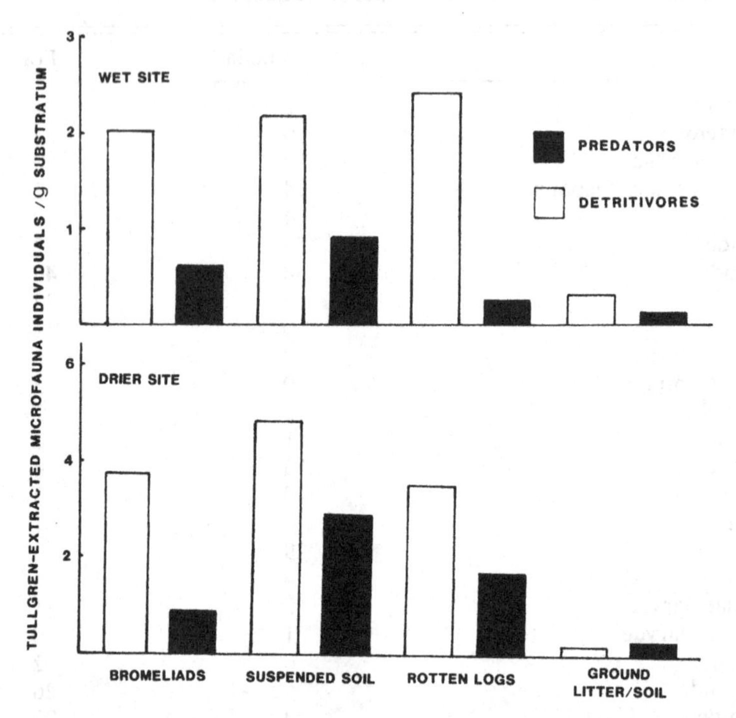

Figure 7.12. Averaged numbers of microfauna (Table 7.5) in numerous samples of epiphytic bromeliad shoots, in suspended humus, and in rotten wood and soil/litter on the ground. The wet site is cloud forest located along the ridge line (ca. 1000 m) at Rancho Grande, northern Venezuela; the dry site is seasonal woodland several hundred meters over the leeward side of the same ridge.

chemistry, or food webs. Location and shoot architecture influence resource flow and biotic diversity within impoundments. Exposed, spreading rosettes often support algal colonies and trophic pyramids dependent on this autotrophic base. Similar subjects from darker sites host heterotrophic communities built on litter (Laessle 1961; Frank 1983). Saprophytes and detritivores are regular inhabitants of tanks; occurrence of phototrophs and vertebrates seems to be much more site-specific.

Relations among the myriad invertebrates native to bromeliad phytotelmata probably fine-tuned shoot form very little if at all, but plant architecture undeniably has an impact on resident fauna. Relatively sedentary but predatory *Toxorhynchites haemorrhoidalis* heavily cropped other immature dipterans when both occupied the single pools contained within tubular *Aechmea nudicaulis* in a Venezuelan forest (Lounibos et al. 1987). Many more potential prey found safety from the larvae of this voracious mosquito in the separate leaf axils of multichambered *A. aquilega* shoots. Vulnerability to foraging birds or reptiles could follow quite different patterns, however. Some dragonflies (Calvert and Calvert 1917), crane flies (Alexander 1912), and syrphids (Knab 1912) show morphologies specialized for life in foliar tanks. Flattening that permits deeper penetration and greater mobility among appressed leaf bases is pronounced in some tank-inhabiting lumbricoids and isopods (M. G. Paoletti, B. R. Stinner, and D. H. Stinner pers. comm.). Intimate tank symbionts also include vertebrates (Fig. 4.16). Tadpoles of several bromeliad frogs are unusually small and slender, presumably in order to negotiate the narrow water columns formed by closely overlapped leaf bases (Rivero 1984). Lizards (e.g., *Abronia*) and snakes (e.g., *Bothrops schlegeli*) are occasional visitors showing no obvious modification for tank habitation.

Of the epiphytes with recognized influence on lower vertebrate evolution, phytotelm bromeliads are the most noteworthy. Tailed amphibians, fundamentally a Laurasian group, made their single substantial incursion into lower latitudes aided by the phytotelmata of cloud-forest Bromeliaceae (Wake 1987). Recent exploration has revealed more than 140 neotropical (80% Mesoamerican) plethodontid salamander species in 11 genera, including many with range restrictions imposed by their narrow temperature tolerances and occurrence in montane habitats. Large bromeliad rosettes filled with moist decomposing debris and abundant invertebrates constitute the specialized microhabitats required by urodeles. A survey conducted along a transect ranging from near sea level to 4000 m in southern Mexico yielded 15 mostly discrete salamander populations. Many of these taxa were heavily dependent upon, if not restricted to, large *Tillandsia* and *Vriesea* shoots. Penetration of neotropical woodlands by Plethodontidae was favored by additional features that distinguish southern from most northern taxa – namely, direct development (no larval stage) and drought tolerance.

Salamanders regularly encountered in bromeliads most often belong to *Dendrotriton, Nototriton,* and *Chiropterotriton.* Form appropriate for life among overlapping leaf bases is evident: Adults are typically small (<50 mm in length) and equipped with long prehensile tails, elongate limbs with widely separated digits, and frontally directed eyes. Other arboreal plethodontids that are active on suspended moss mats rather than in phytotelmata possess shorter appendages and trunks resembling those of terrestrial relatives. Phytotelm dwellers can be quite gregarious, especially during dry weather; up to 34 *Dendrotriton xolocalcae* specimens were counted in a single bromeliad (Wake 1987). As many as half of all sizable rosettes in some Mesoamerican bromeliad populations contain one or more salamanders, and totals often exceed that of all co-occurring vertebrates combined.

Expanding the definition of tropical forest soil

The presence of earthworms and diverse soil-dwelling arthropods in epiphytes and other suspended substrata (Lavelle and Kohlmann 1984) challenges the notion developed by temperate-zone biologists that soil, the quintessential supportive medium, is restricted to the bottom of terrestrial ecosystems. Overlooked are accumulations that appear more and more to be arboreal extensions of the upper humic horizons of subjacent earth soil (Nadkarni 1981, 1984). Both substrata are nutritive, penetrated by roots, and the sites of important soil phenomena like N_2 fixation and humus mineralization. Should much of the resident biota overlap, the case for expanding the definition of soil to accommodate conditions in humid frost-free forests will be even stronger. So far, several invertebrate taxa offer mixed support for the change. For instance, Palaçios-Vargas (1982), working with mites from 22 families in a Mexican forest, observed certain species exclusively in the canopy, others only in earth soil, and a third group thriving in both locations. *Tillandsia* shoots in the same area harbored 18 Collembola taxa, some only during the wet season (Palaçios-Vargas 1981). Thirteen more species never left earth soil. Factors accounting for the broader tolerance of bromeliad-inhabiting populations were not obvious.

The presence of animal inhabitants in the arboreal soils of tropical woodlands also poses interesting questions about evolution and ancestral habitats: If invertebrates in the tree crowns of Venezuelan cloud forest and the Mexican region just described not only occur there but likewise generally achieve densities equal to or greater than those in earth soils, then one can reasonably hypothesize that some of this fauna acquired important characteristics aboveground. Are, in fact, related arboreal and terrestrial soil invertebrates distinguished by traits specific to their native substrata? To what extent are the two groups convergent, or do members share similar morphology and diets through common origins? How much exchange has occurred historically between invertebrate fauna living in soils at opposing regions of the forest profile? How many populations currently span the two compartments? In view of the distribution of soil-like media in many tropical forests, such movements may not be difficult.

Epiphyte phytotelmata

Water chemistry: influence of biota

Much of a phytotelm epiphyte's impact on canopy fauna depends on what happens to animals lured to its tanks. In a water-filled leaf axil functioning as a botanical carnivore's pitcher, only those aquatic visitors immune to a digestive liquor will feed or multiply. At the other extreme, tanks may be especially accommodating to diverse fauna, providing moisture, food, security, and perhaps even dissolved oxygen from adjacent green tissue. Apparently, the real world for most bromeliads falls somewhere between these two possibilities. Laessle (1961) discovered that pH, [O_2], and [CO_2] in tank fluids varied considerably from day to night and from one rosette to another in Jamaica. Concentrations of dissolved gases seemed to be influenced by the kinds and numbers of organisms inhabiting leaf axils, not by the plant itself.

In order to determine whether a bromeliad shoot can exert appreciable influence on the chemistry of impounded fluids, measurements of [CO_2], [O_2], and pH in the central cavity and one lateral leaf axil of *Aechmea bracteata* were made at 6-h intervals for several days (Benzing et al. 1971). This species produces a series of essentially isolated foliar chambers (Fig. 7.5); some of these were gently cleaned and refilled with distilled water; to others, rotting leaf debris, green algae, and/or Odonata nymphs were also added. Small glass beakers filled with the same materials served as controls. Throughout a 12-h photoperiod, PPFD was maintained at about 12% of full sunlight. Results, some of which are depicted in Figures 7.13 and 7.14, indicated that at least in the case of *A. bracteata,* tank water chemistry is influenced most by the organisms present. Illuminated or darkened, both plant chambers and beakers filled with distilled water were never more than about 60% saturated with O_2. During the same runs, [CO_2] showed higher values in tanks, probably through activity of epiphyllous heterotrophs. Only when animals or decomposing vegetable matter were present in illuminated or darkened tanks and beakers did [CO_2] remain high and [O_2] drop markedly. Illumination and algae were required to push [O_2] to near-saturation or above. Tanks were always moderately acid, sometimes more so at night than by day, regardless of their contents or light environment; beakers containing algae, on the other hand, showed pH to be >7.0 during the daytime.

Laessle's results and those from *Aechmea bracteata* suggest that no special intervention by the plant is needed to improve phytotelm quality for animal

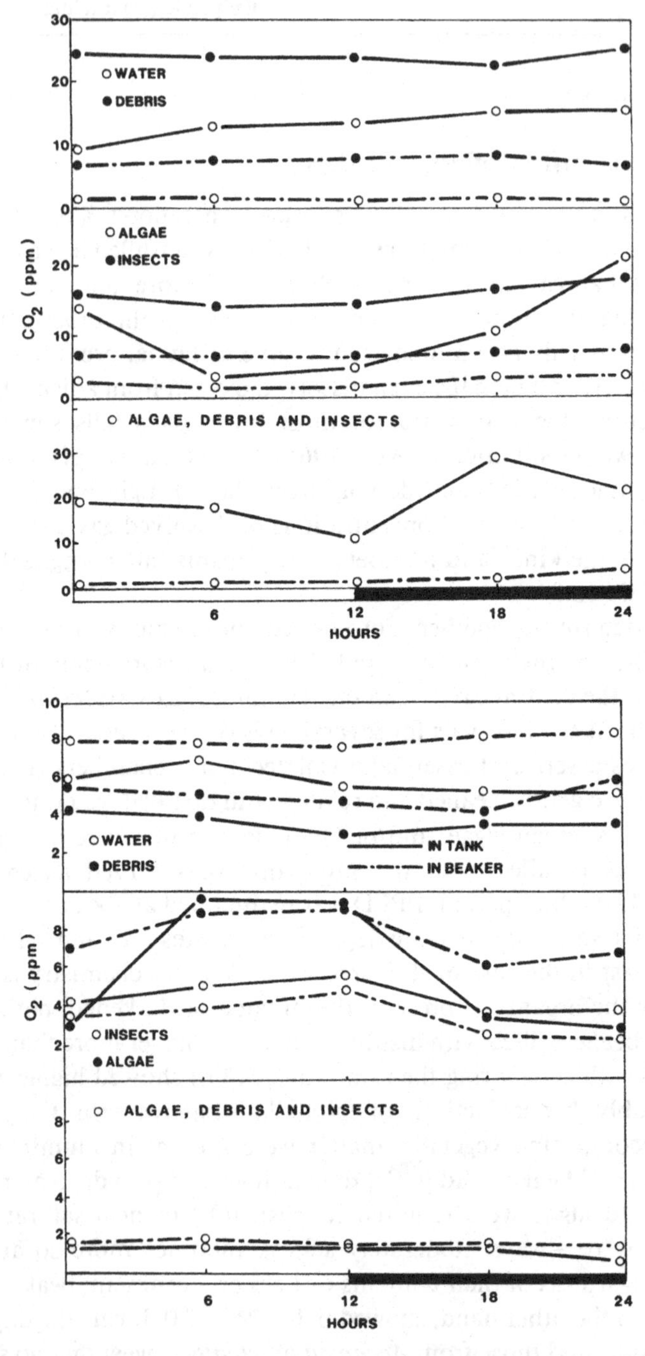

Figures 7.13, 7.14. Diurnal fluctuation in the chemical properties of water [(7.13) dissolved CO_2; (7.14) dissolved O_2] impounded in the central tanks of *Aechmea bracteata* to which had been added algae, insect larvae, and/or decaying leaves. (After Benzing et al. 1971.)

use. Oxygen and CO_2 exchange between leaf axil contents and atmosphere presumably occurs just as if the same fluids and organisms were contained in an opaque, inert receptacle of identical conformation and volume. Resident fauna receive no substantial gaseous or energy inputs from the supporting epiphyte. Nevertheless, shoot features beyond impoundments, such as the ornate pigmentation already described, probably exist primarily to encourage specific kinds of animal use and benefit, although different functions are conceivable (Benzing and Friedman 1981b).

Nutrient processing

Driving another process are microbes and detritivores, sometimes in greater numbers than occur on the forest floor (M. G. Paoletti, B. R. Stinner, and D. H. Stinner pers. comm.; Chap. 4). After about three years (the average life of a tank bromeliad ramet), intercepted vegetable matter is reduced to dark, finely divided, humic soil. Should degradation occur at different rates under a densely bromeliad-infested canopy, effects on element recycling could be substantial. Determining the magnitude and type of processing effected by epiphytes will require considerable study. But events in phytotelmata need not be fully understood to measure important consequences. Outputs can be monitored by examining the chemistry of effluents relative to that of affluents and noting patterns of release. Differences among elements are likely; those in refractory form (e.g., N) will be held longer, perhaps until the shoot dies. Tank biota might outcompete the plant for certain ions, or supplies could so exceed demand that the excess is discharged in overflow. Thought should also be given to the possibility that some tank residents also forage beyond phytotelmata on a regular basis.

The origin and significance of phytotelm form

Abiotic factors must have influenced bromeliad habit, and thus the capacity to intercept and retain various resources. Particular shapes may exist primarily to accommodate specific degrees of drought or light exposure: Deep tanks need replenishment less often than shallow, open ones; the vertical leaves of tubular heliophilic billbergias (Fig. 4.25F) are spared direct-beam insolation at midday; the flat, open rosettes (and therefore shallow axils) of shade-adapted shoots (Fig. 4.25G) maximize photon (and litter) capture. Indeed, tank bromeliads from shady, humid environments tend to produce open, shallow shoots; those from arid habitats are often vase-shaped or tubular (Benzing and Friedman 1981b). Sun- and shade-grown

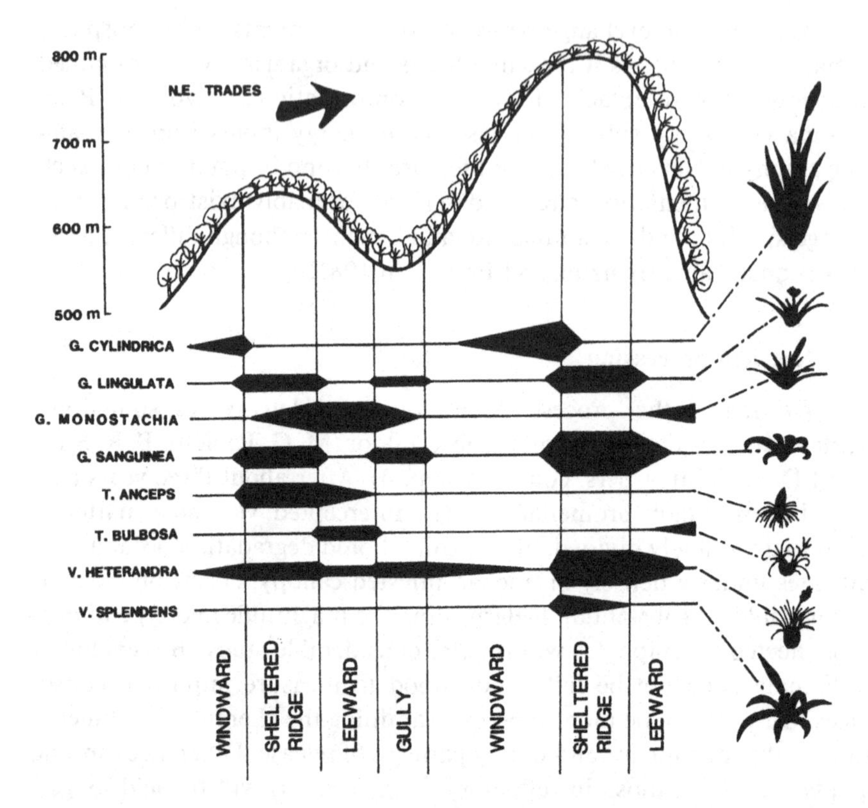

Figure 7.15. Habit and occurrence of eight bromeliads along a ridge–valley system in a northern Colombia cloud forest. (After Sugden 1981.)

Tillandsia utriculata (Figs. 2.11, 2.12) illustrate plasticity in a single population. Wind can provide an additional dimension to the environmental sorting of bromeliads by tank type, according to Sugden (1981).

Eight tillandsioids in allied *Tillandsia* and *Vriesea* grow together along a minor ridge–valley sequence in cloud forest on the Serrania de Macuira in northern Colombia. Here, varied topography and constant wind direction dictate canopy height between about 1 and 10 m. For all but two months each year, moisture is delivered solely as mist, but unevenly. Figure 7.15 illustrates bromeliad occurrence across two adjacent ridges in the system. Progress up the relatively cloud-free leeward slope on the right to ridge top and beyond is marked by changes in canopy flora. Lowest in elevation and comparatively unprotected by dense foliage are heliophilic *Guzmania monostachia* (Fig. 2.15), semibulbous *Vriesea heterandra,* and a few succulent *Tillandsia bulbosa* (Fig. 4.24B) which feature onionlike shoots that often harbor ant colonies. Approaching the summit, the first and last species essentially disappear and more hygrophilic taxa with lax soft rosettes and shallow tanks (*Guzmania lingulata, G. sanguinea,* and *Vriesea splendens*) take their place, reaching maximum density at or near the ridge top. More generally distributed *V. heterandra* also becomes populous in the especially thick ridge–margin forest. Greater cloud flux, denser foliage in taller trees, and coalesced rain drops that are carried over the next windward ridge combine to create wetter conditions along sheltered ridge margins than on leeward slopes. Epiphytes on windward slopes intercept the most moisture of all but also experience stronger desiccating wind. Bromeliads with the shallowest tanks (flattest rosettes) are missing here. Scattered about are only bulbous *V. heterandra* featuring well-insulated leaf axils, and many *Guzmania cylindrica,* a species not often encountered elsewhere in the ridge system. The latter's considerable volume of tank fluid is well protected by upright rigid leaves. Tall trees in the gully below support delicate shade-tolerant *Guzmania lingulata* and *G. sanguinea. Guzmania monostachia* begins to reappear, and *Vriesea heterandra* continues as a fairly common epiphyte. Although cloud contact is low here, a dense canopy and relatively still air reduce evaporative demand.

Bromeliad habits and allied modes of resource securement seem to interact in way that also affect occurrence in more typical forests. Figure 7.16 shows carnivorous, lax humus-based, tubular, and myrmecotrophic habits arrayed along a humidity gradient according to a theoretical response to drought and its impact on productvity. Sunlight is unfiltered across the entire range. The area beneath each diagonal boundary indicates moisture regimens under which each habit is serviceable in the sense that investment can be repaid through photosynthesis during an average shoot lifetime. All four shoot types achieve maximum return on plant input under the wettest conditions. The carnivorous habit, as illustrated by terrestrial *Brocchinia reducta* (Fig. 4.25E), is most restrictive because the tight, overlapping leaves create considerable self-shading, and because production of a fragrance, of copious cuticular substances, and perhaps of tank secretions as well, add to construction and maintenance costs (Givnish et al. 1984). Substantial moisture and irradiance are thus required to sustain the necessary rate of photosynthetic return. The lax mesomorphic shoots of epiphytic *Nidularium bruchellii* (Fig. 4.25G) are too drought-sensitive to be efficacious in any but continuously humid habitats; tanks are shallow, and leaf exposure is highest of any of the impoundment forms. Next in line is the tubular habit of *Billbergia zebrina* (Fig. 4.25F) which shows less foliar overlap than does *Brocchinia reducta,* but much more than *Nidularium bruchellii.* Neither *N. bru-*

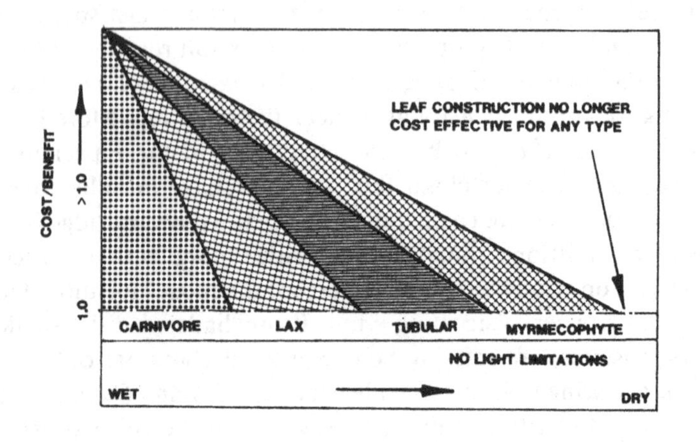

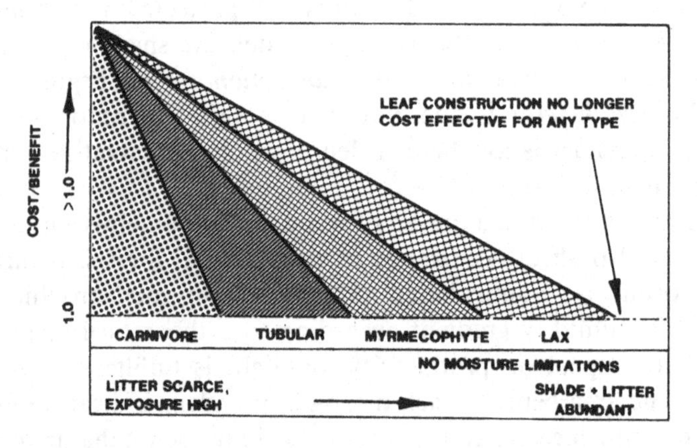

Figures 7.16, 7.17. Graphic economic models predicting the occurrence of bromeliads of various habits relative to moisture supply (7.16) and shading (7.17).

chellii nor *Billbergia zebrina* bears the extra expense associated with carnivory. Ant-fed ant-house forms (Fig. 4.24) are cast as the most flexible of the types comprising this series; cost of production of dry leaf-base chambers is all that is required (in a high exposure area, of course) to gain useful ant products.

Position varies somewhat from that in Figure 7.16 relative to correlated parameters of shade depth and litterfall (Fig. 7.17). Moisture supply is nonlimiting. Carnivorous types are again the least flexible because of substantial light requirements imposed by both high production cost and much self-shading. The danger that phytotelmata will become clogged with enough litter to preclude further trapping may increase risk for the carnivore in dense forests. At the other extreme, lax rosettes remain cost-effective well into deep forest habitats owing to their low light demand. Typically, hanging foliage prevents intercepted debris from blocking too much irradiance. Neither the lax nor the tubular tank types theoretically operate where tree litter is scarce or nonexistent. In fact, tubular forms can tolerate more open sites as access to other nutrient sources (e.g., excrement of periodic vertebrate visitors) increases. Ant-fed ant-house species are less versatile in the second compared to the first context, owing to a compact body that elevates light demands and perhaps to the ant's preference for open microsites.

Both models barely begin to describe the complex interplay between resource supplies (moisture, nutrients, light) and resource use by phytotelm plants of different shoot geometries. For instance, neither model incorporates effects of leaf and shoot longevity on energy return; there is substantial variation in life-spans among bromeliads. The type of photosynthetic pathway is ignored, yet this factor is all-important in determining rates of carbon gain and WUE. Also discounted are inherent differences in vigor that influence mineral ion requirements. Should root systems supply resources in any substantial way, the picture becomes even more complex. Nevertheless, cost–benefit models that include environmental and design constraints are essential to establish performance for a given plant body form in specific circumstances. In this instance, assessment of shoot characters that influence shade and drought tolerance, and the suitability of certain substrates as mineral nutrient sources, should help explain tank bromeliad occurrence. If no other purpose is ever served, study of phytotelm operation will at least illustrate how benefits and liabilities of particular plant forms shift with the environmental context.

Stranglers and other primary hemiepiphytes

When stranglers destroy their hosts, the process constitutes predation. Girdling is reputedly the usual way dicots are dispatched, but shading and eventual competition for earth-soil resources may be equally or more important. Roots of Indian *Ficus religiosa* intrude into moist fissures, then expand and split supporting trunks, but this species is exceptional. On drier sites, infested trees incur much less damage (Galil 1984). Further inquiry is needed to determine how killer species co-opt a benefactor's place in the upper canopy. Less decisive interactions involving weaker-growing primary hemiepiphytes should at most be viewed as competition. In some of these cases, special plant characteristics minimize host injury.

Ficus, the largest predominantly hemiepiphytic genus, contains both true stranglers and others incapable of achieving free-standing status (Todzia 1986). Central American *F. ballenei, F. citriifolia,* and *F. costaricana* can displace sizable hosts, whereas *F. paraensis, F. perforata, F. pertusa,* and most observed nonmoraceous primary hemiepiphytes cannot. True stranglers grow upright and produce thick roots and dense crowns of large leaves (Fig. 1.8). The permanent primary hemiepiphytes fail to support themselves or to confine host trunk expansion very much. Moreover, leaves are usually smaller and crowns more transparent. *Clusia odorata* in Panamanian forests forms a dense spherical crown, but to one side of the host rather than above it, while *Coussapoa panamensis* has an erect habit usually tilted away from the supporting axis. Both attitudes ensure that shade cast by the hemiepiphyte is minimized. Windblown leaves of *F. paraensis* undulate on long petioles, further increasing crown transparency and the probability of host survival (Todzia 1986).

Todzia noted a tendency in certain Panamanian hemiepiphytes to root on particular substrata as another factor influencing potential impact on hosts. *Cosmibuena skinneri, Hautiopsis flexilis,* and *Clusia odorata* (Figs. 7.18–7.20) occurred throughout infected crowns, whereas *Ficus citriifolia, F. paraensis, F. colubrinae, F. obtusifolia, F. trigonata,* and *F. perforata* were restricted to specific sectors. For instance, *Ficus citriifolia, F. trigonata,* and *F. perforata* rooted on trunks only. Only *F. costaricana* occasionally used positions beyond the middle of primary branches and on small peripheral axes. Germination requirements or behavior of dispersal agents may dictate microsite location, which in turn affects the welfare of the epiphyte; smaller limbs quickly succumb to strangling and fall away or simply break under the weight of these relatively large plants. (The smaller species did tend toward the most diffuse distribution.) Shade could also be important. Two strangling figs growing on *Copernicia* palms in the llanos of Venezuela exhibited high light demands; density of juveniles was greatest within or below sparsely foliated crown quadrants averaging 63% exposure. In no instance were subjects receiving less than about 40% of full insolation (Putz and Holbrook 1987).

Effects of epiphytes on associated vegetation

The case for parasitism

Authors of textbooks who consider nonhaustorial epiphytes at all, routinely describe them as plants that grow upon other plants but have no significant effect on hosts. In *A Dictionary of Biology,* Abercrombie, Hickman, and Johnson (1970) described an epiphyte as "a plant attached to

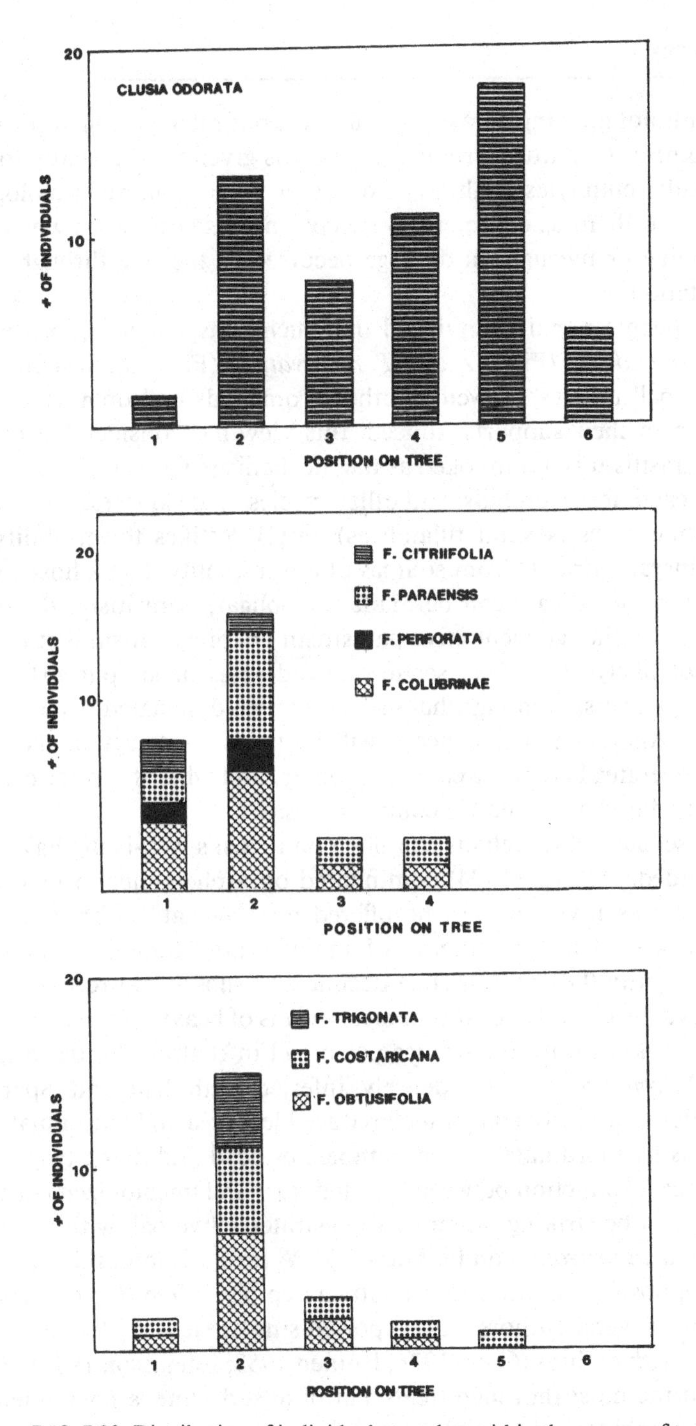

Figures 7.18–7.20. Distribution of individual stranglers within the crowns of trees in a Panamanian humid forest: 1, Below middle of trunk; 2, trunk above middle; 3, crotch in trunk; 4, on primary branch, halfway out; 5, on a primary branch more than halfway out; 6, on small branches at crown periphery. (After Todzia 1986.)

another plant, not growing parasitically upon it but merely using it for support." In point of fact, little serious thought was given until recently to the impact vascular epiphytes might have on other vegetation. Most biologists simply accepted them as commensals except on those occasions when significant shading or mechanical damage occurred to the tree through very heavy infestation.

Many lay people remain convinced that such conspicuous epiphytes as *Tillandsia usneoides* L. (Fig. 7.2) and *T. recurvata* L. (Fig. 7.3), the common Spanish and ball mosses, as well as other bromeliads and orchids, derive sustenance from their supports. Indeed, this view has considerable merit, providing parasitism is not invoked as the mechanism. Occurrence of many bromeliads, cacti, ferns, orchids, and other species on dead trees, rocks, and even telephone wires (several tillandsias) amply testifies to an ability to scavenge mineral nutrients from sources of lower quality than a host's vasculature. Although no case can be made for obligate parasitism, the existence of a direct although facultative parasitism or epiparasitism is still possible, but not likely. This final section considers evidence purporting to demonstrate parasitism among what has so far been designated a free-living canopy flora. Another major concern will be trophic interaction that unequivocally operates between tree and epiphyte, and what the latter can do to supports and perhaps to entire communities.

The adverse effects bromeliads can have on a host's well-being have not gone unrecorded. Billings (1904) commented on a phenomenon in which Spanish moss was involved, but he offered no explanation for its origin except to suggest that experiments of many years' duration would be required to identify the process. Host decline and subsequent recovery following removal by chemical or mechanical means of heavy *Tillandsia* infestations are well known to tree service personnel in central Florida. Native live oak *(Quercus virginiana),* densely infested with ball and Spanish mosses, is often characterized by much reduced leaf area and individual leaf size as well as by inordinately large numbers of dead and dying twigs and larger branches. Distinction between infected trees and uncolonized conspecifics nearby can be striking. Figure 7.2 illustrates a live oak with a heavy load of *Tillandsia usneoides* and scattered *T. recurvata* colonies. Figure 7.4 depicts the opposite condition in a relatively epiphyte-free live oak about 300 m distant. Several authors have reported similar signs of deterioration in trees laden with orchids (Cook 1926; Ruinen 1953; Johansson 1977). Ruinen provided the most thorough treatment of host decline, a phenomenon she labeled "epiphytosis." In each case, orchids were the purported causal agents. Extensive hyphal bridges were demonstrated between orchid roots and adjacent branch vasculature. In no case were host substrates tracked into orchid tissue, however.

Epiphytes are known to aggregate on dead branches, a relationship Johansson (1977) ascribed to orchid use of host substrates in a case involving shootless *Microcoelia* infesting African *Terminalia*. Actually, this association may be typical of many epiphytes even in healthy trees and the impression of parasitism purely illusory. Approximately 80% of the mature ball moss colonies growing in the middle- to lower-crown portions of 10 mature live oaks with normal foliage development in South Florida were anchored to slender twigs in another survey (Benzing 1979). Remaining fruiting specimens were secured to main trunks or its major branches. Of 95 reproducing specimens rooted to small twigs, 70.5% were supported by dead axes. Numerous adjacent stems of apparently similar age and about the same size but harboring no mature bromeliads were also lifeless. Juveniles were abundant on dead and living stems. The best explanation for this occurrence requires knowledge of tree development and architecture.

Although massive and long-enduring, trees even more than many herbs exhibit determinate growth. Beginning with a single axis, the crown expands by stages. Longevity of a particular stem relative to the whole organism's life-span is determined by three factors: (1) place in space and time within the ontogenetic sequence; (2) model-specific form of the species (Hallé, Oldeman, and Tomlinson 1978); and (3) local circumstances that affect the plant's ability to conform to its architectural model. Crown development in most instances begins with formation of a number of primary lateral axes (Fig. 7.21). Some years later, all or most of the lowest ones have died and fallen away. By that time, the remaining laterals located above them have proliferated. Surviving second-order axes undergo a third ramification and so on as the crown continues its expansion. Repeated subdivisions, usually no more than five or six orders in all, yield a sequence of smaller and smaller branch complexes according to model-specific destinies. Not only are successive ramifications progressively miniaturized, but growth potential diminishes apace. Finally, the crown periphery where most of the remaining shoot meristems are located approaches a maximum dimension, eventually lapsing into a kind of dynamic equilibrium. Over a number of seasons, terminal meristems at the periphery of the mature crown elongate slowly while generating leaves and fruiting structures in a fashion not unlike the annual shoot production of many temperate old-field and forest-understory perennials. For a time, branchlets are generated to take the place of those lost through normal attrition, and the outline and density of the crown remain static. At some point, wholesale senescence commences: New meristems are

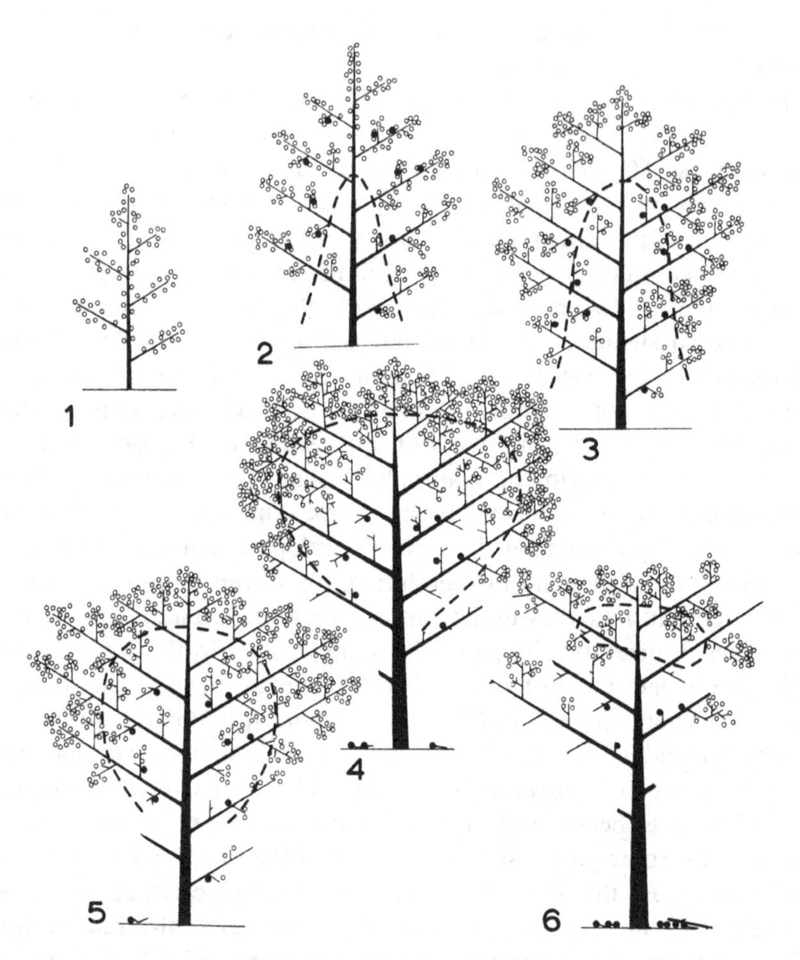

Figure 7.21. The progression of events during phorophyte ontogeny that produces the illusion of parasitism by resident epiphytes. The crown area enclosed in dashed lines represents the self-shaded region where branches will normally die whether or not epiphytes are anchored there. (After Benzing 1979.)

no longer produced at a rate sufficient to replace all of those that die, and the crown thins. Whole branch complexes succumb; finally the organism itself either dies or, in its weakened condition, is toppled by wind or killed by pathogens or predators.

Considered in this perspective, a tree represents an ordered mosaic whose parts differ in time of arrival, growth potential, and life-span. Crown stems fall into distinct and successive but temporally overlapping populations. Major segments – the main axis and large laterals – live longest. But whatever their size, individual crown components must be autotrophic. Death ensues once exposure falls below the light compensation point of attached foliage whether because of self-shading or because of a neighbor's overgrowth. Small lateral shoots whose role was temporary support for expanding leaders are cast off without delay as crowns expand. Displacement inward ensures that, if nothing else, the service life of a particular anchorage site will usually be much shorter than the life of the whole tree. The system is vulnerable to some distortion and, if populated by epiphytes, can give an impression of parasitism whether growth is normal or disturbed.

If a particular branch, for whatever reason, occupies an exposed position longer than usual, that axis will not self-prune on schedule. It will die or become a moribund relic well after similarly programed parts of the same tree fall away at the end of the normal life-span. The more extended its "place in the sun," the more likely that a branch will be successively colonized by epiphytes, and the larger (older) those plants will become before being shed or suppressed by heavy shade. Moreover, the longer a portion of the crown remains alive, the greater the likelihood that the well-established epiphytes located there will appear to be responsible for its eventual death.

Neither the exact exposure requirements nor life histories of epiphytes so far identified as suspected parasites are known. Most assuredly, all need several years to mature and many more seasons to become robust older adults. Assuming that they are also at least moderately intolerant of shade and rain shadows cast by heavy foliage, the highest density of large specimens would be expected on surfaces with the longest history of high exposure. These sites are most common near but not at the center of the crown, exactly where several investigators (Johansson 1975; Catling et al. 1986; Fig. 7.22) documented the highest epiphyte densities in individual trees. Substrata at the crown periphery and deeper within well-foliated regions are likely to be too new or were too quickly shadowed to have nurtured a comparable density of equally vigorous epiphytes. It follows that disproportionate occurrence on dead and dying branches cannot be construed as evidence that epiphytes are parasitic or that they have had any effect whatsoever on a host. Equally plausible, and in fact more likely, is the possibility that this association occurs because the larger axes involved are relics favored with greater durability by extraordinary exposure or, if smaller, were simply programed to die after a normal, brief life-span.

Damage unrelated to mineral nutrition

Epiphytes can injure phorophytes through several mechanisms short of parasitism. Phytotoxins may be released by vascular residents in tree crowns as they are by some epiphyllous lichens (Orús, Estévez, and Vi-

Figure 7.22. Leafless deciduous tree in western Ecuador illustrating the distribution of resident epiphytes in its crown. Most of these plants are tank bromeliads.

cente 1981). Claver, Alaniz, and Caldiz (1983) reported such a phenomenon associated with the rapid spread of *Tillandsia aeranthos* and *T. recurvata* in and around La Plata, Argentina. Defoliation of exotic conifers and broadleaf trees was the principal symptom. Such a phenomenon would favor these two heliophilic bromeliads, but caution must be observed until the origin, chemical identity, and effect of the putative toxin are confirmed. Epiphytes may also damage supports by simply creating conditions conducive to pathogens that attack underlying phorophyte tissue. Some fungi associated with orchid protocorms and roots (Figs. 4.10, 4.12) are more virulent to other plants and under appropriate circumstances may turn epiphytes into reservoirs of infectious disease.

Weight alone injures phorophytes densely colonized by epiphytes, but there are other mechanical effects. Girdling by orchid roots was presumed

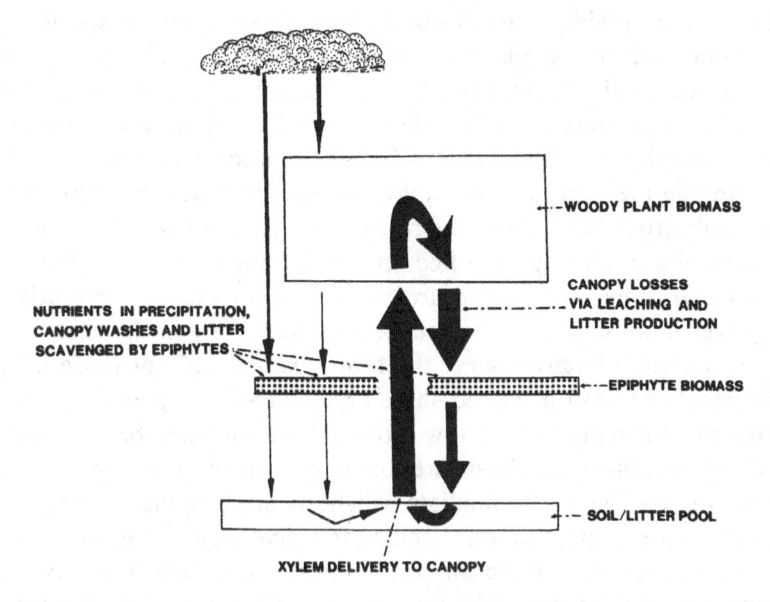

Figure 7.23. Schematic representation of atmospheric input and mineral cycling in a tropical forest, depicting the pirating activities of resident epiphytes.

to cause dieback in some citrus orchards (Cook 1926). Small, wiry bromeliad holdfasts might produce similar effects, but searches in Florida failed to turn up any such evidence (pers. obser.). Shading comes to mind when one observes heavy shrouds of Spanish moss on live oaks and cypress in the southeastern United States (Fig. 7.2); this explanation is less than satisfactory for ball moss, which tends to aggregate inside thin crowns. Spanish moss as well is often restricted to the lower canopy; yet much of the dieback so characteristic of these species' declining supports occurs at the crown periphery. Unusually small, chlorotic leaves, another symptom of host decline, occur throughout the canopy, providing further evidence that stress has a systemic rather than a localized cause.

Nutritional piracy

Epiphytes may either create problems for phorophytes or intensify existing deficiency by acting as nutritional pirates (Benzing and Seemann 1978). In order to appreciate the suitability of this label, one must first consider how plant nutrients cycle through forest ecosystems (Fig. 7.23). Once absorbed, required substances circulate (except for fractions immobilized in such inert and durable tissue as wood); ions of each chemical species flux (if phloem-mobile) among ephemeral organs as they age and between plant and adjacent earth. Substantial quantities are brought back to the ground whenever litter is produced, to be released by decomposition. Various ions leached outright from intact organs by precipitation return to roots; movement through the plant and utilization are then repeated. Efficient recycling is especially crucial on porous infertile earth where much of the ecosystem's most critical nutrient capital is tied up in plant tissue (Jordan 1985). Failure to recover lost ions under these circumstances may exact a substantial price in vigor, or if scarcity is great enough, promote severe nutritional stress.

Botanists traditionally reserve the term "parasitism" for those instances where benefactor and benefited share organic continuity; without this sort of integrity, nutrients cannot flow directly from one member of the pair to the other. Mistletoes and hosts are joined by haustoria, whereas epiparasitism involves a fungal intermediate which, by bridging the two individuals, obviates need for direct union. Lacking invasive organs, true epiphytes are neither parasites of, nor typical competitiors with, their hosts. But like a parasite, an epiphyte does take nutrients from a phorophyte. It does so by absorbing essential ions before the support has a chance to recover or, for that matter, even lose them to the understory.

In essence, intermixed forest-floor plants compete for mineral ions; epiphytes, however, enjoy special access to them when they arrive from the atmosphere and subsequently circulate among community members. Nonimpounding epiphytes extract ions from rainfall, leachates, and dry deposition, whereas trash-basket and phytotelm species harvest them from impounded litter as well. Whatever the mode of interception, epiphytes thus deprive their hosts of nutrients with the same effect as if they had acted as direct parasites or epiparasites. Moreover, slow growth, long life, and modest litter production combine to ensure that epiphyte nutrients remain immobilized for extended periods. Canopy-based vegetation is thus well suited by scavenging capacity, growth characteristics, and position in the forest (Fig. 7.23) to influence community-wide nutrition and all related phenomena.

Two sites were chosen for quantifying pirating activities of bromeliads on *Quercus virginiana* in South Florida; an instantaneous rather than a long-term analysis was used (Benzing and Seemann 1978). One study plot was located in a coastal strand community dominated by dwarf live oak just a few kilometers north of Naples, Florida. Also present in this community were two other large but less abundant species, *Pinus clausa* and *Serenoa repens. Quercus* crowns supported dense Spanish and ball moss infestations

Table 7.6. **Soil fertility data (mean ppm) and pH (mean) in two sites in Florida**

Site	Available Ca	Exchangeable K	Available Mg	Total N	Available P	pH
#1						
Dwarf oak	322 ±22	23.3 ± 0.8	41.8 ±3.4	2270 ± 428	2.7 ±0.3	5.03 ± 0.16
#2						
Vigorous oak	477 ± 124	54.5 ± 5.7	63.3 ± 10.8	9392 ± 2234	25.8 ± 7.8	4.82 ± 0.27

Source: Benzing and Seemann 1978.

(Fig. 7.3); *Tillandsia balbisiana, T. fasciculata,* and *T. utriculata* occurred in smaller numbers. A sparse, patchy cover over white, acidic, sandy earth-soil was provided by fruticose lichens, poikilohydrous *Selaginella arenicola,* and scattered forbs. Key nutrient pools were deficient in the ground (Table 7.6) and in vegetation as well (Table 7.7).

Materials harvested from two dwarf oak crowns revealed that 35–57% of the total N, P, and K located in phorophyte shoots, subtending twigs, and attached bromeliads was present in the epiphytes (Table 7.8). The poor nutrition of *Quercus* and attached *Tillandsia usneoides* at the strand site was demonstrated by comparison with live oak foliage and Spanish moss collected from vigorous trees on more fertile ground near Tampa, Florida (Table 7.7). Direct inferences from these data are not possible, but had the N, P, and K present in *Tillandsia* tissue been retained or initially intercepted by *Quercus,* one can reasonably question whether dwarfing would have been so pronounced at the strand. The nutrient capital co-opted, plus light also captured by epiphytes, would have been available to create additional oak biomass.

More direct information on the pirating capacities of epiphytes and the effect on broader cycling will be costly to obtain, but some preliminary observations are encouraging. Nadkarni (1984) reported relatively rich throughfall (in Ca, K, Mg, N, P) beneath *Clusia alata* branches bearing heavy epiphyte loads in a Costa Rican cloud forest during pairwise tests. In contrast to wetter months, dry season rainfall lost ionic strength with pas-

Table 7.7. **Mineral nutrient concentrations in leaves of dwarf and vigorous *Quercus virginiana* and the shoots of their respective *Tillandsia usneoides* colonists**

	Percent dry weight						Parts per million (ppm)					
	N	P	K	Ca	Mg	Na	Mn	Fe	B	Cu	Zn	Mo
Leaves from canopies of dwarf oaks	1.43	0.236	0.826	0.752	0.205	0.0512	128.1	145.5	20.0	13.8	37.0	1.30
Leaves from canopies of vigorous oaks	1.88	0.286	0.846	0.629	0.234	0.0436	204.7	116.0	21.2	16.2	38.2	1.36
Shoots of *T. usneoides* on dwarf oaks	0.945	0.140	0.463	0.700	0.197	0.130	57.3	471.3	16.3	10.3	27.0	1.66
Shoots of *T. usneoides* on vigorous oaks	1.19	0.133	0.520	0.587	0.153	0.130	114.0	457.3	17.0	10.0	57.0	1.40

Source: Benzing and Seemann 1978.

Table 7.8. **Mineral nutrient capital in two dwarf *Quercus virginiana* hosts: percentage of total found in the epiphyte load**

	N	P	K	Ca	Mg	Na	Mn	Fe	B	Cu	Zn	Mo
Specimen #1	35.4	53.4	50.2	41.4	76.4	60.8	44.0	77.1	36.2	55.2	62.4	62.1
Specmen #2	35.9	33.9	57.2	43.4	43.9	69.9	55.8	28.6	39.0	49.4	60.5	50.2

Source: After Benzing and Seemann 1978.

sage through the crown vegetation. Local sampling must be extensive and include litter fall in order to extrapolate whole-system behavior with this approach, however. Forest canopies and even individual tree crowns are patchworks of different successional stages. Limbs supporting epiphyte colonies younger or older than the modest number sampled by Nadkarni lose or gain solutes at changing rates depending on age, just as do whole regenerating ecosystems.

Broader effects of epiphyte nutrition on the forest

Epiphytes influence element apportionment and related aspects of plant community performance according to local circumstances. On stable sites heavily dependent on atmospheric deposition to balance outflow, a massive epiphyte–epiphyll presence may enhance the forest's storage and nutrient-capturing and -retaining capacities without depriving any community members (Nadkarni 1981). Here the negative connotation of "piracy" renders that term less appropriate for describing the epiphyte's role in nutrient flux. Rather than continuously accumulating minerals, as any expanding biomass must, a mature flora suspended in the canopy probably more nearly approaches a nutritional steady state (Fig. 7.24), losing essential elements about as fast as they can be intercepted. Nadkarni's Central American primary cloud forest may have been close to that steady-state condition. She unquestionably discovered that extended systems of canopy roots produced by trees further promote nutritional equity on some very moist sites supporting mature forest. Whether such a well-developed epiphyte community pays a photosynthetic return commensurate with the resources preempted and host foliage it displaces is an open but important question.

At present, no values are available on standing crops of epiphylls (Fig. 7.6). Only a few studies provide data on vascular epiphytes, and all of these deal with moderately to very wet systems (Fittkau and Klinge 1973;

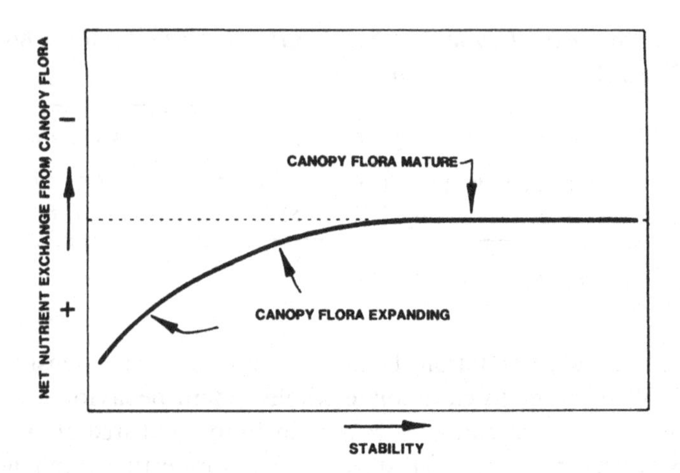

Figure 7.24. A graphic model illustrating the relationship between state of development and nutrient exchange by epiphytic flora.

Edwards and Grubb 1977; Golley, Richardson, and Clements 1978; Tanner 1980; Grubb and Edwards 1982). Most report that canopy-based flora constitute only a few percent of the total aboveground biomass, a statistic which obscures the fact that this compartment figures more prominently in aggregate leaf surface area. Edwards and Grubb (1977) did calculate epiphyte weight at about half of that for tree leaf biomass in a New Guinea lower montane rain forest. Tanner (1977) observed values up to 35% in Jamaica. Nonvascular and vascular epiphytes constituted much of the green canopy biomass in a Costa Rican elfin forest (Nadkarni 1984; Fig. 7.25), and the percentage of a given mineral element contributed by these plants to the total in the ecosystem foliage reached as high as 45% (Fig. 7.26). Epiphytes (bryophytes and ferns) contained more Chl than did understory herb and shrub layers (11.3% vs. 10.1% and 1.0%, respectively) in a western Himalayan forest dominated by *Quercus floribunda* (Singh and Chaturvedi 1982); the actual amount of Chl was 2080 mg m^{-2}, a quantity sufficient to drive photosynthesis at an estimated 0.027 J $m^{-2}s^{-1}$. No survey published to date provides assessments of the physiology of epiphyte as compared to adjacent phorophyte foliage.

Effects of epiphytic vegetation on forest productivity probably vary from additive to suppressive, according to climate and its influence on terrestrial versus canopy-based vegetation. Mineral use efficiency should shift along the same gradient. Because trees have access to more continuous moisture supplies than do most of the plants growing in their crowns, nutrient

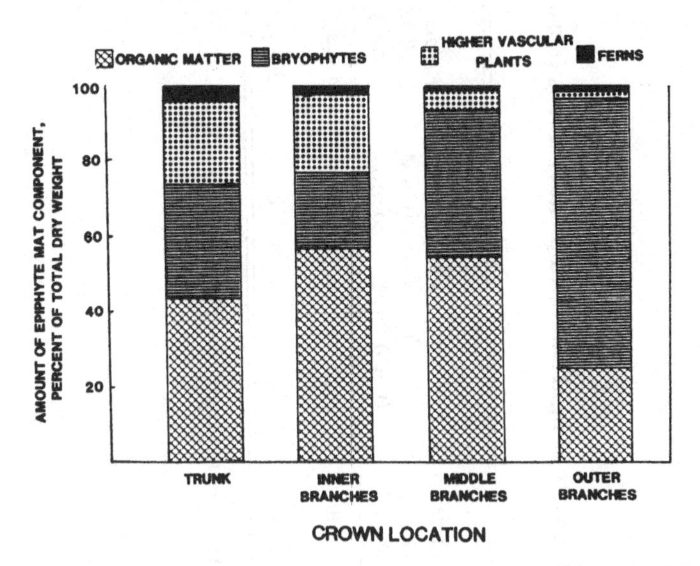

Figure 7.25. Epiphyte biomass present on *Clusia alata* in a Costa Rican cloud forest. (After Nadkarni 1984.)

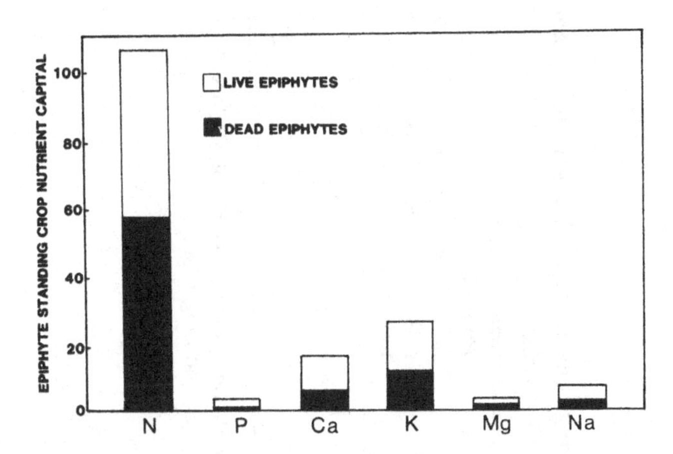

Figure 7.26. Nutrient elements present in epiphyte biomass on *Clusia alata*. (After Nadkarni 1984.)

resources that the two allot to foliage should yield different rates of return. Little is known about epiphylls, but epiphytes are generally modest producers. Adapted for drought to a greater degree than their supports, these plants probably achieve higher WUE but lower instantaneous returns on nutrient

Table 7.9. **Gas exchange characteristics of selected C_3 and CAM epiphytes at light saturated photosynthesis (during phase IV for the CAM species)**

	Pyrrosia lanceolata (CAM fern)	*Aglaomorpha heracleum* (C_3 fern)	*Platycerium grande* (C_3 fern)	*Anthurium hookeri* (C_3 aroid)	*Drymoglossum piloselloides* (CAM fern)	*Pyrrosia longifolia* (CAM fern)	*Kalanchoe uniflora* (CAM)
CO_2 uptake (μmol $m^{-2}s^{-1}$)	2.1	3.2	3.3	3.4	0.8	1.4	0.6–4.2
Transpiration (mmol $m^{-2}s^{-1}$)	28	14	22	26	21	23	20–95

Note: A_{max} of more xerophytic epiphytes with typical CAM (e.g., atmospheric *Tillandsia*) are lower still.
Source: Lüttge, Ball, Kluge, and Ong 1986a.

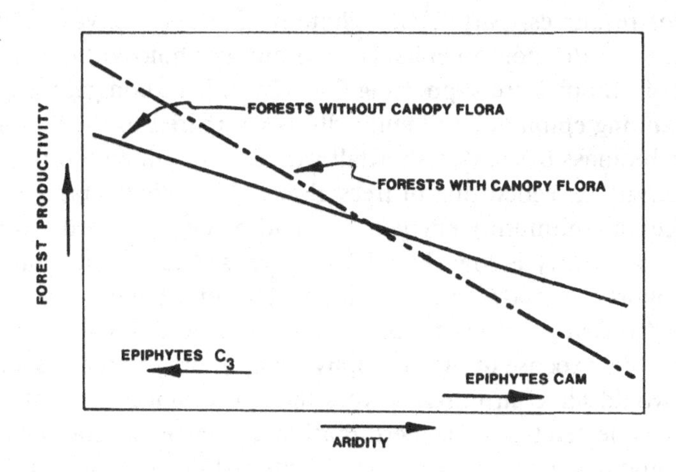

Figure 7.27. A graphic illustration of the contribuion made by epiphyte photosynthesis to total canopy productivity in forests of differing humidity levels. Note that epiphytes in humid forest tend to be C_3 types, whereas those on drier sites utilize CAM.

investments. Nitrogen is more concentrated in C_3 than in CAM foliage, and A_{max} (including that of some epiphytes; Table 7.9) is proportionally even higher (Larcher 1980). Given that the leaves of most epiphytes are evergreen, however, values for integrated MUE are no doubt closer. Patterns of gas exchange and mineral deployment in epiphytes and supporting trees (Fig. 7.27) should converge with progression from dry to wet forests. Table 3.2 compares a single phorophyte with a group of co-occurring epiphytes utilizing a variety of water-balance mechanisms. The best performer of the lot in terms of both A_{max} and instantaneous N and P use efficiency was the tree. Returns on invested nutrients recorded for the well-watered C_3 epiphytes were fairly similar to that for *Psidium.* Year-round integration would have revealed much different performances, however. Lacking access to ground moisture, the epiphytes' leaves are either shed *(Microgramma polypodioides)* or moisture-stressed enough to slow or eliminate carbon gain through much of the dry season. Note that only one of the epiphytes exhibited net CO_2 uptake before the seasonally dry supporting mat was irrigated.

One could reason that not only do sizable epiphyte–epiphyll loads on equable sites markedly increase nutrient storage and retention and promote N_2 fixation but additionally, canopy productivity could be greater than would be possible in their absence. An extensive epiphyte–epiphyll presence of the sort encouraged by high humidity increases genetic diversity in wet forests (e.g., Fig. 7.10), perhaps to the extent of promoting a more effective partitioning of resources, particularly photons. Certainly, investments in wood and bark will do more to enhance community photosynthesis if leafless portions of canopies are supporting flora. Even if carbon gain is somewhat slower among epiphytes and epiphylls as compared to their supports on an area or biomass basis, that shortfall would be eliminated if the combined productivity exceeded that of trees alone. Conversely, epiphyte–epiphyll influence on community energetics should be negative in dry forests, especially where fertility is low (Fig. 7.27). Here, scarce mineral ions preempted to support the modest photosynthetic rates of arboreal CAM plants and poikilohydrous vegetation (in this case, mostly lichens) would, to some extent, come at the expense of the phorophyte's far more productive foliage. Disturbance would have an impact across the entire range of forest types. Communities subjected to enough perturbation to prevent substantial nutrient buildup could never achieve a mature nutritional steady state (Fig. 7.24); their epiphyte–epiphyll loads would be constantly regenerating and accumulating mineral ions at the expense of supports. Broadly based inquiries into the participation of canopy flora in the nutritional dynamics of a variety of tropical woodlands are much needed; without them, knowledge of tropical-forest structure and function will remain incomplete.

8 Epiphyte occurrence

This final chapter concerns epiphyte occurrence in three contexts: global, taxonomic, and ecological. First, global and taxonomic patterns and the question of why epiphyte floras are unevenly developed throughout the tropics are addressed. Second, the effects of climate, topography, and soil fertility on species range and abundance are considered. Finally, hypotheses are offered to explain why certain plant lineages have been more successful than others in forest canopy habitats.

Distribution: taxonomic and geographic

At the higher taxonomic levels, epiphytes are diverse; excluding the mistletoes, 84 families, including 69 in Magnoliophyta, contain qualifying taxa. But rather few major clades account for most of the species; just 23 families harbor about 98% of the total flora in 87% of the epiphytic genera (Tables 1.1, 1.2). Fifteen families include but a single epiphyte; 52% of the 871 epiphytic genera contain five or fewer species, and about half of those contain only one. Heaviest contributors are Araceae, Bromeliaceae, Ericaceae, Gesneriaceae, Melastomataceae, Piperaceae, Orchidaceae, Rubiaceae, and several fern families. Forty-three genera each contain more than 100 epiphytic species (Table 1.2): of the 43, 22 are orchids, 8 are ferns, 4 are bromeliads, 3 are from Aracaceae, and the remaining 6 are contributions from five additional families.

Geographic asymmetry is also considerable, especially in the more advanced taxa (Madison 1977). Of the 86 canopy-adapted fern genera listed by Madison, approximately two-thirds are pantropical; the remainder are divided about equally between the two subregions – 16 exclusively neotropical and 14 paleotropical. Only *Pyrrosia* of the sizable, predominantly epiphytic, genera is restricted to a single geographic zone. *Lycopodium* and *Psilotum* are cosmopolitan (the former also contains temperate species); *Tmesipteris,* like *Pyrrosia,* is Australasian. Relatively broad distribution of polypodiaceous fern genera in tree crowns may reflect high dispersibility or great age – in effect, pronounced evolutionary stasis.

A comparison of Old versus New World distribution shows that contributing angiosperm families approach parity (42 vs. 43; Gentry and Dodson 1987b) and that the genera are more insular. Families (Table 8.1) with epiphytes in just one major zone (33) outnumber those in two or more zones (23). Of the 32 seed plant families with five or more epiphytic species, 26 are represented in canopies of the neotropics, 25 in those of Australasia, and just 14 in Africa where cool, dry Pleistocene climate had an unusually devastating impact on humid forest flora. Some of these families contribute epiphytes to only one zone: 18 do so to Australasia; 15 to the neotropics; and one to Africa (Costaceae), and that group barely qualifies because closely allied counterparts belonging to Zingiberaceae occupy tree crowns in tropical Asia. Others colonize forest canopies in two zones but not the third: For example, nine families at least are epiphytic in both the neotropics and Australasia, but not in Africa. In contrast to angiosperms as a whole, neotropical and Australasian epiphyte floras exhibit greater affinity with each other than with African counterparts.

Only 30 angiospermous genera containing epiphytes are found in more than one of the three major geographic regions; just 13 of these inhabit tree crowns in both Africa and Australasia. Orchidaceae account for five of the truly pantropical genera (less than might be expected); the others are *Begonia, Ficus, Myrsine, Peperomia, Rhipsalis, Schefflera, Utricularia,* and *Vaccinium.* Terrestrialism outweighs epiphytism in about half of these taxa. Canopy life is sporadic or rare in *Begonia, Burmannia, Gaultheria, Lycianthes, Myrsine, Pilea, Piper, Psychotria, Senecio, Solanum, Utricularia* and *Vaccinium.* Diverse biology, including dispersal mechanisms, obscure reasons why 30 disparate lineages have penetrated widely distributed forest canopies more than have all others. About the only unifying feature is occurrence in continuously humid rather than seasonal locations.

At the species level, epiphytic flowering plants in the neotropics outnumber those in the paleotropics (ca. 15,500 vs. 12,600; Madison 1977), even though the latter region is larger and more fragmented. Tree crown habitats are rather thinly colonized throughout much of the Old World tropics. Australia, a largely frost-free but relatively arid land mass, supports only about 380 canopy-adapted species representing just 1.4% of its total vascular flora (Wallace 1981). Greater diversity is present at higher taxonomic levels, however; of all the families containing epiphytes, more than a third (33) include Australian species. Africa is somewhat better stocked with species than is Australia; western Africa alone contains 239 epiphytic orchids (Sanford 1969), and the entire continent harbors about 2400 species (Madison 1977).

By comparative land area, New Guinea and Borneo are still better supplied with epiphytes, particularly Orchidaceae. Orchids are generally useful for geographic comparisons; Dressler's (1981) compilation of species by region parallels pretty closely the overall pattern of epiphyte distribution (Fig. 8.1).

By virtue of sheer bulk and abundance, Bromeliaceae dominate the neotropical canopy-adapted flora, especially where insolation and humidity are high. Although often outnumbered by co-occurring orchid and fern species at these locations, tank forms of the commonly more prolific phytotelm bromeliads subordinate the rest of the arboreal vegetation to the human eye. For instance, Kelly et al. (1988) reported that eight bromeliad species, seven with phytotelmata, constituted the majority of compact epiphytes in a Jamaican rain forest, even though 29 pteridophytes, 20 orchids, 6 dicots, and an aroid population shared the same canopy. *Asplenium, Drynaria,* and *Platycerium* produce similar silhouettes in some paleotropical forests, but never do they fill as much space as is occupied by bromeliad rosettes at the peak of their development in American woodlands. Old World epiphyte communities are usually best described by orchid/fern components (Table 1.2, 8.2); nonorchid monocots and dicots are underrepresented there, although such shrubby forms as *Rhododendron* or tuberous Hydnophytinae sometimes contribute most to epiphyte biomass.

Reasons for the neotropical bias

Epiphyte diversity follows a general pattern: Neotropical angiosperm species of all habits outnumber those in the paleotropics (90,000 vs. 60,000; Raven 1976), but special causes and perhaps even some coincidence may account for the uneven occurrence of arboreal forms. Bromeliaceae and Cactaceae – the two largest, essentially endemic, New World families – contain many epiphytic species (Table 8.1). One could argue that both lineages succeeded in tree crowns in part because of a tendency toward xerophytism – and just happened to have neotropical origins. The presence of Bromeliaceae is particularly significant because it allowed a remarkable opportunity to be utilized better in tropical America than anywhere else. Plants of no other family create equally elaborate phytotelmata in lieu of resource-rich rooting media, although there is no obvious reason why Liliaceae (particularly *Astelia*), Commelinaceae, Costaceae, Pandanaceae, and Zingiberaceae – families that all possess some tank development or shoot form readily modifiable for impoundment – have not become more successful bromeliad analogs.

Table 8.1. **Abundance and range of epiphytic species among angiosperm and pteridophyte families**

Strictly neotropical			Pantropical			Strictly paleotropical	
Family	No. of epiphytic spp.	Range assignment[a]	Family	No of neotropical/ paleotropical epiphytic spp.[b]	Range assignment[a]	Family	No. of epiphytic spp.
I. Angiosperms							
Agavaceae	1	U	Araceae	684/166	NA	Amaryllidaceae	1
Bromeliaceae	919	NA	Araliaceae	5/8 (60)	NA	Aquifoliaceae	1
Cactaceae	133	Gondwanan	Asclepiadaceae	2/133	U	Celastraceae	2
Campanulaceae	18	NA	Balsaminaceae	(5)	L	Costaceae	4
Commelinaceae	3	U	Begoniaceae	(10)	NA	Cunoniaceae	2
Compositae	3	NA	Bignoniaceae	2/1	A	Elaeocarpaceae	1
Cyclanthaceae	31	NA	Burmanniaceae	1/1	Guayanan	Myrtaceae	7
Dulongiaceae	1	U	Clusiaceae	7/0 (85)	U	Nepenthaceae	6
Gentianaceae	1	L	Crassulaceae	2/2	L	Pandanaceae	4
Marcgraviaceae	94	NA	Ericaceae	263/193(23)	NA	Pittosporaceae	5
Onagraceae	3	SA	Gesneriaceae	430/126	NA	Poaceae	2
Rapateaceae	6	Guayanan	Griseliniaceae	(3)	U	Potaliaceae	20
			Lentibulariaceae	(12)	U	Ranunculaceae	1
			Liliaceae	4/13	L	Rosaceae	3
			Melastomataceae	137/346	NA	Vitaceae	4
			Moraceae	20/1 (500)	A	Winteraceae	1

	Myrsinaceae	11/5	(12)	NA		
	Orchidaceae	15,000		NA		
	Philesiaceae	1/1		U		
	Piperaceae	700/0	(10)	NA		
	Rubiaceae	52/157	(7)	NA		
	Scrophulariaceae	3/3		L		
	Solanaceae	15/2	(15)	SA		
	Urticaceae	1/20	(20)	NA		
II. Pteridophytes						
	Aspidiaceae	254			Schizaeaceae	2
	Aspleniaceae	400				
	Davalliaceae	185				
	Hymenophyllaceae	500				
	Lycopodiaceae	150				
	Ophioglossaceae	2				
	Polypodiaceae	1027				
	Psilotaceae	8				
	Selaginellaceae	5				
	Vittariaceae	142				

[a]Range assignments: A, Amazonian-centered; NA, northern Andean-centered; SA, southern Andean-centered; L, Laurasian; U, unassigned.
[b]For pantropical families, ratios denote # neotropical/# paleotropical species; parentheses enclose numbers of additional species in genera whose neotropical/paleotropical ratios are not given in Madison (1977).
Sources: After Madison 1977; Gentry 1982b.

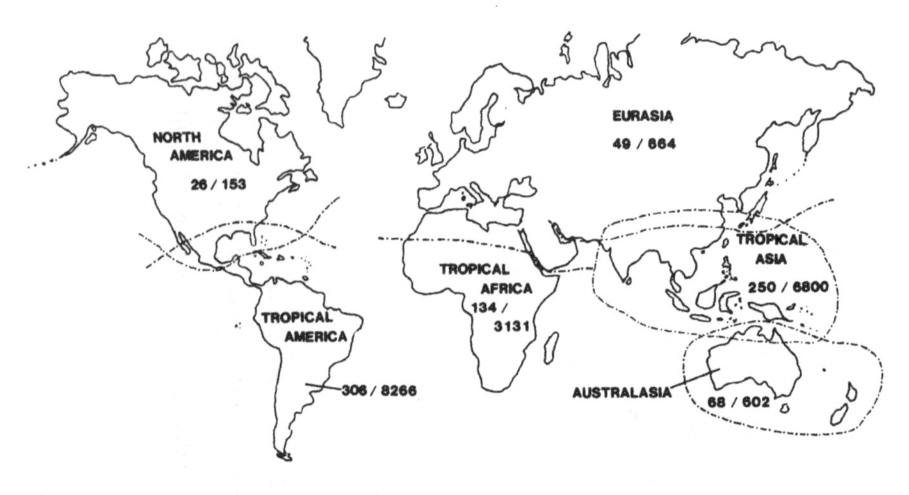

Figure 8.1. Worldwide occurrence of orchid species and genera. (After Dressler 1981.)

Imbalance within several families containing New and Old World epiphytes points to additional area-specific reasons for the neotropical bias. More than a third of all American epiphytes are represented by the three exclusively New World orchid subtribes Maxillariinae, Pleurothallidinae, and Oncidiinae. New World *Anthurium* and *Philodendron* are much larger than any of seven African and Australasian aroid genera containing epiphytes. Gesneriad epiphytism is best developed in the Americas where subfamily Gesnerioideae and particularly taxa like *Columnea, Dalbergaria, Drymonia,* and *Trichantha* have radiated extensively. Pantropical *Peperomia* is particularly diverse in tropical America. Only one of the families that contain sizable numbers of epiphytes (Ericaceae) is about equally represented in New and Old World forest canopies. Asclepiadaceae have populated tree crowns in the paleotropics far more extensively than in the neotropics, but with relatively few species. Among the larger widespread groups of related epiphytes, Melastomataceae and Rubiaceae exhibit the strongest paleotropical emphasis, the latter largely because of the success of myrmecophytic *Hydnophytum, Myrmecodia,* and several smaller allied genera.

Disproportionate neotropical radiations are further documented in Table 1.2. Of the 21 nonorchid genera containing the largest epiphytic contingents, just three (*Medinilla, Rhaphidophora,* and *Pyrrosia*), containing about 400 species, are paleotropical, whereas eight others (ca. 2150 species) include only American species. Six of the seven nonorchid monocot genera are neotropical; *Rhaphidophora* is the lone exception. Four of the remaining six

Table 8.2. **Taxonomic distribution of epiphytic species at New and Old World locations**

Location and extent of area represented	Ferns and allies	Nonorchid monocots	Orchids	Dicots
Entire continent of Australia[a]	169	10	152	61
Dwarf cypress forest in South Florida[b]	2	8	3	0
Multiple plots on four islands in Micronesian rain forest[c]	61	5	52	19
Several plots in lowland wet forest near Veracruz, Mexico[d]	32	51	87	42
Lowland wet forest (1000-m^2 sample) at Río Palenque, Ecuador[e]	26	61	81	67
Dry rain forest (1000-m^2 sample) at Juaneche, Ecuador[e]	4	15	33	11
Dry forest (100-m^2 sample) at Capeira, Ecuador[e]	0	2	5	0
New Zealand[f]	36	2	7	5
Dry forest in West Africa at Makokou, Gabon[e]	26	10	21	2
Lowland wet forest on Barro Colorado Island, Panama[e]	0	2	5	0

Sources: [a]Wallace 1981; [b]Benzing pers. obser.; [c]Hosokawa 1955; [d]Valdivia 1977; [e]Gentry and Dodson 1987b; [f]Oliver 1930.

taxa contain natives of Old as well as New World forests; *Columnea* and *Drymonia* are strictly American. Neotropical *Blakea* (Melastomataceae) and *Cavendishia* (Ericaceae) fall just short of inclusion in Table 1.2. The same pattern holds for the orchids: Five of 10 neotropical genera with 100 species or more include over 500 members each; mean size for the lot tops 450. Of the strictly paleotropical genera with more than 100 members, just two exceed 500 species, and average size is below 300. The single sizable pantropical genus. *Bulbophyllum,* is largely Australasian. Wholly Australasian *Thrixspermum* is another that almost qualifies for Table 1.2. All taxa considered, epiphyte success in New versus Old World forests is based not so much on the number of times that the epiphytic habit has evolved as on the number of sizable radiations that followed these events.

Topography

At least three additional abiotic phenomena beyond coincidence have contributed in various measure to the numerical superiority of American epiphytes: topography, ethology, and paleoclimate. Neotropical topography has provided ideal conditions for epiphyte radiation. All but the most stress-tolerant plants in tree crowns are unusually sensitive to climate because they lack access to the more stable resource pools - particularly moisture - of many terrestrial soils. Epiphytic populations growing along roads through montane regions bear witness to this fact; species quickly appear and disappear when patterns of temperature, and especially of humidity, change abruptly. Experienced collectors know that clouds channeled through valleys promote greatest epiphyte development along narrow, wet, ridge lines. Adjacent slopes and nearby hill crests tend to support much sparser canopy floras. Founder events must abound where evaporative demand, the most decisive ecoclimatic determinant of epiphyte presence, varies sharply over short distances. Numerous Andean orchid genera show evidence of speciation aided by this mechanism; the *Stanhopea jenishiana* complex in the Cauca valley of Colombia is a well-documented example. Two other distinct but ecologically comparable species (*S. embreei* and *S. frymirei*) are clearly derived from *S. jenishiana;* they occupy narrow ranges in central Ecuador (Gentry and Dodson 1987b) and are almost certainly products of long-range dispersal across territory featuring hostile growing conditions. Epiphytic American Melastomataceae, most of which are midmontane, generally exhibit similar restricted distribution (Renner 1986), very likely reflecting similar migrations. Predominantly midaltitude, epiphytic *Anthurium* also includes numerous confined populations; for instance, 85 of 150 Panamanian species are endemic to that small mountainous nation (Croat 1986).

Moist montane habitats occur in New and Old World Tropics, but less widely in the latter. The Andes is by far the most extensive mountain range at low latitudes. In effect, relatively recent orogeny has created an immense land archipelago extending from Mexico to temperate South America that has favored epiphyte speciation and dense packing of resulting taxa. The Andes have not been uniformly conducive to radiation among canopy flora, however. Epiphytes in most neotropical angiosperm families with a sizable canopy-adapted membership are clustered in the younger northern region; only two (Onagraceae and Solanaceae), with less than 20 epiphytic species between them, exhibit a southerly bias. Climate more than geologic history may provide an explanation. Southern and central Andean areas achieved their current height (Zeil 1979) in the mid-Tertiary, but perhaps too far south to create the most favorable habitat for epiphytes. Most of the North Andean uplift has taken place during the past five million years, still enough time for orchids and several other lineages to generate the largest of all canopy-adapted floras, an exuberant speciation which continues to this day. Gentry and Dodson (1987b) cite colliding offshore Pacific cold and warm currents at low latitudes as a major cause of the microclimatic variety that has helped foster the unparalleled development of Andean epiphytism. The resulting "much finer niche partitioning" supposedly favors high alpha diversity (densely packed species), whereas dissected topography promotes evolution of ecologically similar (in terms of growing conditions), allopatric (at least initially) species.

Ethology

Neotropical fauna have contributed much to cladogenesis in numerous plant taxa, including several with a strong bias toward epiphytism. Bromeliads commonly attract hummingbirds, as do many co-occurring gesneriads, ericads, and mistletoes. Nowhere else are avian pollinators as numerous as in northern Andean forests. Euglossine bees, an exclusively neotropical group whose foraging behavior has become almost legendary, maintain sexual isolation among many American Orchidaceae (Dodson et al. 1969; Dressler 1981; Williams and Whitten 1983). Emergence of those two derived, isolated *Stanhopea* populations mentioned above involved a switch from pollination by *Euglossa* species to *Eulaema bomboides* (Gentry and Dodson 1987b). All three conspecifics elaborate methyl cinnamate accompanied by distinct sets of scent-modifying compounds. An isolated *Stanhopea jenishiana* population in southern Ecuador beyond the range of *Eulaema bomboides* remains little changed, including its utilization of *Euglossa* species.

The ethological explanation may be less applicable elsewhere, however. Pleurothallidinae (about 3800 species) is perhaps the best neotropical candidate for a taxon that belies the widely held view that specialized pollinators are the most important reason why Orchidaceae is so large and, by extension, why epiphytes are so numerous. Fetid rather than pleasant odors characterize many species' blossoms, and bizarre shapes are common. Flowers are often too small to attract strong fliers. Moreover, except for some pseudocopulating tachinids attracted to neotropical *Trichoceros,* and perhaps *Telipogon* and *Stellilabium* as well (Dressler 1981), there is no evidence that dipterans suspected of servicing pleurothallids (e.g., members of the large, diverse fly genus *Bradesia*) are as constant, and therefore as effective, as male euglossines in producing either isolation among co-occurring populations or fruit set by widely dispersed conspecifics. Quite likely, beetle-pollinated *Anthurium* and *Peperomia* have been less affected by ethological factors than some other groups and instead owe their large size to aspects of substratum and topography. Definitive judgments on all three of these relatively obscure angiospermous taxa require fuller knowledge of their reproductive biology, but one case is now unequivocal. Pollinators have played no role in pteridophyte evolution, yet the rich Costa Rican fern flora is approximately 70% epiphytic (Wagner and Gómez 1983), whereas less than a third of pteridophytes overall are canopy-adapted.

Refuge theory

Efforts to explain numerical disparities between New and Old World forest floras have taken note of Pleistocene climatology and humid refuges (Haffer 1969, 1978; Richards 1973; Hammen 1974; Simpson and Haffer 1978). Islands of residual tropical forest created by aridification during glacial advances at higher latitudes were supposedly most extensive in the New World. Thus, their purported twofold effect on biotic diversity was especially marked in the neotropics where mesic plant taxa were most often (1) conserved and (2) encouraged to undergo allopatric speciation. African floras, and perhaps to an even greater extent those of some tropical lands to the east, incurred proportionally greater extinction during recent global coolings. Today, for instance, 80% of Australia's epiphytes remain confined to pockets of humid forest occupying a small fraction of the continent's total land area. A canopy-adapted flora of such modest size probably also reflects recent rifting from the south. Stocking of late from more speciose, originally tropical, New Guinea and smaller islands to the north is suggested by low endemism – just eight monotypic genera. Only *Sarcochilus* has speciated extensively, having produced 21 Australian species (Wallace 1981).

Refuges may be overplayed in analyses of neotropical diversity. Gentry's (1982b) appraisal of South and Central American floristics, combined with palynological data, indicates that less plant speciation can be attributed to shifting paleoclimate than recent theories propose. Humid-refuge theory is especially inadequate to explain the current phytogeographic and systematic status of some neotropical synusiae, including epiphytes. Amazonian-centered taxa, a subset of the tropical American flora Gentry identifies as composed predominantly of canopy trees and lianas, show the most convincing evidence of refuge influence among New World Gondwanan plant groups (Table 8.1). Closely related populations tend to be geographically isolated, and genera are relatively small. Humid lowlands are common habitats. Andean-centered taxa – the other Gondwanan group – exhibit a different profile. Genera are larger and frequently contain sympatric species, as noted earlier. Shrub, palmetto, and epiphyte habits characterize much of its membership, and ranges are concentrated in premontane and low montane humid forests where refugia were less developed.

Lineages originating in Laurasia have made only minor contributions to plant diversity in South America except at relatively high elevations, and have shaped local epiphytic flora even less. Almost half of the Laurasian species are wind-pollinated or otherwise poorly suited for sympatric speciation (Gentry 1982b) or evolution of epiphytism. Also contributing to the limited distribution and size of this group is its recent introduction to the deep neotropics. Most Laurasian taxa were probably denied southward passage before the Isthmus of Panama closed in the Pliocene, producing the first continuous post-Cretaceous land bridge between Mesoamerica and South America (Keigwin 1978). Whatever the reason, Amazonian- and Andean-centered groups together still constitute more than 78% of all tropical South American species (Gentry 1982b).

Epiphytes, compared to most ground-rooted flora, challenge the humid refuge model in another respect. The fact that members of this synusia are quite dispersible and that certain specialized forms (admittedly a minority of the species) tolerate and even require periodic drought suggests that at least the more desiccation-resistant taxa were less subject than trees to isolation by changes in rainfall patterns. Quite a few of the most stress-tolerant populations also anchor on rocks. Dry rather than moist refuges possibly had the most telling effect on atmospheric bromeliads and comparably moisture-sensitive forms. Several predominantly xeric *Tillandsia* subgenera with separate South American centers of diversity may owe their beginnings to arid refugia (Gilmartin and Brown 1986). Granville (1982) proposed that localized dry zones provided sanctuary for certain epiphytes in French Guyana during the wettest parts of glacial interphases. *Aechmea setigera* (Bromeliaceae), a dry-growing *Epidendrum nocturnum* segregate (Orchidaceae), and *Topobea parasitica* (Melastomataceae) are cited as probable beneficiaries of arid-zone insulation. Nevertheless, the continuity of Pleistocene forest must account in part for the unparalleled botanical variety of epiphytes and terrestrials alike in regions like the currently pluvial Colombian Chocó.

A more pervasive impetus for radiation

Certain aspects of substratum point to an area-nonspecific reason why fully 10% of all vascular species are epiphytic. Mechanisms center on the nature and distribution of rooting media and operate on coarse and fine scales. Epiphyte habitat is discontinuous everywhere, even in primary forest, but not to the same degree. It is especially so in low to midmontane regions where composition of the phorophyte community, and therefore of the epiphytes' substratum, is fragmented, a condition conducive to speciation (Templeton 1981). Where climate, elevation, and type of vegetation are more monotonous (e.g., Amazonia), epiphyte species tend to be wide-ranging but not very numerous.

Conditions beyond those favoring proliferation of taxa are necessary to build communities, especially if the component populations have similar growth requirements, as do many epiphytes. Studies of other communities indicate how habitat at the finer scale promotes co-occurence; they may also provide insight into how combinations of certain epiphytes – for instance, those miniature orchids native to citrus or guava twigs (Catling et al. 1986; Chase 1987) – exist. Instability is decisive in each case (Connell 1978); if appropriately distributed in time and space, disturbance allows populations with the requisite reproductive characteristics to share anchorages. Aggregations comprised of organisms as disparate as barnacles, coral reef fish, and moist tropical forest trees where epiphyte loads may help promote gaps (Strong 1977) all maintain associations in part through lottery-like rotations that prevent competitive exclusion. Dense packing is possible because none of the resident populations is sufficiently mobile or fecund to prevent coexistence. Communities persist because vacant sites (regenerative niches) are usually filled by the first propagule to arrive, an event that does not favor one parent species over another.

A constant rain of dislodged epiphytes attached to broken twigs, bark fragments, and (less often) whole fallen trees documents a steady progression of small and larger scale disturbances even in a stable forest. Overgrowth of one plant by another is exceptional; instability seems to be too great and living space too fragmented to allow even the most aggressive epiphyte to match the expansion over large areas achieved by many a terrestrial via seeds or ramets. Thus epiphyte synusiae are, perhaps more than some others, shaped by disturbance and patchiness rather than by competition. Statistical evidence for randomness as a prime determinant of epiphyte community structure is admittedly scanty. Hazen (1966) failed to demonstrate spatial patterns among canopy associates forming dense tank bromeliad colonies in wet forest near Turrialba, Costa Rica, but his analysis could have been more rigorous. Tree crowns on drier sites feature greater epiphyte dispersion because low vigor enforced by stress magnifies the inhibitory effect of disturbance and patchiness on population growth (Benzing 1978a,b).

The disproportionate proliferation of epiphytes compared to plants of several other habits makes yet another case for the promotive effects of tree crowns as sites for cladogenesis. Radiation of neotropical epiphytes on humid sites has been exceptional compared to that of co-occurring trees and lianas (Fig. 8.2). In a wet forest at Río Palenque, Ecuador (elev. 150–220 m), about 35% of all vascular species are canopy-adapted according to one common definition. Other statistics are even more revealing. Epiphyte flora is quite homogeneous taxonomically, a fact lending credence to Gentry's (1982b) impression that evolution continues to be exceptionally active in this synusia. At Río Palenque, there are 21 flowering-plant genera containing seven or more species. Eleven of the 21 are fully or largely epiphytic, and only one contains trees. On Barro Colorado Island, 12 genera include at least 10 species; five genera are mostly epiphytic; none are arborescent. Congeners can be numerous in drier sites as well. Seven species of *Tillandsia* grow on stunted cypress in southwestern Florida along with *Catopsis berteroniana, Psilotum nudum,* a few ferns, and several orchids (Table 1.1). It is not at all unusual to encounter a single tree harboring three or more congeneric orchids in New and Old World forests (e.g., Sanford 1969; Wallace 1981).

Gentry's (1982b) observation that the New World vascular epiphyte synusia continues to expand while its woody associates appear closer to saturation says as much about the habitability of tree crowns as it does about neotropical epiphyte diversity. Earth soil is not equivalent to bark on many counts, and should not necessarily accommodate similar botanical variety. Within a forest, total bark surface greatly exceeds that of ground area and can be more densely packed with plants. Rooting media in canopies are also diverse, although whether more or less so than earth soil is unclear. In effect, tree crowns may be especially permissive habitats that foster dense species packing for vascular and nonvascular plants alike. Pócs (1982) recorded approximately 100, 60, and 50 canopy-adapted bryophytes in single localities in Vietnam, East Africa, and Cuba, respectively, suggesting that small body size further increases density in canopy communities; individual leaves supported 10–20 different epiphylls. Many vascular epiphytes, particularly orchids, are small, sometimes weighing only a few grams at flowering. Dodson and Gentry (pers. comm.) determined that 11 of the 17 commonest

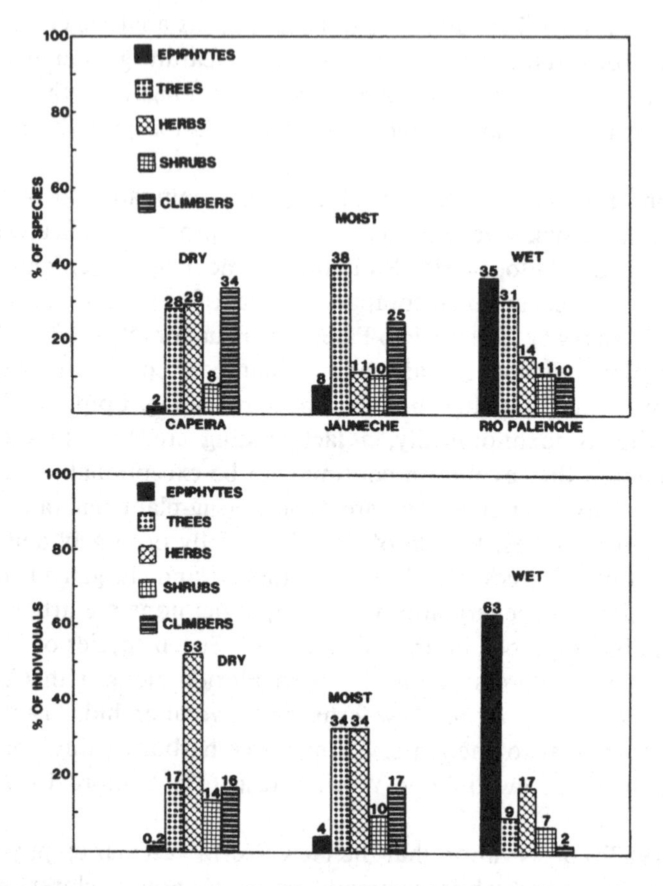

Figure 8.2. Occurrence of species (top) and individuals (bottom) representing different habits in dry, moist, and wet Ecuadoran forest. (After Gentry and Dodson 1987b.)

vascular species in the forest at Río Palenque were canopy-adapted. Among the remainder, just one was a dominant tree; another was a large shrub; and the rest were terrestrial herbs. Small-scale disturbance that continuously affects crown surfaces probably provides an especially effective vehicle for rotation of epiphytes. That is to say, the carrying capacity of forest vegetation for epiphytes probably exceeds that of subjacent earth for tree species (although not usually for all terrestrial taxa combined, as noted below). Very likely, much of the epiphyte synusia is not yet saturated with either biomass or taxa, even in tropical America; perhaps insufficient stock limits colonization everywhere, but especially in the paleotropics. Too few lineages have evolved adequate stress tolerance for a broader epiphyte presence in seasonal woodlands.

Table 8.3. **Climatic data from six Australian forest types where epiphyte diversity, abundance, and life form were studied**

Site	Temperature (°C) Summer Maximum	Temperature (°C) Winter Minimum	Approximate annual rainfall (mm)	Type of climate	Type of canopy
Dorrigo	31–35	1–1.5	2000	Moist subtropical/ warm temperate; mists common	Relatively dense and even; 18–30 m aboveground
Shelly Beach	30.5–39	1–3	1500	Moist subtropical; mists uncommon	Cover 50–80%, strong maritime influence; salt spray probably significant
Long Point	32–39	0–3.5	1100	Moderately moist subtropical; (dry rain forest); night mists relatively common except in summer	Canopy uneven and low (3–10 m); with numerous light breaks
Humber Hill	26–33	−1 to −1.5	1600	Moist, warm temperate; mists common	Cover 60–85% with large light breaks
Wrights Lookout	25–30	0–12.5	1800	Moist, cool temperate; mists common	Cool, temperate rain forest; 50–80% cover, 15- to 30-m height
Leo Creek	<35	>12	2000	Monsoon: wet season Dec.–Apr. but significant rain May–Nov. with much cloud cover	Monsoon rain forest (semievergreen); dense, continuous at 10- to 15-m height

Source: After Wallace 1981.

Regional and smaller-scale determinants of distribution

Climate, elevation, and latitude

Epiphytes occur along a broad moisture continuum ranging from pluvial woodlands to cactus savannas and microphyllous forests. Biomass, diversity (Fig. 1.21), and habit all vary along this axis, with each parameter usually increasing as seasonality and hence aridity diminish. Wallace (1981) recorded decreasing numbers of epiphytic species across increasingly dry forests (Tables 8.3, 8.4) in eastern Australia. Sanford (1974) documented a similar sequence in his survey of epiphytic orchids along a rainfall gradient in Nigeria. Length of dry season (about one to four months) and WVPD

Table 8.4. **Extent of colonization of phorophytes by epiphytes, and the diversity and abundance of the latter in sample plots at six locations in Australia**

Site	Number of phorophyte species	Percentage of phorophytes colonized	Number of epiphytic species	Number of individual epiphytes per hectare
Dorrigo	29	21.5	28	2744
Shelly Beach	17	14.7	19	296
Long Point	16	80.0	15	8217
Humber Hill	11	61.8	20	1776
Wrights Lookout	8	62.5	13	1136
Leo Creek	30	45.6	43	3304

Source: After Wallace 1981.

were more important determinants of floristic richness than was total rainfall. A second set of three Australian sites chosen by Wallace to illustrate low-temperature effects, including occasional frost at the extreme, yielded 28, 20, and 13 epiphytic species. Gentry and Dodson (1987b) thoroughly demonstrated the effects of climate on epiphyte success (Figs. 8.3, 8.4). Florulas were developed at three sites supporting dry, moist, and wet forests; the number of resident species increased by a factor of 17.5 across the three communities. Contributing families numbered just two at Capeira; 18 were recorded at Río Palenque. Epiphytic species constituted 2%, 8%, and 35% of the total indigenous forest flora, respectively. What promises to be the highest tally yet – perhaps 50% – is being recorded for a still wetter Ecuadoran montane forest where the list of orchids alone has already reached 322 species. Figures for lowland forests in Santa Rosa, Costa Rica and Barro Colorado, Panama fall within the sequence as predicted (Fig. 8.4). Trees and shrubs contributed more consistently to the Ecuadoran communities, whereas numbers of herb and climber species fluctuated moderately, with lower diversity and abundance on wetter sites (Fig. 8.2).

Kelly et al. (1988) report figures from central and eastern Jamaica that parallel those of Gentry and Dodson, Sanford, Wallace, and others, further

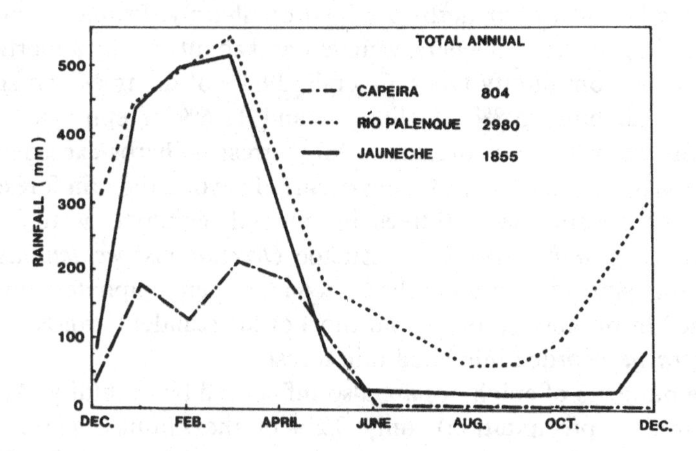

Figure 8.3. Annual distribution of rainfall at three sample sites in Ecuador. (After Gentry and Dodson 1987b.)

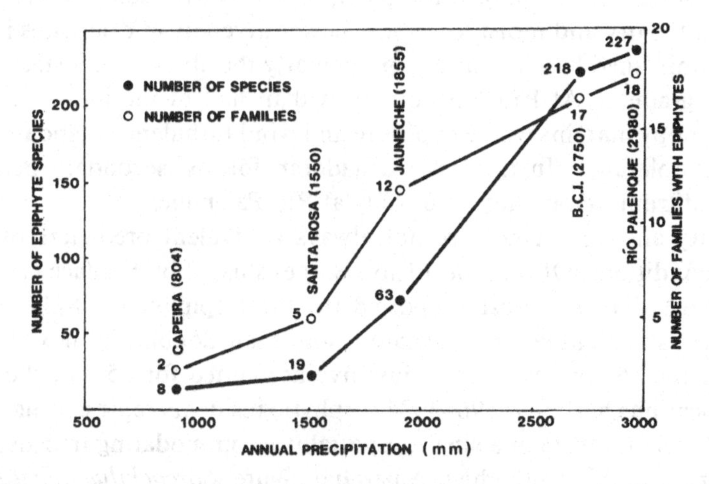

Figure 8.4. Relation of annual rainfall to occurrence of epiphyte taxa at five locations in tropical America. (After Gentry and Dodson 1987b.)

underscoring differences in ecological amplitude between epiphytes and co-occurring life forms. Three sites located over calcareous bedrock and chosen to illustrate the effects of increasingly seasonal climate (ca. 3800, 1600, 1000 mm/yr; corresponding dry season: 0, 4, 9 months) were sampled. Numbers of compact epiphyte species and their representation in the local flora were 73 (26.1%), 37 (15.0%), and 7 (5.4%). Tree, climber, and shrub species constituted fairly consistent or increasing proportions of the total flora as aridity increased, while forest floor herbs dropped off sharply. Primary hemiepiphytes, including several stranglers, were few and about equally apportioned among the three community types. Overall, 29.3% of the trees and shrubs taller than 2 m, plus 25.8% of climbers and 13.6% of epiphytes, were observed on two or three of the sites. Only terrestrial herbs exceeded epiphytes in insularity; just 10.7% of them occurred beyond the rain forest site. Present in all three sites were 10 trees, but the only epiphytes were facultative *Anthurium grandifolium,* the mistletoe *Oryctanthus occidentalis,* the strangler *Ficus pertusa,* and a climber. Secondary hemiepiphytes were not distinguished from other climbers, but most of the scandent aroids and the only *Marcgravia* recorded inhabited rain forest.

Absolute numbers of epiphytes are also influenced by humidity. At Capeira (804 mm/yr precipitation), only 0.2% of the sampled plants were anchored in the canopy; at Jauneche (1855 mm/yr), 4%; and at Río Palenque (2980 mm/yr), 63%. At this last site, no month receives less than about 50 mm of rainfall (Fig. 8.3). Little precipitation is available for five to six months at the dry and moist locations. Lower diversity at drier sites in part reflects diminished habitat variety, particularly the absence of a steep light/humidity gradient. At Río Palenque, less than half of the local epiphytes occupy canopy margins where exposure and wind turbulence mandate superior stress tolerance. In the three Ecuadoran forests, secondary hemiepiphytes and trunk sciophytes occur only at Río Palenque.

Diversity and abundance are not always equivalent predicters of local growing conditions. Of six plots (Table 8.4) evaluated by Wallace (1981) in eastern Australia, the driest supported the most epiphytes – 8217 mostly orchidaceous individuals per hectare, more than double the next highest value. Of the 15 species present, just five accounted for 85% of the total. Wetter locations harbored 296–3304 epiphytic residents representing 13–43 species. Large numbers of a single, especially accommodating tree also contributed measurably to the high Australian figure. *Sarcochilus falcatus* and *S. hillii* on *Backhausia sciadophora* accounted for 73% of the total epiphyte flora at the Long Point site. Basal area of all phorophytes was relatively low here and canopy exposure correspondingly high. A similar pattern prevailed in certain parts of the relatively arid Big Cypress Forest and some coastal stands of dwarf *Quercus virginiana* (Fig. 7.3) in South Florida. Trees are too extensively festooned with bromeliads *(Tillandsia recurvata, T. usneoides, T. balbisiana, T. paucifolia, T. utriculata,* and *T. fasciculata)* to allow accurate counts. Dense epiphyte populations in these two rather homogeneous Florida forest communities are fostered by several factors, including frequent morning fog in winter, the stability of oak and cypress bark, the durability of small twigs, and the stress-induced transparency of hosting crowns.

Epiphyte habit is quite sensitive to climate. Shrubby forms, like vining hemiepiphytes, grow in warm humid canopies; herbs are much more versatile except for ferns, which constitute a disproportionate share of the most drought-sensitive taxa. Hemiepiphytic climbers seem to replace lianas in some extremely wet Andean sites (Gentry and Dodson 1987b). Climbers and trash-basket epiphytes were most abundant in the wettest and warmest of the same six sites examined by Wallace (1981), whose study focused particularly on growth forms (Table 8.5). Note that only 14 species were present at the driest site, and none were debris collectors or hemiepiphytes; most were compact CAM orchids. The two unusually prolific tangle-root types were encouraged by frequent mists and the moderating effects of nearby ocean waters.

Moisture is not always conducive to epiphytism; in fact, heavy rain in humid lowlands seems to discourage it. Sugden and Robins (1979) concluded that sites subjected to frequent torrential storms offer poor colonizing opportunity for these plants, dislodging propagules before anchorage can take place. The presence of fewer fern species than expected without similar reduction in those of Araceae, Ericaceae, Melastomataceae, and arillate Guttiferae prompted Gentry and Dodson (1987b) to cite the importance of heavy seeds for establishment in some neotropical, pluvial, forest canopies. Epiphytism is best developed at midelevation, where moisture supply is usually continuous but not necessarily plentiful. Trees support the heaviest burdens in cool cloud forests. Diversity shifts with elevation, first increasing to a maximum with decreasing evaporative demand and then dropping off as temperature minima fall below about 18^{0}C. In Peru's Huascaran National Park (3500–5000 m), only seven species or 1% of the flora is epiphytic (Gentry and Dodson 1987b).

Another survey by these authors of three moist Ecuadoran sites at roughly 200, 600, and 1000 m revealed greatest diversity at the intermediate elevation. Here 337 species in 20 families representing 35% of the total vascular flora were present (Table 8.6). Contrary to the trend along a moisture gradient, proportional contributions from families changed little with altitude. Orchidaceae, Araceae, and ferns, in that order, were best represented at every location. Piperaceae *(Peperomia)* and Bromeliaceae were next, with the number of bromeliads remaining remarkably constant. Only Moraceae among important families evidenced a strong altitudinal bias, reflecting the general absence of stranglers in any but warm and at least reasonably wet

Table 8.5. **Diversity and abundance of epiphytic growth forms in sample plots within six Austrialian forests**

Site	Number of epiphytic species	Number of epiphytic individuals	Epiphytic growth forms (# species/total # of individuals)							
			Compact habit, restricted anchorage	Long creeping and mat formers	Tangle-root	Nest formers	Nest invaders	Hemi-epiphytes	Semi-epiphytic climbers	Accidental epiphytes
Dorrigo	31	458	12/177	3/44	1/25	3/64	4/34	1/2	4/107	3/5
Shelly Beach	16	41	5/16	1/1	—	2/9	3/5	2/4	—	3/6
Long Point	14	928	7/702	1/60	2/152	1/6	—	1/2	—	2/6
Humber Hill	18	314	7/140	2/40	1/34	2/56	1/2	—	3/38	2/4
Wrights Lookout	8	137	2/68	2/48	1/3	—	—	—	2/16	1/2
Leo Creek	41	528	19/269	6/101	—	2/50	10/90	3/17	—	1/1

Source: After Wallace 1981.

Table 8.6. **Number of species in families of epiphytic floras at three different altitudes in Ecuadoran humid forests**

Family	Río Palenque (200 m)	Centinela (600 m)	Tenafuerste (1000 m)
Orchidaceae	81	133	68
Araceae	35	52	26
Ferns and allies	26	38	28
Piperaceae	19	19	11
Bromeliaceae	18	23	18
Moraceae	13	10	—
Gesneriaceae	12	16	8
Cyclanthaceae	8	5	3
Marcgraviaceae	5	3	2
Guttiferae	4	9	3
Cactaceae	3	2	1
Ericaceae	3	9	9
Araliaceae	2	4	—
Bignoniaceae	2	2	—
Melastomataceae	1	4	2
Polemoniaceae	1	1	—
Solanaceae	1	2	1
Urticaceae	1	1	1
Acanthaceae	—	1	—
Rubiaceae	—	1	—
Total	227	337	181
Percentage of total regional flora	22	35	31

Source: Gentry and Dodson 1987b.

communities. The species list for the 1000-m site and a second pluvial location at that altitude near Mera remain incomplete but may eventually include higher percentages of epiphytes than any recorded so far. Interestingly, families that extended into the driest lowland sites, particularly Bromeliaceae and Orchidaceae, were also among the last to drop out along the elevational gradient. In the northern Andes, epiphyte diversity peaks at 1000–2000 m and abundance at about 2000–2500 m; both peak at somewhat lower elevations in Central America (Gentry and Dodson 1987b). Working in West Africa, Schnell (1952) discovered that 31–33% of the vascular flora was epiphytic in sampled montane forests but only 4–16% in subjacent lowlands. Utilization of trees (≥10 cm dbh) comprising evergreen forest up the slope of an ultrabasic mountain in Sabah, Malaysia, further illustrated the relationship between elevation and epiphyte success (Proctor et al. 1988). Just 5.5% of all boles supported arboreal flora at 280 m; 75.9% were colonized at 610 m. One or more epiphytes resided on every available trunk at 780 m near the often cloud-enshrouded summit. More data are needed to determine whether the epiphytes' seeming deviation from decreasing diversity with increasing elevation (seen in most other tropical flora) is real.

The location of so many epiphytic taxa in montane habitats is well illustrated by subfamily Tillandsioideae of Bromeliaceae (Fig. 8.5, 8.6) and neotropical Melastomataceae. Six South American tillandsioid genera range over much of the highlands rimming Amazonia, with greatest diversity in the Andean foothills (Smith and Downs 1977). Only *Tillandsia* extends through the vast area between. Within this single genus of almost 450 species, the pattern is repeated; all but one of its six South American subgenera are essentially upland. *Tillandsia* and *Vriesea* exhibit similar geographic distribution in Mexico and Central America, respectively. Montane sites are heavily colonized, but large subgenus *Tillandsia* is once again concentrated primarily in what are often arid lowlands (e.g., *T. concolor,* Fig. 3.1C; *T. paucifolia,* Fig. 1.11; *T. ionantha*). Mexican microphyllous forest sometimes supports these and related bromeliads at very high density. Epiphytic American melastomes are most abundant at midelevations, particularly in the superhumid Colombian Chocó region and in Andean forests (Renner 1986). Of 227 species, 28% occur only below 500 m, 53% between 500 and 3000 m, and 19% from sea level to 2000 m. Just 5% of the 235 Central Amazonian melastome species are epiphytes.

The relationship between latitude and epiphyte occurrence is not consistent on both sides of the equator. Compared to north temperate forests, those in the more frost-free south feature a sizable canopy-based flora complete with some endemic genera. Just two vascular epiphytes inhabit Caverns State Park in Florida (30°50′N), whereas 15 such species are recorded for Puyehue National Park at 41°S in Chile (Gentry and Dodson 1987b). Northernmost epiphytes in the New World are mostly outlying species representing large tropical genera. Ecologically equivalent taxa restricted to temperate South America include *Synammia* (a fern), *Pfeiffera* (Cactaceae), and *Luzuriaga* (Liliaceae), each with a single species. The monotypic fern *Anarthropteris, Collospermum* (Liliaceae), and the only epiphytes in Cunoniaceae occur on South Pacific islands. Chile and New Zealand constitute the entire range of largely canopy-adapted Griseliniaceae. An epiphytic grass *(Microlaena)* is another feature of New Zealand's unique epiphyte flora. Similarly exceptional in the north are epiphytic Himalayan members of *Cosmopolliton, Ilex, Euonymous, Thalictrum,* and some other

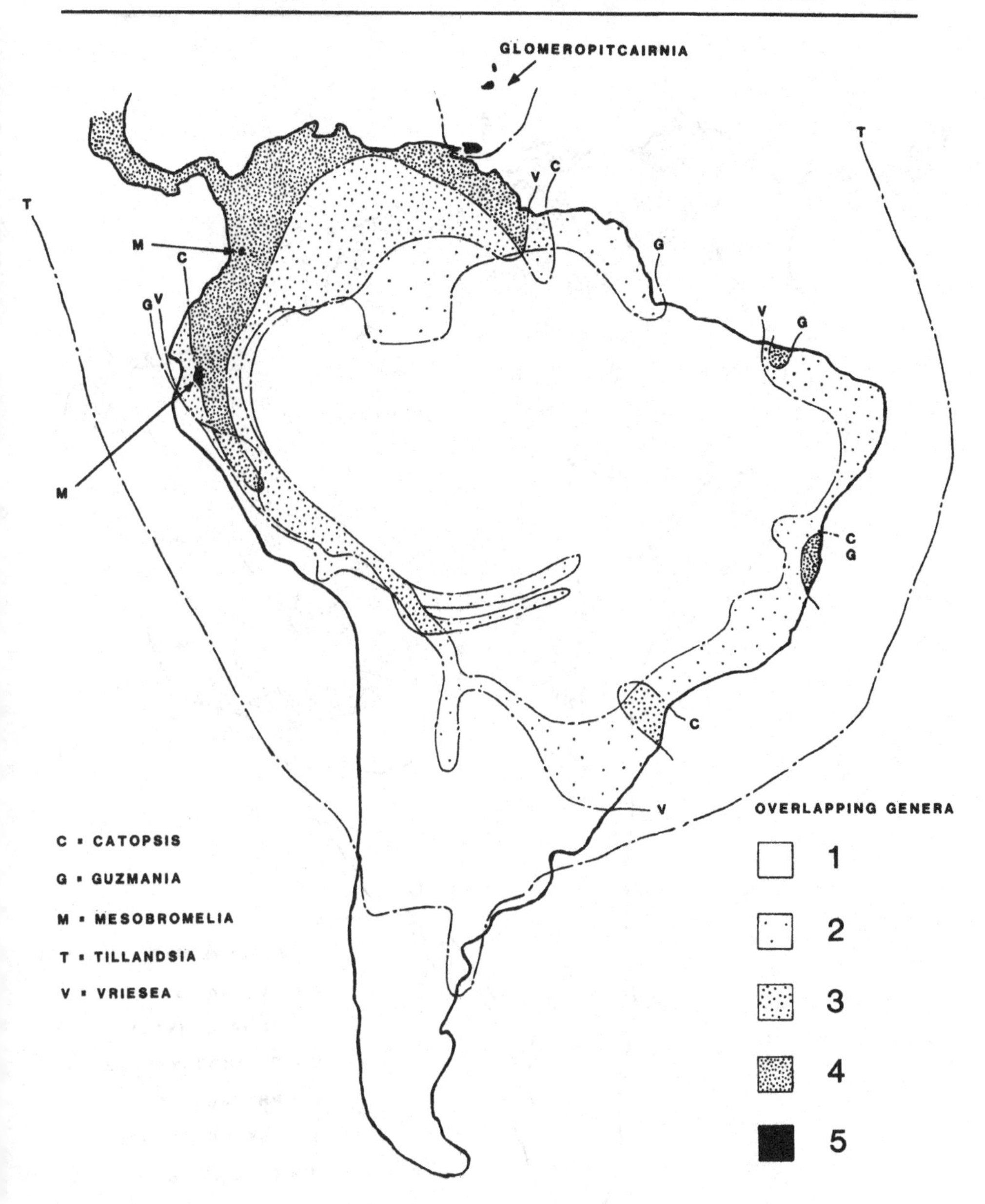

Figure 8.5. Range of six genera in subfamily Tillandsioideae (Bromeliaceae) in South America. (After Smith 1934.)

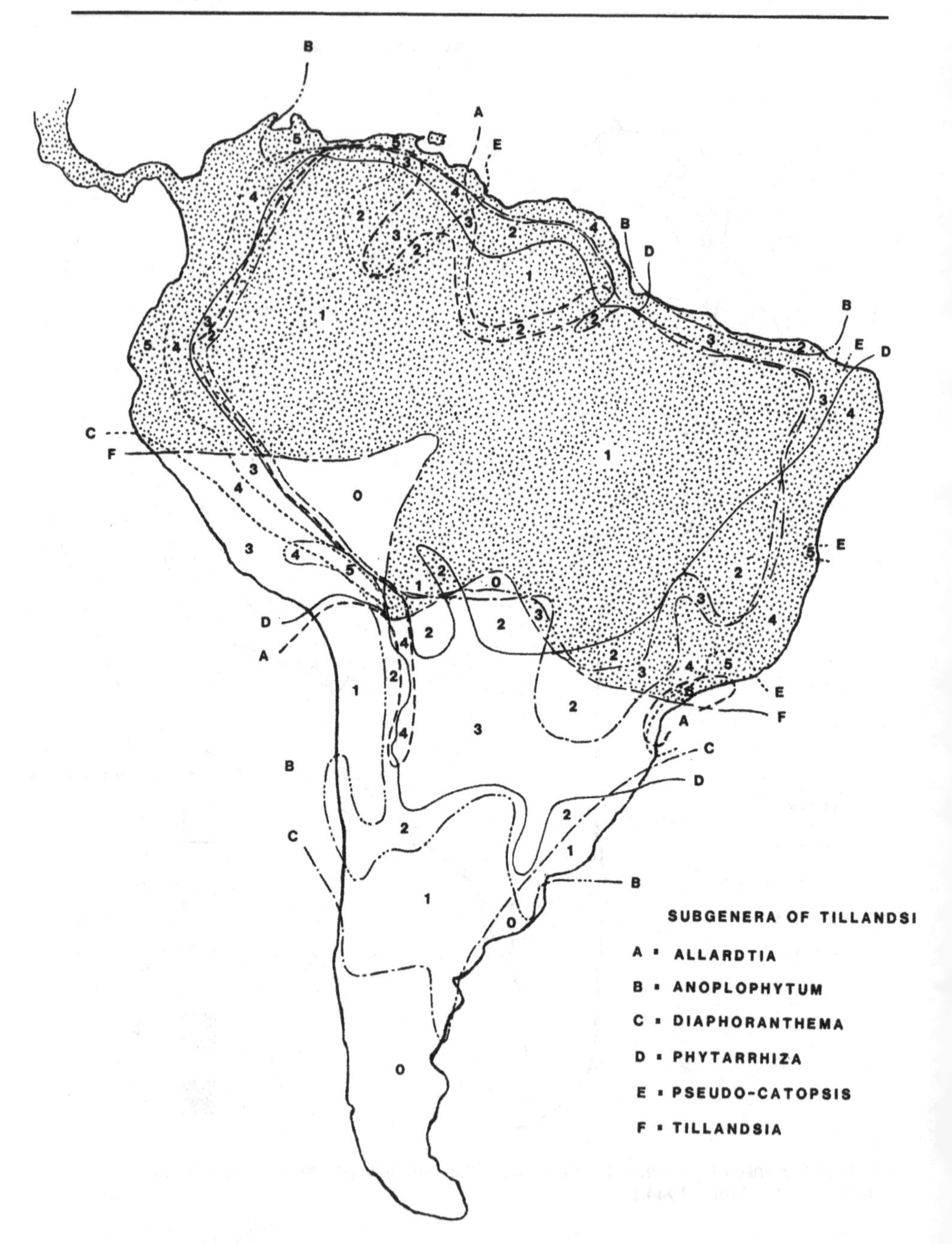

Figure 8.6. Range of subgenera of *Tillandsia* in South America. Numbers indicate overlap of subgenera. (After Smith 1934.)

equally unlikely candidates for life in a forest canopy (Gentry and Dodson 1987b).

Ground soil conditions

Just one effort has been made to document the effect of earth-soil type or fertility on epiphyte development (Gentry and Emmons 1987), but there have been additional comments on the subject (Janzen 1974; Benzing and Seemann 1978; Gentry and Dodson 1987b). In fact, the situation probably varies, depending on the type of epiphyte and where it anchors. Because so many species are heliophilic by nature, opaque canopies promoted by moist, fertile substrata should limit colony growth in many instances. On the other hand, heavily leached or acidic soil, if it leads to sparser phorophyte foliage, should allow bark to be more fully exposed to light and precipitation. Amazonian forests over extremely impoverished white sands can support substantial epiphyte communities (Anderson 1981), but these may be especially stress-tolerant forms or species with access to rich, unconventional nutrient pools (e.g., ant nests). Scattered short trees in parts of the very nutrient-depleted Gran Sabana of eastern Venezuela accommodate dense populations of heliophilic *Catopsis berteroniana* (Fig. 4.11) and *Tillandsia flexuosa* (Fig. 4.25B). Both of these bromeliads, however, may supplement precipitation and litter-based input with nutrients provided by animals. *Catopsis berteroniana* is supposedly carnivorous, whereas *Tillandsia flexuosa* is sometimes ant-inhabited although generally not considered myrmecotrophic. For the same reason, rubiaceous myrmecophytes may be able to build up sizable populations over exceptionally sterile substrata supporting low-diversity "Kerangas" in northern Borneo.

Janzen (1977) has offered what seems to be an opposing, as yet untested, view to the shade-limiting argument: that epiphytes without access to supplemental resources are, in effect, "starved off" dipterocarps and other phorophytes throughout much of the paleotropics. He suggested that neotropical supports, because they most often grow on richer, younger volcanic soils, produce much more harvestable food. The resulting better-developed avifauna sustained by this largesse create a "rain of nutrients" that presumably promotes luxuriant epiphytic growth. Frugivores are certainly abundant enough to provide adequate seed transport in paleotropical forests (Madison 1977). Except for Gesneriaceae, whose fleshy-fruited neotropical epiphytes have mostly wind-dispersed Old World counterparts, similarly differentiated canopy-adapted flora exhibit no consistent area-specific use of wind or animals as dispersers. At this point there is no evidence that concentrations of N and P in tree foliage differ between neotropical and Old World forests or that canopy leachates or other substrates are more conducive to sound epiphyte nutrition in the Americas. Nor is there reason to presume that other media in tree crowns should generally be more nutrient-laden in one part of the globe than another.

Mechanisms described earlier in this chapter to account for neotropical epiphyte diversity seem more persuasive than Janzen's relatively narrow trophic explanation, although nutrients in ground substrata have some effect as demonstrated by Gentry and Emmons (1987). Earth-soil fertility and occurrence of shade-tolerant understory epiphytes (particularly Araceae) were definitely interrelated throughout six neotropical sites; more than twice as many species and individuals were encountered in forests over lateritic and alluvial substrata compared to communities based on impoverished white sands. Lianas and terrestrial herbs – but not trees – followed suit. The fact that all sampled locations were wet enough to support vertically stratified epiphyte floras probably explains why rainfall pattern had no effect on the development of these well-insulated epiphytes. Mechanisms could vary in upper-canopy regions, however, depending on supplies of light, moisture, and key nutritive ions. The extent of colonization at mid- and upper-canopy levels in the same woodlands inspected by Gentry and Emmons would be instructive. Available data and extensive field experience have already convinced Gentry and Dodson (1987b) that epiphyte diversity and abundance are highly sensitive to the land's fertility, perhaps more so than are trees and lianas.

Too many factors influence geographic occurrence to cite any single cause as predominant. Montane climate, the various factors mentioned above that fragment gene pools, and fertile substrata all combine to create the most species-rich epiphyte floras, although not necessarily the greatest standing crops. Additional second- and higher-order mechanisms come into play under special circumstances involving specific types of epiphytes and environments. Highest densities of Amazonian ant nest-garden species, for instance, occur in seasonally inundated forests and scattered through adjacent, better-drained communities as disturbance permits (Davidson 1988). Already cited as important to occurrence have been drying effected by the strong insolation purportedly required by ant colonies, the heliophilic nature of most ant plants, and the rewards presented to ants by phorophytes. Trees bearing the heaviest ant nest-garden loads, including members of *Gauzuma, Hasseltia, Inga, Pithecellobium, Sapium,* and *Senna,* do often produce extrafloral nectar and regularly maintain abundant phloem-feeding

Homoptera. Production of these and other expensive ant resources encourages nest building but also requires sites on moist, fertile land; here, the most heavily used supports plus their ants and the products of the seeds they plant do indeed tend to congregate.

Hydrology may affect ant–epiphyte occurrence through trophic and associated factors beyond those arising directly from the usual high fertility of alluvial substrata. If seasonal inundation reduces the number of terrestrial nesting sites for ants more than it diminishes their food resources, epiphytic myrmecophytes that can protect brood in plant cavities or secure carton with their roots could increase in proportion. Studies on demand for nest-garden space and specifications for acceptable domatia could be done using ersatz plant organs. Artificial chambers fitted with nylon mesh sleeves containing appropriate exchange resins can provide additional information on plant benefits, the full value of which cannot be determined from occupancy rates alone.

Finally, quirks of history cannot be ignored in attempts to explain epiphyte distribution in particular types of habitat. Dry forests would surely support richer canopy-based floras if *Tillandsia* extended beyond tropical America. Ant ranges may have been important in determining where certain myrmecotrophic conditions could develop (e.g., ant-fed ant-house epiphytism in tropical America and Australasia but not Africa). *Crematogaster* is the major epiphyte associate in peninsular Malaysia; *Iridomyrmex* engages in the most specialized Australasian relationships; *Azteca* fills similar niches in the neotropical forest canopy as does *Camponotus.* Of course, the cosmopolitan ranges of *Camponotus, Crematogaster,* and some other key groups militate against important historical coincidence unless unrecognized variations exist within these ant taxa.

Systematic occurrence

Epiphytism has probably evolved independently in every qualifying family of seed plants and most families of ferns. Moreover, separate events permitted canopy colonization of each continent by a number of pantropical angiosperm lineages (Gentry and Dodson 1987b). Most of the epiphytic gesneriads belong to subfamilies Cyrtandroideae (paleotropics) or Gesnerioideae (neotropics). Similarly, Rhododendroideae and Vaccinioideae account for much of Old and New World ericaceous epiphytes. Epiphytic members of Central American *Cynanchum* are only distantly related to the more numerous paleotropical canopy-dwelling asclepiads. So it seems that bias toward arboreal life is rather fundamental in some major clades. Epiphytism probably also arose at different locations in some wide-ranging genera (e.g., *Begonia, Gaultheria, Pilea, Solanum, Utricularia*). In others (e.g., *Clusia, Coprosma, Ficus, Griselinia, Luzuriaga,* and *Peperomia*), long-range dispersal occurred almost certainly after founders were able to grow in tree crowns. Except for the ferns, strangler figs, *Peperomia, Rhipsalis,* and a few orchid genera, the epiphyte floras of different continents were seemingly derived independently (Gentry and Dodson 1987b).

Most diverse on taxonomic grounds are the epiphytes that colonize suspended soils in humid forests. Virtually every family containing epiphytes includes at least one such humiphile; most contain no other type. Where aridity is pronounced, few higher taxa are represented, although these are sometimes species-rich (Benzing 1978a). Neotropical ant nest-gardens provide substrata for a modest myrmecochorous flora which often offers food to ants. Perhaps additional traits as yet unrecognized that might enable rooting in ant nest-gardens favor this mutualism in Araceae, Bromeliaceae (specifically subfamily Bromelioideae), Cactaceae, Gesneriaceae, Marcgraviaceae, Orchidaceae, and Piperaceae. Some Asclepiadaceae, Melastomataceae, and Rubiaceae engage in similar but less well-defined nest-garden symbioses in Australasia (Janzen 1974). Those myrmecotrophs that exhibit unequivocal modification for ant occupancy all belong to Asclepiadaceae, Bromeliaceae, Melastomataceae, Orchidaceae, Polypodiaceae, Rubiaceae, and Solanaceae (Huxley 1980). Rubiaceae seem to be most specialized for ant-fed ant-housing (Figs. 1.4, 4.24). At least 20 dicot families contain primary hemiepiphytes (Putz and Holbrook 1986). Stranglers – about 300 in all – come primarily from Moraceae (mostly *Ficus* subg. *Urostigma;* Fig. 1.8). Species reputedly capable of killing hosts also belong to *Schefflera* (Araliaceae), *Posoqueria* (Rubiaceae), and *Metrosideros* (Myrtaceae). Secondary hemiepiphytes have vining habits (Fig. 1.9), and indeed most of them belong to groups with a scandent tendency (e.g., Araceae, Cyclanthaceae, Marcgraviaceae). Bromeliaceae account for most of the tank formers (Fig. 1.2). Trash-basket species are scattered throughout several families (Figs. 1.5, 1.19). The PS epiphytes are almost exclusive to Bromeliaceae (numerous *Tillandsia,* a few *Vriesea*) and Orchidaceae (many subtribes).

Both parallelism and convergence produced much redundancy among the epiphytes. Parallelism is illustrated by CAM, the key feature of which (CO_2 fixation via β-carboxylation) is fundamental to intracellular pH regulation and osmotic balance. Embellishments required for CAM, including stepped-up PEPc activity, a tonoplast capable of high malic acid traffic, and hypervacuolate cells, have emerged in close to 35 families and more than once among some of the larger, heavily epiphytic, lineages (e.g., Bromeliaceae,

Fig. 2.4; Orchidaceae). In canopy- and soil-based vegetation alike, extended longevity of both the whole plant and its individual leaves, along with stress-compatible physiology, probably reflects similar selective pressures on common potential. Less widespread potential fostered evolution of impoundment and ant occupany (Fig. 4.24). In effect, a common genetic base underlies machinery governing resource use by epiphytes. But a greater variety of less homologous mechanisms are required to cope with the special problems of resource acquisition in tree crowns.

Predisposition and phylogenetic constraints

Compared to vertebrates, angiosperms have exhibited great adaptive plasticity. Quite commonly, species with distinct habits and substrata (e.g., earth soil, tree crowns) belong to the same genus (e.g., *Epidendrum, Senecio, Schefflera, Smilacina, Vaccinium*). Such broad radiations have been fostered by the higher plant's greater physiological plasticity aided by continuous turnover of both vegetative and reproductive organs. Selection may occur among ramets – perhaps even among tissues within individual organs – during the life of a single clone (Walbot and Cullis 1985). Homoeosis (the transference of traits from one type of organ to another during evolution – e.g., leaf to cladophyll) can be pronounced, possibly in part because only a small number of genes need be involved to alter structure, function, and ecological tolerance (Gottlieb 1984; but see Coyne and Lande 1985). Evolutionary tempo is reputedly rapid enough to produce species in as little as 10^4–10^5 years – or perhaps a few decades! – in epiphytic Orchidaceae (Gentry and Dodson 1987b).

Some angiosperm lineages are, however, quite conservative, as if for them entry into certain adaptive zones has restricted access to others. Frequently, invariant ecological themes can typify entire, albeit small, specialized families (e.g., Sarraceniaceae, Lamnaceae). More impressive evidence for constraints on direction but not on speciation in plant evolution is provided by greater redundancy in other families. Examples include halophytism in Chenopodiaceae, ruderalism in Brassicaceae, xerophytism in Cactaceae, and of course, epiphytism in Orchidaceae. No single life-style characterizes the entire clade in any of these examples, but particular themes turn up too often to deny underlying family-wide dispositions.

Ecological syndromes, including the various expressions of epiphytism, are comprised of structural, functional, and phenological components that together promote survival in proscribed environments. Ruderals succeed through their capacity to create seed mass rapidly, because time is the major constraint affecting success in their resource-rich, disturbed habitats. Components of this well-defined syndrome are habits with minimal mechanical and root tissue; vigorous photosynthesis; small, long-lived, often light-sensitive seeds; and self-compatibility. Here, such traits as low shoot/root ratios, woodiness, and extensive succulence are never found because they are inconsistent with an abbreviated life cycle. Although the habits of vascular halophytes range from large trees (mangroves) to succulent terrestrial and submerged herbs, the common threat of desiccation mandates a single countermeasure. Osmotic compensation with borrowed and manufactured osmotica provide the only mechanism for water balance. Toxic ions, taken in from the environment to lower tissue water potential, must in every case be sequestered in vacuoles and balanced across the tonoplast by "compatible" osmotica of plant origin. Salt balance can be fine-tuned further by taxon-specific features; chenopod halophytism is aided by multicelled vesicular hairs which excrete excess Na^+ and Cl^-; succulence provides for ion dilution. The presence of organic solutes, including free proline, in other salt-tolerant taxa instead of the betaines found in Chenopodiaceae reveals that physiological convergence among natives of saline environments is not complete. Likewise, the malic acid synthesized by all CAM plants is usually decarboxylated through only one of the three possible reactions: mediation by NAD- or NADP-dependent malic enzyme or PEP carboxykinase.

On balance, taxonomic diversity among the plant species comprising an ecological category reflects the equability of, and access to, the type of habitat they occupy. Fertile, moist soils and strong irradiance have fostered wide convergence; ruderals are a good example. Weeds in crop fields belong to many families, reflecting easy access to a collection of required character states. Tropical forest canopies have been similarly colonized but only to a certain degree. Whereas many lineages have evolved traits for growth on the suspended histosols of everwet forests, few have invaded more demanding zones, as demonstrated by narrower taxonomic participation in PS epiphytism (Benzing 1978a; Fig. 1.21).

Inherited – in effect, phyletic – constraints determined which ancestors of modern families could generate epiphytic derivatives. Potential to express key components of a required mechanism was not sufficient to ensure its establishment, however. Widespread occurrence of CAM-like function, for instance, did not guarantee adoption of CAM by all xerophytes or by all aquatic macrophytes in the softwater lakes where nocturnal fixation is favored by limited carbon supply (Keeley 1981; Richardson et al. 1984). Each component of a complex mechanism must be free of unalterable relationships with incompatible traits; for instance, plants with vigorous cambia may rarely employ CAM because costly woodiness is not sustainable without greater capacity for carbon gain. An exception among epiphytes is the strangler *Clusia rosea* (Ting et al. 1985b), but substantial C_3 activity ($\delta^{13}C$ = about −18‰) continues in this species. Moraceous stranglers resist desiccation without CAM, as do a number of *Clusia* species with comparable habits (Popp et al. 1987; Ting et al. 1987).

Plant evolution has been, and is now, constrained by aspects of acquisition and use of photons, water, and N (Raven 1985); major limiting factors are shade, drought, and infertile substrata, respectively. Type of N source imposes different demands depending on where (in which organ) processing takes place, how much water and energy is needed per unit of product, and the environmental context (is light or moisture scarce or abundant?). Calculation of comparative costs must extend beyond input for chemical synthesis; transport and pH regulation should be included. Excess protons must be eliminated by users of NH_4^+ and N_2; resultant OH^- is consumed or excreted by NO_3^- assimilators. Although the ammonium-to-protein pathway is least expensive in terms of energy consumption, overriding factors may still dictate another choice even where NH_4^+ is the form of N in greatest environmental supply. Soils are the usual sinks for H^+ generated by NH_4^+ use. Indeed, owing to the immobility of protons in phloem, terrestrial plants process most of their acquired NH_4^+ in roots, a constraint with special relevance for epiphytism. If this is the universal rule, then what compensation, if any, accompanied root system reduction in advanced Bromeliaceae? Evidence indicates that NH_4^+ is the predominant form of N in at least some tropical forest canopies, albeit with variation depending on which fluids are tapped (e.g., Curtis 1946; Table 4.3). Conceivably, the absence of similar morphological diminution in nonbromeliad lineages is in part related to their less flexible N metabolism, although the presence of bromeliad foliar trichomes as replacements for roots cannot be ignored in such comparisons. Perhaps slow-growing plants like the atmospheric bromeliads metabolize N at such low rates that complications are avoided. Either the internal biochemical pH-stat is adequate for disposal or excess protons are dumped while shoots are wetted. Raven's (1988) choice of *Tillandsia paucifolia* to illustrate unusual solutions for potential problems of plant metabolism underscores the need for comprehensive cost accounting as well as a thorough knowledge of functional incompatibilities and phylogenetic constraints when interpreting epiphyte history. It is at these levels of performance that the trade-offs, economies, and accommodations necessary for plant evolution occur.

Historical basis for canopy dwelling

There is now no way to explain fully why one lineage developed epiphytism while another did not. Partial answers are available in some cases, however; several of the more notable ones are discussed below. Three questions provide focus: Why are proportionally more ferns than seed plants epiphytic? Why do so many monocots, particularly orchids, inhabit tree crowns? Why have several families of dicots with no obvious advantage by basic habit or water balance mechanism succeeded so widely as epiphytes?

Ferns

Small diaspores are an important attribute for pteridophytic epiphytism (Table 8.7) as are spongelike masses of finely divided, durable roots (e.g., *Campyloneurum angustifolium;* (Fig. 1.19). Unique water and carbon balance mechanisms limit occurrence of many species to specific conditions, however. Poikilohydry is pronounced in exceptional taxa (e.g., *Polypodium polypodioides;* Fig. 1.6), but many others also exhibit desiccation tolerance superior to that of most seed plants. A fern's xerophytism, unlike that of most CAM plants, seems more serviceable in shade than in fuller irradiance. Greater exposure might be tolerable, and upper as well as lower strata colonized equally, but for the trade-off associated with desiccation tolerance. Resurrection is effective for life under the occasional drought, but not for the repeated dehydration that the epiphytic fern would have to survive on most rooting media. Recall that, when poikilohydrous foilage dries too often, carbon balance tends to become negative (Alpert and Oechel 1985). Raven (1985) cites the rates of photosynthesis and particularly respiration which lie at the low ends of the ranges reported for tracheophytes as a reason why ferns are so well-equipped to inhabit shady, drought-prone locations.

Nevertheless, ferns have accomplished a modest invasion of drier sunny sites either by drought avoidance via seasonally deciduous foliage (e.g., *Phlebodium aureum)* or by enduring dry periods with photosynthetic organs intact. Occurrence in some arid Australasian sites is possible for evergreen *Pyrrosia* and its equally coriaceous relatives through that odd juxtaposition of structural and physiological characters reminiscent of both poikilohydry and desiccation-resistant (CAM) xerophytism previously described. Field assays indicated that CAM allowed *Drymoglossum piloselloides* to conserve much respired CO_2 but ΔH^+_{max} and WUE were far below that of angiospermous epiphytes (Kluge et al. 1989). A thorough examination of ferns with regard to microclimate, substratum, and water–carbon relations in both gametophyte and sporophyte stages is needed to place discussion of the evolution of pteridophytic epiphytism on a firmer foundation.

Liliopsida as a whole

Orchids account in large measure for the immense number of epiphytic species, but even without this family, monocots would be overrepresented in tree crowns (Table 1.1). Bromeliaceae and Araceae rate second and third. Over 30% of all monocots are epiphytic; a mere 2% of dicots qualify. Although Araceae, Bromeliaceae, and Orchidaceae contain the most arboreal species, there is no common adaptive theme (Table 8.7). Two photosynthetic pathways in many variations, tank and trash-basket impoundments, myrmecotrophism, foliar trichomes, velamentous roots, and virtually all the dispersal modes enabling life aboveground occur in Liliopsida. A peculiar body plan, shared to some extent with the higher ferns but less so with dicots, may have offered class-wide opportunity.

Shoots of vascular plants, but particularly those of Liliopsida, are serialized into relatively independent physiological units roughly corresponding to phytons (IPUs; sensu Watson and Casper (1984) – in effect, to nodes with associated leaves, buds, and adventitious roots. Axillary meristems tend to receive photosynthate primarily from subtending foliage. Partitionment into vertical compartments (more a feature of dicots) is evidenced by movement of labeled photosynthate among leaves and associated buds forming longitudinal series (orthostichies). The prototypical rhizomatous monocot body with its reticulate "atactostele" helps distinguish the angiosperm classes and seemingly has imparted distinct evolutionary opportunity through novel plasticity. For instance, remote sources and sinks of certain spreading grasses change identity and interdependence over time to allow optimum harvesting of patchy resources (e.g., Callaghan 1984; Welker, Rykiel, Briske, and Goeschl 1985). Another aid, if not impetus, to the development of monocot epiphytism was the capacity to modify shoot segments into connectors and leafy floriferous regions as a system for episodic growth. Serial production of sympodial shoots consisting of one (e.g., *Dendrobium ultissimum*) to many (e.g., *Catasetum* spp.; Fig. 1.15) phytons each year is a recurrent theme in Orchidaceae responsible for many epiphytic habits serving thousands of species. Bromeliaceae offer a parallel, with *Tillandsia usneoides* (Fig. 4.25D) as the most reduced form. Other architectures that foster repeated production of closely placed, self-sufficient shoots exist but are less common in epiphytic Magnoliopsida (e.g., some Cactaceae, Gesneriaceae, and *Lycopodium;* (Fig. 1.18).

Modification of individual organs further promoted monocot over dicot epiphytism. Differentiation of roots into feeder and holdfast types (Fig. 1.5), a useful division of labor for the vine or epiphyte, is also overrepresented in Liliopsida, as is an important aspect of root anatomy. Recent studies

Table 8.7. **Preliminary tabulation of predominant and less common vegetative features underlying the epiphytic habit in angiospermous families containing more than 50 canopy-adapted species**

Group or family (# epiphytic species/# parent genera)	Habitat humidity	Most pervasive adaptive features	Less pervasive adaptive features	Common ecological types	Minor ecological types
Ferns 2388/90	Wet to moderately dry	Dust-size propagules; poikilohydrous tendency; shade tolerance; diverse habits	Macroimpoundment; brood chambers for ants; CAM; pronounced resurrection capacity; absorbing foliar trichomes	General humus-rooted, sciophytic epiphytes	Trophic myrmecophytes; resurrection forms; drought-enduring, CAM forms; trash-basket epiphytes
Araceae 1349/13	Wet	Vining habit; macroimpoundment; microimpoundment (velamen); heterophylly		Secondary hemiepiphytes; trash-basket and general humus-rooted epiphytes	Nest-garden epiphytes
Araliaceae 78/9	Wet	Versatile root growth and function		Woody hemiepiphytes	General humus-rooted, shrubby epiphytes
Asclepiadaceae 137/8	Wet	Vining habit; CAM; xeromorphy; various ant associations		Vining, often humus-rooted epiphytes	Trophic myrmecophytes
Bromeliaceae 1144/26	Wet to dry	Macroimpoundment; microimpoundment (foliar trichomes); CAM; vegetative reduction; xeromorphy	Carnivory; brood chambers for ants; deciduousness	Tank epiphytes; PS epiphytes (atmospherics)	Nest-garden and general humus-rooted epiphytes; trophic myrmecophytes; drought avoiders
Cactaceae 150/18	Wet to dry	CAM; xeromorphy		Secondary hemiepiphytes; general humus-rooted epiphytes	

Clusiaceae 85/6	Wet	Versatile root growth and function	CAM; xeromorphy	Stranglers	General humus-rooted, shrubby epiphytes
Cyclanthaceae 86/7	Wet	Vining habit		Secondary hemiepiphytes	
Ericaceae 672/36	Wet	Mycorrhizas (?); propensity for acid, organic substrata		General humus-rooted, shrubby epiphytes	
Gesneriaceae 560/30	Wet to moderately dry	CAM; xeromorphy; various ant associations; diverse habits		General humus-rooted epiphytes	Ant nest-garden epiphytes
Marcgraviaceae 89/7	Wet	Vining habit; heterophylly		Secondary hemiepiphytes	
Melastomataceae 648/33	Wet	Propensity for acid, organic substrata		General humus-rooted, shrubby epiphytes	Hemiepiphytes
Moraceae 552/4	Wet to moderately dry	Versatile root growth and function; strangling habit		Stranglers	
Orchidaceae 13,951/440	Wet to dry	Microimpoundment (velamen); CAM; vegetative reduction; microsperms; xeromorphy; fungus-assisted juvenile nutrition; mycorrhizas (?); diverse habits Photosynthetic root	Macroimpoundment; brood chambers for ants; deciduous foliage	General humus-rooted epiphytes; PS epiphytes (drought-enduring)	Trophic myrmecophytes; nest-garden epiphytes
Piperaceae 710/2	Wet to moderately dry	CAM; xeromorphy; small size		General humus-rooted epiphytes	
Rubiaceae 223/25	Wet	CAM; xeromorphy; brood chambers for ants		Myrmecophytes; humus-rooted epiphytes	

(Peterson 1988) have revealed Casparian strips, hence an exodermis, in the outer root cortex of most flowering plants, but only in monocots is an adjacent velamen ever present. There are parallels in the shoot (M. Madison pers. comm.). Sparse branching, a common feature of Liliopsida and less so of dicots, mandates durable leaves and hence preadaptation for climatic stress. Aiding performance in arid environments are expressions of CAM that grant water economy unexceeded elsewhere.

Nonorchid monocots

Bromeliaceae, with far fewer species and almost exclusively neotropical distribution, nevertheless rival Orchidaceae for variety of epiphytic mechanisms (Table 8.7) and vastly exceed the latter's biomass in tropical American forests. Tank habits have evolved independently in two bromeliad subfamilies, and in all three if sometimes-epiphytic *Brocchinia* (Fig. 4.25E) is correctly assigned to Pitcairnioideae (Benzing et al. 1985). A rosulate shoot was required for each transference of absorptive role from root to foliage. Ancestry was apparently mesic in both Tillandsioideae and Pitcairnioideae (Fig. 2.4; Benzing and Renfrow 1971a; Medina 1974; Benzing et al. 1985 – but see Pittendrigh 1948); tank shoots show C_3 photosynthesis in each subfamily. Bromelioideae, with about 500 species capable of creating substitutes for ground soil in leaf bases, are fundamentally CAM plants that probably acquired nocturnal CO_2 fixation and impoundment as terrestrials in arid habitats. Specialization for PS epiphytism – in effect for greater stress tolerance – has proceeded farthest in the derived atmospheric forms of Tillandsioideae (Benzing et al. 1985), where absorbing trichomes (Fig. 3.21) are perfected to the highest degree while the vegetative apparatus is reduced to simplest form (Fig. 4.25A–D).

Early Bromeliaceae were clearly predisposed to epiphytism by the presence of a suitable epidermal appendage and habit. Here, life in the canopy is based on a modified shoot with the foliar trichome as its keystone feature (Table 8.7). Leaf bases need not be highly specialized to tap tanks continuously filled with moist humus and the organisms required to reduce it to plant nutrients, but atmospheric epiphytism requires an extraordinary foliar indumentum (Benzing and Pridgeon 1983). Myrmecophytism, and a single case of epiphytic carnivory (*Catopsis berteroniana;* Givnish et al. 1984), are also associated with modified shoot surfaces and tubular rosettes (Benzing 1970b). Hypotheses concerning how the bromeliad foliar epidermis may have acquired its current function and importance are described elsewhere

(Pittendrigh 1948; Benzing et al. 1985). (Briefly: Contrary to Pittendrigh's proposition that absorptive function would emerge only under drought selection, Benzing et al. posited a mesic, infertile, ancestral habitat and a foliar epidermis and habit that assumed their unusual characteristics to promote utilization of impounded humus or perhaps animal prey.) Bromeliad seeds are disseminated by birds (Bromelioideae; Fig. 5.2G) or wind (Pitcairnioideae and Tillandsioideae; Fig. 5.2K). Pollination syndromes are diverse.

Aroid, by comparison with bromeliad or orchid, epiphytism is neither as advanced nor as versatile, although a capacity for life in tree crowns has originated at least three times in the family (Madison 1977). There are no reports of CAM here (Table 2.1) although perhaps CAM-cycling exists (Ting et al. 1985b), and overlapping foliage that might mitigate drought lacks the watertight quality of inflated bromeliad leaf bases. Trash-basket catchments (Fig. 1.5) sometimes trap falling litter but little moisture. Roots fail to produce velamina as elaborate as those of the most drought-tolerant Orchidaceae, nor is there any indication that these organs can contribute substantially to the plant's carbon budget as can those of some orchids. Seasonally deciduous leaves on green or tuberous stems occur in *Philodendron* and *Remusatia,* respectively, but these are minor themes represented by few species. Arboreal existence in Araceae is based predominantly on two mechanisms, both humus-based (Table 8.7): impoundment, seen in short-stemmed *Anthurium* and some *Philodendron;* and secondary hemiepiphytism (Fig. 1.9), a more widespread phenomenon often encountered in *Anthurium, Philodendron,* and *Rhaphidophora.* Velamentous roots, and vining habits truncated by progressive stem decay are responsible for aroid hemiepiphytism. Both sympodial (e.g., *Philodendron*) and monopodial (e.g., *Pothos*) habits are involved. Water and nutrient balance exhibit no obvious modifications for arboreal life, but they have not been examined closely. Ant nest-gardens are utilized by some *Anthurium* and *Philodendron.* Baccate fruit is an integral part of the aroid epiphytic syndrome, but it occurs throughout the family without habitat restriction. Pollinators range from beetles to euglossines. Specialized pollen vectors, including coleopterans that appear to be unexpectedly faithful to specific flowers (C. Dodson pers. comm.), may have prompted radiation of *Anthurium* and possibly other genera.

Cyclanthaceae, the only other nonorchid monocot family with a sizable epiphyte contingent (largely *Asplundia;* Table 1.1), mostly utilize the forest canopy as permanently ground-rooted climbers and secondary hemiepiphytes. True epiphytism occurs in *Sphaeradenia* and *Stelestylis* (G. Wilder pers. comm.). Stems and internodes are shorter than those of related hemiepiphytes.

Orchids

Orchidaceae owe their numerical superiority among epiphytes to an exceptionally propitious set of vegetative and reproductive features, including pollen conveyance by specialized insects lured by novel floral syndromes (Benzing and Atwood 1984). Vegetative mechanisms vary tremendously according to the taxon's native substratum and microclimate (Table 8.7), but there are several important attributes common to all canopy-adapted family members that, in some form, predisposed early stock for arboreal life. The specialized roots of epiphytic orchids vary in photosynthetic performance and water balance, depending on structure and metabolism; uptake is enhanced in all cases by a nonliving velamen (Fig. 3.19) which imbibes precipitation containing solutes for subsequent sorption through transfer cells in an underlying exodermis. This same mantle effectively retards desiccation. Hyperovulate gynoecia and aggregated pollen characterize most of the family. Microspermy – up to millions of tiny, lightly provisioned seeds per capsule – requires fungal intervention for germination and fosters high fecundity without frequent pollination. As a result, specialization for high-fidelity, long-range, but often inefficient pollinators is sustainable and has led to much ethological isolation and exuberant speciation (Benzing in press).

Pre-epiphytic orchid stock probably possessed velamentous roots, as do most extant terrestrial family members and some other nonepiphytic monocots (e.g., Amaryllidaceae). Microspermy and associated mycotrophic nutrition were also acquired in a terrestrial context, as suggested by the habitats in which all other such heterotrophic plant symbioses occur (e.g., Monotropaceae, achlorophyllous Gentianaceae). The stage was set for migration into forest canopies (including many that are uninhabitable for less stress-tolerant epiphytes) by the orchid's body plan, production of numerous tiny diaspores (Fig. 5.2B,D), ability to subsist on transitory resource supplies, and maintenance of high water and nutrient use efficiencies. Evolution of pheromone-like fragrances and specialized floral morphology tightened relationships with specific pollen vectors and ensured extensive proliferation of several clades that happened to be canopy-adapted. Large clusters of related species among taxa seemingly utilizing smaller short-range vectors with no known propensity for exclusive foraging suggest that additional factors have been more important to cladogenesis elsewhere in the family.

Dicots

Magnoliopsida are, on the whole, poorly disposed to epiphytism (Table 1.1), but there are exceptional taxa. Success in a clade is often predicated on a single theme; no other families incorporate the diverse mechanisms for resource procurement or stress tolerance exhibited by epiphytic Bromeliaceae and Orchidaceae. Except in Marcgraviaceae (<150 species), dicot terrestrials always outnumber confamilial epiphytes. In tree crowns, *Peperomia* ranks first in size among successful dicot genera and even families – a statistic fostered by pantropical distribution, the presence of CAM variations (Sipes and Ting 1985; Nishio and Ting 1987; Patel and Ting 1987), and high-volume production of small adhesive fruit (Fig. 5.2N). Habits range from shrubby to minute and creeping. Assignment to Piperales along with Chloranthaceae, a family characterized by pollen similar to that contained in early middle Cretaceous sediments, further highlights *Peperomia* as possibly one of the first angiospermous lineages to develop epiphytism.

Moraceae also owe much of their major epiphytic presence to a single genus *(Ficus)* with a similarly broad range; here the strangler habit underlies success in forest canopy habitats. Rampant speciation within a relatively narrow adaptive mode has again been encouraged by circumtropic range and host-specific pollinators – in this case, the fig wasps. According to Ramírez (1977), the moraceous strangling habit evolved "as a response to lack of light at forest level," a rather cryptic observation saying little about mechanisms. Requirements for success also included the presence of viscid hyaline coats covering seeds that would germinate only on moist humus; long aerial roots; adequate WUE in seedlings; and dispersal by winged vertebrates. Marcgraviaceae and Clusiaceae are additional single-theme families, emphasizing secondary and primary hemiepiphytism respectively.

Most epiphytic Asclepiadaceae belong to closely related, succulent, vining *Dischidia* and *Hoya.* Flasklike leaves of *D. rafflesiana* (Fig. 4.24C) provide housing for ants; ants disperse seeds for many more congenerics. Forms with less specialized foliage regularly root in or grow against ant debris, suggesting how ant-fed relatives evolved. Ovoid, dome-shaped leaves of *D. collyris* (Fig. 4.24D) are pressed against bark, providing shelter for *Iridomyr-*

mex colonies (Huxley 1980). Photosynthesis involves CAM and/or CAM-cycling (Kluge and Ting 1978).

Cactaceae were poised for canopy life through previous drought selection in terrestrial habitats. Fleshy, small-seeded fruit and a climbing habit aided by adventitious rooting would additionally favor epiphytism, but some reversals occurred during transition. Originally aphyllous stems of the most advanced epiphytic forms, which happen to be natives of humid forests (e.g., *Zygocactus,* Fig. 3.1A; *Rhipsalis*), have lost their armature and become much flattened or narrowed if still terete (e.g., *Hatiora*), presumably to improve photosynthetic performance in shade. Family-wide CAM is probably present in relatively muted form as well. Despite the extreme drought tolerance of such terrestrial relatives as *Mammillaria* and *Ferrocactus,* epiphytic Cactaceae never colonize the most severe bark and twig exposures. Long-lived roots usually tap more or less continuous supplies in tree fissures, suspended humus, or earth soil (secondary hemiepiphytes – e.g., *Hylocereus*).

Less obvious is the basis for high epiphyte success in Ericaceae, Gesneriaceae, Melastomataceae, and nonmyrmecophytic Rubiaceae. Most canopy-adapted members in all four families grow exclusively on humus mats in humid forests. Woody habits and sclerophyllous foliage, sometimes complemented by storage tubers (e.g., *Macleania, Psammisia*) characterize Ericaceae and Melastomataceae. Many rubiaceous epiphytes (e.g., *Hydnophytum, Myrmecodia*) supplement mineral nutrition and store moisture via swollen ant-inhabited hypocotyls (Figs. 1.4, 4.24A). Basically herbaceous Gesneriaceae feature an extraordinary variety of growth forms. Like some *Peperomia,* these gesneriads exhibit trilayered mesophyll (Fig. 3.1E; Nishio and Ting 1987; Patel and Ting 1987) that signals the unusual photosynthesis discussed in Chapter 2. Substrata are more diverse in this family, ranging from ant nest-gardens to less defined humus. *Codonanthe* and related genera, along with some scandent cacti, are probably the best drought-insulated of the dicot epiphytes. Baccate fruit provides seed mobility in most cases, although *Rhododendron,* a few gesneriad genera, and large proportions of Melastomataceae and Rubiaceae ripen wind-borne seeds (Fig. 5.2). Ant-associated species are myrmecochorous.

Representation in canopy habitats varies among these families, ranging from 4% to 35% of all genera in Rubiaceae and Ericaceae, respectively (Madison 1977). The ericad statistic is all the more impressive in view of the family's modest size and numerous temperate taxa. The three larger families are exclusively moist-tropical or nearly so, hence have had greater access to tree crowns. Breadth and depth of specialization for life in canopies are further indicated by comparing the number of exclusively epiphytic genera containing two or more species with the number that includes terrestrial species as well. Gesneriaceae is most canopy-adapted by this measure with 13 genera meeting each criterion, whereas Ericaceae is least so with only four terrestrial-free genera out of 22 containing epiphytes; the largest of the four contains only eight species. *Vaccinium* is especially noteworthy for its wide range throughout Old and New World boreal to equatorial zones and diverse habitats. The ratios of epiphytic-only to mixed genera for Melastomataceae and Rubiaceae are 8:12 and 5:14, respectively.

Chemical peculiarities of substrata or unusual mycorrhizas may be responsible for uneven epiphyte development among higher monocot and dicot taxa. Ericaceae, Melastomataceae, and Orchidaceae exhibit family-wide affinity for acidic, usually moist, infertile organic soils. Substrata in humid forests where most epiphytes live tend to be sodden, at least moderately acid, and certainly organic. Use of NH_4^+ rather than oxidized N by plants native to such substrata may have been a predisposing character for epiphytism. Some ericads are notably deficient in nitrate reductase, a sign of long utilization of reduced N. Ericaceae was perhaps especially well positioned for canopy invasion via a type of mycorrhiza seen in some extant terrestrials that mobilize N and P from sterile organic soil (Stribley and Read 1975; St. John, Smith, Nicholas, and Smith 1985). Terrestrial Orchidaceae are also strongly mycorrhizal, but the advantage, if any, of fungi to canopy-adapted adults remains little studied and largely controversial (Ruinen 1953; Hadley and Williamson 1972; Sanford 1974; Benzing and Friedman 1981a). Broad surveys of epiphyte roots and N-cycle microbes in canopy substrata could prove rewarding. This effort should be part of a broader one designed to expand and refine information of the type presented in Table 8.7 and Figure 1.22.

Envoi

Too few aspects of epiphyte biology were well enough known by January 1989 to permit more than preliminary treatment as this first attempt at an overview goes to press. Numbers of species and their taxonomic and geographic distributions are fairly accurate now, but not so impressions of how plant form, physiology, and life history are tailored for existence in tree crowns. Even less refined is knowledge of animal–epiphyte interactions and effects on phorophytes and the entire community. A quick glance at the literature reveals how soon this text can be revised. Of all the writings cited here, nearly 40% are no more than five years old. Second, as heretofore, explorations of epiphytism will profit from knowledge gathered elsewhere. Appreciation of the many similarities between canopy-based and much earth-rooted vegetation will continue to mount as additional plant data on water, carbon, and mineral balance accumulate. Nevertheless, unique combinations of sometimes novel characteristics will always justify the special recognition and status accorded the epiphytes by monographic treatment.

References

Abercrombie, M., Hickman, C. J., and Johnson, M. L. 1970. *A dictionary of biology*. Harmondsworth: Penguin Books.

Ackerman, J. D. 1986. Coping with the epiphytic existence: pollination strategies. *Selbyana* 9: 52–60.

Ackerman, J. D., Ontalvo, A. M., and Vera, A. M. 1989. Epiphyte–host specificity of *Encyclia krugii*, a Puerto Rican endemic orchid. *Lindleyana* 4: 74–77.

Adams, W. W. 1988. Photosynthetic acclimation and photoinhibition of terrestrial and epiphytic CAM tissues growing in full sunlight and deep shade. *Aust. J. Plant Physiol.* 15: 123–134.

Adams, W. W., and Martin, C. E. 1986a. Heterophylly and its relevance to evolution within the Tillandsioideae. *Selbyana* 9: 121–125.

1986b. Morphological changes accompanying the transition from juvenile (atmospheric) to adult (tank) forms in the Mexican epiphyte *Tillandsia deppeana* (Bromeliaceae). *Am. J. Bot.* 73: 1204–1214.

Alexander, C. P. 1912. A bromeliad-inhabiting crane fly (Tipulidae, Diptera). *Entomol. News* 23: 415–417.

Alosi, M. C., and Calvin, C. L. 1985. The ultrastructure of dwarf mistletoe (*Arceuthobium* spp.) sinker cells in the region of the host secondary vasculature. *Can. J. Bot.* 63: 889–898.

Alpert P., and Oechel, W. C. 1985. Carbon balance limits the microdistribution of *Grimmia laevigata*, a desiccation-tolerant plant. *Ecology* 66: 660–669.

Anderson, A. B. 1981. White sand vegetation of Brazilian Amazonia. *Biotropica* 13: 199–210.

Atsatt, P. R. 1983. Mistletoe leaf shape: a host morphogen hypothesis. Pp. 259–276 in *The biology of mistletoes*, eds. D. M. Calder and P. Bernhardt. New York: Academic Press.

Avadhani, P. N. 1976. Carbon dioxide fixation in orchids. Pp. 412–413 in *Proc. 8th World Orchid Conf.*, Frankfurt (1975). Frankfurt: German Orchid Soc.

Avadhani, P. N., and Goh, C. J. 1974. Carbon dioxide fixation in the leaves of *Bromheadia finlaysoniana* and *Arundina graminifolia* (Orchidaceae). *Singapore Nat. Acad. Sci.* 4: 1–4.

Avadhani, P. N., Goh, C. J., Rao, A. N., and Arditti, J. 1982. Carbon fixation in orchids. P. 177 in *Orchid biology–reviews and perspectives II*, ed. J. Arditti, Ithaca: Cornell University Press.

Avadhani, P. N., Khan, I., and Lee, Y. T. 1978. Pathways of carbon dioxide fixation in orchid leaves. Pp. 1–12 in *Proc. Symp. Orchidology*, ed. E. S. Teoh. Singapore: Orchid Soc. S.E. Asia.

Banister, P. 1989. Nitrogen concentration and mimicry in some New Zealand mistletoes. *Oecologia* 79: 128–132.

Barkman, J. J. 1958. *Phytosociology and ecology of cryptogamic epiphytes.* Assen, Netherlands: Van Gorcum.

Barlow, B. A. 1983. Biogeography of Loranthaceae and Viscaceae. Pp. 19–46 in *The biology of mistletoes,* eds. D. M. Calder and P. Bernhardt. New York: Academic Press.

Barlow, B. A., and Wiens, D. 1977. Host–parasite resemblance in Australian mistletoes: the case for cryptic mimicry. *Evolution* 31: 69–84.

Barthlott, W., and Capesius, I. 1975. Mikromorphologische und funktionelle Untersuchungen am Velamen radicum der Orchideen. *Ber. dtsch. bot. Ges.* 88: 379–390.

Bawa, K. S., Bullock, S. H., Perry, D. R., Coville, R. E., and Grayum, M. H. 1985. Reproductive biology of tropical lowland rain forest trees. II. Pollination systems. *Am. J. Bot.* 72: 346–356.

Beckman, K. M. 1964. The influence of light on germination of western dwarf mistletoe (*Arceuthobium campylopodum* cf. *campylopodum*). *Phytopathology* 54: 1431–1432.

Bennett, B. C. 1984. A comparison of the spatial distribution of *Tillandsia flexuosa* and *T. pruinosa. Florida Sci.* 47: 141–144.

1987. Spatial distribution of *Catopsis* and *Guzmania* (Bromeliaceae) in southern Florida. *Bull. Torrey Bot. Club* 114: 265–271.

Bentley, B. L., and Carpenter, E. J. 1984. Direct transfer of newly-fixed nitrogen from free-living epiphyllous microorganisms to their host plant. *Oecologia* 63: 52–56.

Benzing, D. H. 1970a. An investigation of two bromeliad myrmecophytes: *Tillandsia butzii* Mez, *T. caput-medusae* E. Morren and their ants. *Bull. Torrey Bot. Club* 97: 109–115.

1970b. Foliar permeability and the absorption of minerals and organic nitrogen by certain tank bromeliads. *Bot. Gaz.* 131: 23–31.

1978a. The life history profile of *Tillandsia circinnata* (Bromeliaceae) and the rarity of extreme epiphytism among the angiosperms. *Selbyana* 2: 325–337.

1978b. Germination and early establishment of *Tillandsia circinnata* Schlecht. (Bromeliaceae) on some of its hosts and other supports in southern Florida. *Selbyana* 2: 95–106.

1979. Alternative interpretations for the evidence that certain orchids and bromeliads act as shoot parasites. *Selbyana* 5: 135–144.

1980. Pp. 87, 104 in *The biology of the bromeliads.* Eureka, CA: Mad River Press.

1981a. The population dynamics of *Tillandsia circinnata* (Bromeliaceae): cypress crown colonies in southern Florida. *Selbyana* 5: 256–263.

1981b. Bark surfaces and the origin and maintenance of diversity among angiosperm epiphytes: an hypothesis. *Selbyana* 5: 248–255.

1983. Vascular epiphytes: a survey with special reference to their interactions with other organisms. Pp. 11–24 in *Tropical rain forest: ecology and management,* eds. S. L. Sutton, T. C. Whitmore, and A. C. Chadwick. Oxford: British Ecological Society.

1984. Epiphytic vegetation: a profile and suggestions for future inquires. Pp. 155–172 in *Physiological ecology of plants of the wet tropics,* eds. E. Medina, H. A. Mooney, and C. Vázquez-Yánes. The Hague: Junk.

1986. Foliar specializations for animal-assisted nutrition in Bromeliaceae. Pp. 235–256 in *Insects and the plant surface,* eds. B. E. Juniper and T. R. E. Southwood. London: Edward Arnold.

1987a. Vascular epiphytism: taxonomic participation and adaptive diversity. *Ann. Mo. Bot. Gdns.* 74: 183–204.

1987b. The origin and rarity of botanical carnivory. Pp. 364–369 in *Trends in ecology and evolution II.* New York: Elsevier Scientific.

In press. The evolution of orchids. *Trends in ecology and evolution.* New York: Elsevier Scientific.

Benzing, D. H., and Atwood, J. T. 1984. Orchidaceae: ancestral habitats and current status in forest canopies. *Syst. Bot.* 9: 155–165.

Benzing, D. H., Bent, A., Moscow, D., Peterson, G., and Renfrow, A. 1982a. Functional correlates of deciduousness in *Catasetum integerrimum* (Orchidaceae). *Selbyana* 7: 1–9.

Benzing, D. H., and Davidson, E. 1979. Oligotrophic *Tillandsia circinnata* Schlecht. (Bromeliaceae): an assessment of its patterns of mineral allocation and reproduction. *Am. J. Bot.* 66: 386–397.

Benzing, D. H., Derr, J., and Titus, J. 1971. Factors affecting the water chemistry of microcosms associated with the epiphytic bromeliad *Aechmea bracteata. Am. Mid. Nat.* 87: 60–70.

Benzing, D. H., and Friedman, W. E. 1981a. Mycotrophy: its occurrence and possible significance among epiphytic Orchidaceae. *Selbyana* 5: 243–247.

1981b. Patterns of foliar pigmentation in Bromeliaceae and their adaptive significance. *Selbyana* 5: 224–240.

Benzing, D. H., Friedman, W. E., Peterson, G., and Renfrow, A. 1983. Shootlessness, velamentous roots, and the pre-eminence of Orchidaceae in the epiphytic biotope. *Am. J. Bot.* 70: 121–133.

Benzing, D. H., Givnish, T. J., and Bermudes, D. 1985. Absorptive trichomes in *Brocchinia reducta* (Bromeliaceae) and their evolutionary and systematic significance. *Syst. Bot.* 10: 81–91.

Benzing, D. H., Henderson, K., Kessel, B., and Sulak, J. 1976. The absorptive capacities of bromeliad trichomes. *Am. J. Bot.* 63: 1009–1014.

Benzing, D. H., and Ott, D. W. 1981. Vegetative reduction in epiphytic Bromeliaceae and Orchidaceae: its origin and significance. *Biotropica* 13: 131–140.

Benzing, D. H., Ott, D. W., and Friedman, W. E. 1982b. Roots of *Sobralia macrantha* (Orchidaceae): structure and function of the velamen–exodermis complex. *Am. J. Bot.* 69: 608–614.

Benzing, D. H., and Pockman, W. 1989. Why do nonfoliar green organs of leafy orchids fail to exhibit net photosynthesis? *Lindleyana* 4: 53–60.

Benzing, D. H., and Pridgeon, A. 1983. Foliar trichomes of Pleurothallidinae (Orchidaceae): functional significance. *Am. J. Bot.* 70: 173–180.

Benzing, D. H., and Renfrow, A. 1971a. Significance of the patterns of CO_2 exchange to the ecology and phylogeny of the Tillandsioideae (Bromeliaceae). *Bull. Torrey Bot. Club* 98: 322–327.

1971b. The significance of photosynthetic efficiency to habitat preference and phylogeny among tillandsioid bromeliads. *Bot. Gaz.* 132: 19–30.

1971c. The biology of the atmospheric bromeliad *Tillandsia circinnata* Schlecht. I. The nutrient status of populations in South Florida. *Am. J. Bot.* 58: 867–873.

1974a. The mineral nutrition of Bromeliaceae. *Bot. Gaz.* 135: 281–288.
1974b. The nutritional status of *Encyclia tampense* and *Tillandsia circinnata* on *Taxodium ascendens* and the availability of nutrients to epiphytes on this host in South Florida. *Bull. Torrey Bot. Club* 101: 191–197.
1980. The nutritional dynamics of *Tillandsia circinnata* in southern Florida and the origin of the "air plant" strategy. *Bot. Gaz.* 141: 165–172.
Benzing, D. H., and Seemann, J. 1978. Nutritional piracy and host decline: a new perspective on the epiphyte–host relationship. *Selbyana* 2: 133–148.
Benzing, D. H., Seemann, J., and Renfrow, A. 1978. The foliar epidermis in Tillandsioideae (Bromeliaceae) and its role in habitat selection. *Am. J. Bot.* 65: 359–365.
Bermudes, D., and Benzing, D. H., in press. Fungi in neotropical epiphyte roots.
Bernhardt, P. 1983. The floral biology of *Amyema* in south-eastern Australia. Pp. 87–100 in *The biology of mistletoes,* eds. M. Calder and P. Bernhardt. New York: Academic Press.
Bhatnagar, S. P., and Johri, B. M. 1983. *Biosystems* Pp. 47–67 in *The biology of mistletoes,* eds. M. Calder and P. Bernhardt. New York: Academic Press.
Billings, F. H. 1904. A study of *Tillandsia usneoides. Bot. Gaz.* 38: 99–121.
Bloom, A. J., Chapin, F. S., and Mooney, H. A. 1985. Resource limitation in plants–an economic analogy. *Annu. Rev. Ecol. Syst.* 16: 363–392.
Böttger, M., Soll, H., and Gasché, A. 1980. Modificaton of the external pH by maize coleoptiles and velamen radicum of *Vanilla planifolia* Andr. *Z. Pflanzenphysiol.* 99: 89–93.
Bradshaw, W. E. 1983. Interaction between the mosquito *Wyeomyia smithii,* the midge *Metriocnemus knabi,* and their carnivorous host *Sarracenia purpurea.* Pp. 161–189 in *Phytotelmata: terrestrial plants as hosts for aquatic insect communities,* eds. J. H. Frank and L. P. Lounibos. Medford, NJ: Plexus.
Bradshaw, W. E., and Holzapfel, C. M. 1988. Drought and the organization of treehole mosquito communities. *Oecologia* 74:507–514.
Brasell, H. M., and Sinclair, D. F. 1983. Elements returned to forest floor in two rainforest and three plantation plots in tropical Austrialia. *J. Ecol.* 71:367–378.
Bronstein, J. L., and Hoffman, K. 1987. Spatial and temporal variation in frugivory at a neotropical fig, *Ficus pertusa. Oikos* 49: 261–268.
Burgeff, H. 1936. *Samenkeimung der Orchideen.* Jena: G. Fischer Verlag.
Burt, K. M., and Benzing, D. H. 1969. The absorption of nutrients by leaves and roots in *Billbergia chlorosticta. Bromeliad Soc. Bull.* 19: 5–10.
Calder, D. M. 1983. Mistletoes in focus: an introduction. Pp. 1–18 in *The biology of mistletoes,* eds. D. M. Calder and P. Bernhardt. New York: Academic Press.
Callaghan, T. V. 1984. Growth and translocation in a clonal southern hemisphere sedge, *Uncinia meridensis. J. Ecol.* 72: 529–546.
Calvert, A. M., and Calvert, P. P. 1917. *A year of Costa Rican natural history.* New York: Macmillan.
Casper, S. J. 1987. On *Pinquicula liqnicola,* an epiphytic heterophyllic member of the Lentibulariaceae in Cuba. *Pl. Syst. Evol.* 155: 348–354.
Catling, P. M., Brownell, V. R., and Lefkovitch, L. P. 1986. Epiphytic orchids in a Belizean grapefruit orchard: distribution, colonization, and association. *Lindleyana* 1: 194–202.
Catling, P. M., and Lefkovitch, L. P. 1989. Guilds of small orchids and the application of the density-independent model to orchid diversity. *Biotropica* 21: 35–40.

Chapin, F. S., Bloom, A. J., Field, C. B., and Waring, R. H. 1987. Plant responses to multiple environmental factors. *BioScience* 37: 49–57.

Chase, M. W. 1987. Obligate twig epiphytes in the Oncidiinae and other neotropical orchids. *Selbyana* 10: 24–30.

Chazdon, R. L., and Fetcher, N. 1984a. Photosynthetic light environments in a lowland tropical rain forest in Costa Rica. *J. Ecol.* 72: 553–564.

1984b. Light environments of tropical forests. Pp. 27–36 in *Tasks for vegetation science 12; Physiological ecology of plants of the wet tropics,* eds. E. Medina, H. A. Mooney, and C. Vázquez-Yánes. The Hague: Junk.

Chazdon, R. L., and Pearcy, R. W. 1986. Photosynthetic responses to light variation in rainforest species. II. Carbon gain and photosynthetic efficiency during lightflecks. *Oecologia* 69: 524–531.

Clark, J., and Bonga, J. M. 1970. Photosynthesis and respiration in black spruce *(Picea mariana)* parasitized by eastern dwarf mistletoe *(Arceuthobium pusillum), Can. J. Bot.* 48: 2029–2031.

Clarkson, D. T., Kuiper, P. J. C., and Lüttge, U. 1986. II. Mineral nutrition: sources of nutrients for land plants from outside the pedosphere. Pp. 81–96 in *Progress in botany,* vol. 48. Berlin: Springer.

Claver, F. K., Alaniz, J. R., and Caldíz, D. O. 1983. *Tillandsia* spp.: epiphytic weeds of trees and bushes. *Forest Ecol. Mgmt.* 6: 367–372.

Clay, K., Dement, D., and Rejmanek, M. 1985. Experimental evidence for host races in mistletoe *(Phoradendron tomentosum). Am. J. Bot.* 72: 1225–1231.

Cockburn, W., Goh, C. J., and Avadhani, P. N. 1985. Photosynthetic carbon assimilation in a shootless orchid, *Chiloschista usneoides* (DON)LDL: a variant on crassulacean acid metabolism. *Plant Physiol.* 77: 83–86.

Cole, L. C. 1954. The population consequences of life history phenomena. *Q. Rev. Biol.* 29: 103–137.

Connell, J. H. 1978. Diversity in tropical rain forests and coral reefs. *Science* 199: 1302–1310.

Connor, J. J., and Shacklette, H. T. 1984. *Factor analysis of the chemistry of Spanish moss: preliminary report.* Washington, D.C.: U.S. Dept. of Interior Geological Survey.

Cook, M. T. 1926. Epiphytic orchids, a serious pest on citrus trees. *J. Dept. Agric. Puerto Rico* 10: 5–9.

Coyne, J. A., and Lande, R. 1985. The genetic basis of species differences in plants. *Am. Nat.* 126: 141–145.

Croat, T. B. 1986. The distribution of *Anthurium* (Araceae) in Mexico, Middle America and Panama. *Selbyana* 9: 94–99.

Curtis, J. T. 1946. Nutrient supply of epiphytic orchids in the mountains of Haiti. *Ecology* 27: 264–266.

1952. Outline for ecological life history studies of vascular epiphytic plants. *Ecology* 33: 550–558.

Davidson, D. W. 1988. Ecological studies of neotropical ant gardens. *Ecology* 69: 1138–1152.

Davidson, D. W., and Epstein, W. W. in press. Epiphytic associations with ants. In *Phylogeny and physiology of epiphytes,* ed. U. Lüttge. Berlin: Springer.

Davidson, D. W., Foster, R. B., Snelling, R. R., and Lozada, P. W. in press. Variable composition of some tropical ant–plant symbioses. In *Herbivory: tropical and temperate perspectives,* ed. P. W. Price. New York: Wiley.

DeSanto, A. V., Alfani, A., and DeLuca, P. 1976. Water vapour uptake from the atmosphere by some *Tillandsia* species. *Ann. Bot.* 40: 391–394.

Docters van Leeuwen, W. M. 1929. Einige Beobachtungen ueber das Zusammenleben von *Camponotus quadriceps* F. Smith mit dem Ameisenbaum *Endospermum formicarum* Becc. Aus Neu-Guinea. *Treubia* 10: 431–437.

1954. On the biology of some Javanese Loranthaceae and the role birds play in their life-histories. *Beaufortia, Misc. Publ.* 4: 105–207.

Dodson, C. H., Dressler, R. L., Hills, H. G., Adams, R. M., and Williams, N. H. 1969. Biologically active compounds in orchid fragrances. *Science* 164: 1243–1249.

Dodson, C. H., and Gentry, A. H. 1978. Flora of the Río Palenque Science Center. *Selbyana* 4: xix.

Dolzmann, P. 1964. Elektronenmikroskopische Untersuchungen an den Saughaaren von *Tillandsia usneoides* (Bromeliaceae). I. Feinstruktur der Kuppelzelle. *Planta* 60: 461–472.

1965. Electron microscopic investigations on the absorptive hairs of *Tillandsia usneoides* II. Observations on the fine structure of plasmodesmids. *Planta* 64: 76–80.

Dressler, R. L. 1981. *The orchids.* Cambridge, MA: Harvard University Press.

Dudgeon, W. 1923. Succession of epiphytes in the *Quercus incana* forest at Landour, western Himalayas. *Prelim. note: J. Indian Bot. Soc.* 3: 270–272.

Dueker, J., and Arditti, J. 1968. Photosynthetic $^{14}CO_2$ fixation by green *Cymbidium* (Orchidaceae) flowers. *Plant Physiol.* 43: 130–132.

Dycus, A. M., and Knudson, L. 1957. The role of the velamen of the aerial roots of orchids. *Bot. Gaz.* 119: 78–87.

Earnshaw, M. J., Winter, K., Ziegler, H., Stichler, W., Cruttwell, N. E. G., Kerenga, K., Cribb, P. J., Wood, J., Croft, J. R., Carver, K. A., and Gunn, T. C. 1987. Altitudinal changes in the incidence of crassulacean acid metabolism in vascular epiphytes and related life forms in Papua New Guinea. *Oecologia* 73: 566–572.

Edwards, P. J., and Grubb, P. J. 1977. Studies of mineral cycling in a montane rain forest in New Guinea. I. The distribution of organic matter in the vegetation and soil. *J. Ecol.* 65: 943–969.

Ehleringer, J. R., Cook, C. S., and Tieszen, L. L. 1986. Comparative water use and nitrogen relationships in a mistletoe and its host. *Oecologia* 68: 279–284.

Ehleringer, J. R., Schulze, E. -D., Ziegler, H., Lange, O. L., Farquhar, G. D., and Cowan, I. R. 1985. Xylem-tapping mistletoes: water or nutrient parasites? *Science* 227: 1479–1481.

El-Sharkawy, M. A., Cock, J. H., and Hernandez, A. P. 1986. Differential response of stomata to air humidity in the parasitic mistletoe *(Phthirusa pyrifolia)* and its host, mandarin orange *(Citrus reticulata). Photosynth. Res.* 9: 333–343.

Emmart, E. W. 1940. *The Badianus Manuscript.* Baltimore: Johns Hopkins University Press.

Epstein, E. 1972. P. 63 in *Mineral nutrition of plants: principles and perspectives.* New York: Wiley.

Erwin, T. L. 1983. Beetles and other insects of tropical forest canopies at Manaus, Brazil, sampled by insecticidal fogging. P. 73 in *Tropical rain forest: ecology and management,* eds. S. L. Sutton, T. C. Whitmore, and A. C. Chadwich. Oxford: British Ecological Society.

Farquhar, G. D., and Sharkey, T. D. 1982. Stomatal conductance and photosynthesis. *Annu. Rev. Plant Physiol.* 33: 317–345.

Feehan, J. 1985. Explosive flower opening in ornithophily: a study of pollination mechanisms in some Central African Loranthaceae. *Bot. J. Linn. Soc.* 90: 129–144.

Field, C., and Mooney, H. A. 1986. The photosynthesis–nitrogen relationship in wild plants. Pp. 25–55 in *On the economy of plant form and function,* ed. T. J. Givnish. Cambridge: Cambridge University Press.

Fineran, B. A., and Hocking, P. J. 1983. Features of parasitism, morphology and haustorial anatomy in loranthaceous root parasites. Pp. 205–227 in *The biology of mistletoes,* eds. D. M. Calder and P. Bernhardt. New York: Academic Press.

Fish, D. 1976. Structure and composition of the aquatic invertebrate community inhabiting bromeliads in South Florida and the discovery of an insectivorous bromeliad. Ph.D. thesis. Gainesville: University of Florida.

1983. Phytotelmata: flora and fauna. Pp. 1–28 in *Phytotelmata: terrestrial plants as hosts for aquatic insect communities,* eds. J. H. Frank and L. P. Lounibos. Medford, NJ: Plexus.

Fish, D., and Hall, D. W. 1978. Succession and stratification of aquatic insects inhabiting the insectivorous pitcher plant *Sarracenia purpurea. Am. Mid. Nat.* 99: 172–183.

Fisher, B. L., and Zimmerman, J. K. 1988. Ant/orchid associations in the Barro Colorado National Monument, Panama. *Lindleyana* 3: 12–16.

Fisher, J. T. 1983. Water relations of mistletoes and their hosts. Pp. 161–184 in *The biology of mistletoes,* eds. D. M. Calder and P. Bernhardt. New York: Academic Press.

Fittkau, E. J., and Klinge, H. 1973. On biomass and trophic structure of the central Amazonian rain forest ecosystem. *Biotropica* 5: 2–14.

Frank, J. H. 1983. Bromeliad phytotelmata and their biota, especially mosquitoes. Pp. 101–128 in *Phytotelmata: terrestrial plants as hosts for aquatic insect communities,* eds. J. H. Frank and L. P. Lounibos. Medford, NJ: Plexus.

Frank, J. H., and O'Meara, G. F. 1984. The bromeliad *Catopsis berteroniana* traps terrestrial arthropods but harbors *Wyeomyia* larvae *(Diptera culicidae). Florida Entomologist* 67: 418–424.

Frei, Sister John Karen, and Dodson, C. H. 1972. The chemical effect of certain bark substrates on the germination and early growth of epiphytic orchids. *Bull. Torrey Bot. Club* 99: 301–307.

Fritz-Sheridan, R. P., and Portécop, J. 1987. Nitrogen fixation on the tropical volcano, La Soufrière (Guadeloupe): 1. A survey of nitrogen fixation by blue-green algal microepiphytes and lichen endophytes. *Biotropica* 19: 194–199.

Galil, J. 1984. *Ficus religiosa*–the tree splitter. *Bot. J. Linn. Soc.* 88: 185–203.

Garth, R. E. 1964. The ecology of Spanish moss *(Tillandsia usneoides):* its growth and distribution. *Ecology* 45: 470–481.

Gentry, A. H. 1982a. Patterns of neotropical plant species diversity. Pp. 1–84 in *Evolutionary biology,* eds. M. K. Hecht, B. Wallace, and G. T. Prance. New York: Plenum Press.

1982b. Neotropical floristic diversity: phytogeographical connections between Central and South America, Pleistocene climatic fluctuations, or an accident of the Andean orogeny? *Ann. Mo. Bot. Gdns.* 69: 557–593.

Gentry, A. H., and Dodson, C. H. 1987a. Contribution of nontrees to species richness of a tropical rain forest. *Biotropica* 19: 149–156.

1987b. Diversity and biogeography of neotropical vascular epiphytes. *Ann. Mo. Bot. Gdns.* 74: 205–233.

Gentry, A. H., and Emmons, L. H. 1987. Geographical variation in fertility, phenology, and composition of the understory of neotropical forests. *Biotropica* 19: 216–227.

Gessner, F. 1956. Wasserhaushalt der Epiphyten und Lianen. Pp. 915–950 in vol. 3, *Handbuch der Pflanzenphysiology,* ed. W. Ruhland. Berlin: Springer.

Gilmartin, A. J. 1983. Evolution of mesic and xeric habits in *Tillandsia* and *Vriesea* (Bromeliaceae). *Syst. Bot.* 8: 233–242.

Gilmartin, A. J., and Brown, G. K. 1985. Cleistogamy in *Tillandsia capillaris* (Bromeliaceae). *Biotropica* 17: 256–259.

1986. Cladistic tests of hypotheses concerning evolution of xerophytes within *Tillandsia* subg. *Phytarrhiza* (Bromeliaceae). *Am. J. Bot.* 73: 387–397.

Giovannetti, M., and Mosse, B. 1980. An evaluation of techniques for measuring vesicular arbuscular mycorrhizal infection in roots. *New Phytol.* 84: 489–500.

Givnish, T. J., Burkhardt, E. L., Happel, R., and Weintraub, J. 1984. Carnivory in the bromeliad *Brocchinia reducta,* with a cost/benefit model for the general restriction of carnivorous plants to sunny, moist, nutrient-poor habitats. *Am. Nat.* 124: 479–497.

Glazner, J. T., Devlin, B., and Ellstrand, N. C. 1988. Biochemical and morphological evidence for host race evolution in desert mistletoe, *Phoradendron californicum* (Viscaceae). *Plant Syst. Evol.* 161: 13–21.

Godschalk, S. K. B. 1983. Mistletoe dispersal by birds in South Africa. Pp. 117–128 in *The biology of mistletoes,* eds. D. M. Calder and P. Bernhardt. New York: Academic Press.

Golley, F. B., Richardson, T., and Clements, R. G. 1978. Elemental concentrations in tropical forests and soils of northwestern Colombia. *Biotropica* 10: 144–151.

Gottlieb, L. D. 1984. Genetics and morphological evolution in plants. *Am. Nat.* 123: 681–709.

Gottsberger, G. 1986. Some pollination strategies in neotropical savannas and forests. *Plant Syst. Evol* 152: 29–45.

Granville, J. -J. de 1982. Rain forest and xeric flora refuges in French Guiana. Pp. 159–181 in *Biological diversification in the tropics,* ed. G. T. Prance. New York: Columbia University Press.

Griffiths, H. 1988. Carbon balance during CAM: an assessment of respiratory CO_2 recycling in the epiphytic bromeliads *Aechmea nudicaulis* and *Aechmea fendleri. Plant Cell Environ.* 11: 603–611.

Griffiths, H., Lüttge, U., Stimmel, K. -H., Crook, C. E., Griffiths, N. M., and Smith, J. A. C. 1986. Comparative ecophysiology of CAM and C_3 bromeliads. III Environmental influences on CO_2 assimilation and transpiration. *Plant Cell Environ.* 9: 385–393.

Griffiths, H., and Smith, J. A. C. 1983. Photosynthetic pathways in the Bromeliaceae of Trinidad: relations between life-forms, habitat preference and the occurrence of CAM. *Oecologia* 60: 176–184.

Griffiths, H., Smith, J. A. C., Lüttge, U., Popp, M., Cram, W. J., Diaz, M., Lee, H. S. J., Medina, E., Schäfer, C., and Stimmel, K. -H. 1989. Ecophysiology of xerophytic and halophytic vegetation of a coastal alluvial plain in northern Venezuela. IV. *Tillandsia flexuosa* S. W. and *Schomburgkia humboldtiana* Reichb., epiphytic CAM plants. *New Phytol.* 111: 273–282.

Grubb, P. J., and Edwards, P. J. 1982. Studies of mineral cycling in a montane rain forest in New Guinea. III The distribution of mineral elements in the aboveground material. *J. Ecol.* 70: 623–648.

Grubb, P. J., Lloyd, J. R., Pennington, T. D., and Whitmore, T. C. 1963. A comparison of montane and lowland rain forest in Ecuador. *J. Ecol.* 51: 567–601.

Guralnick, L. J., Ting, I. P., and Lord, E. M. 1986. Crassulacean acid metabolism in the Gesneriaceae. *Am. J. Bot.* 73: 336–345.

Haas, N. F. 1975. ^{32}P, ^{22}Na, und ^{99m}Tc in Versuchen über den Wassertransport in Luftwurzeln von *Vanda tricolor* Lindl. *Z. Pflanzenphysiol.* 75: 427–435.

Haberlandt, G. F. J. 1914. *Physiological plant anatomy.* London: Macmillan Press.

Hadley, G. 1982. Orchid mycorrhiza. Pp. 83–118 in *Orchid biology: reviews and perspectives II,* ed. J. Arditti. Cambridge: Cambridge University Press.

1984. Uptake of [^{14}C]glucose by asymbiotic and mycorrhizal orchid protocorms. *New Phytol.* 96: 263–273.

Hadley, G., and Williamson, B. 1972. Features of mycorrhizal infection in some Malayan orchids. *New Phytol.* 71: 1111–1118.

Haffer, J. 1969. Speciation in Amazonian forest birds. *Science* 165: 131–137.

1978. Distribution of Amazon forest birds. *Bonn. Zool. Beitr.* 29: 38–78.

Hall, P. J., Badenoch-Jones, J., Parker, C. W., Letham, D. S., and Barlow, B. A. 1987. Identification and quantification of cytokinins in the xylem sap of mistletoes and their hosts in relation to leaf mimicry. *Aust. J. Plant Physiol.* 14: 429–438.

Hallé, F., Oldeman, R. A. A., and Tomlinson, P. B. 1978. Pp. 133–250 in *Tropical trees and forests.* New York: Springer-Verlag.

Hammen, T. van der 1974. The Pleistocene changes of vegetation and climate in tropical South America. *Am. J. Biogeogr.* 1: 3–26.

Harris, J. A. 1918. On the osomotic concentration of the tissue fluids of phanerogamic epiphytes. *Am. J. Bot.* 5: 490–506.

Harris, J. A., and Lawrence, J. V. 1916. On the osmotic pressure of the tissue fluids of Jamaican Loranthaceae parasitic on various hosts. *Am. J. Bot.* 3: 438–455.

Hawksworth, F. G. 1983. Mistletoes as forest parasites. Pp. 157–158 in *The biology of mistletoes,* eds. D. M. Calder and P. Bernhardt. New York: Academic Press.

Hawksworth, F. G., and Wiens, D. 1972. *U.S. Dept. Agric., Agric. Handb.* 401, 234 pp.

Hazen, W. E. 1966. Analysis of spatial pattern in epiphytes. *Ecology* 47: 634–635.

Hellmuth, E. O. 1971. Eco-physiological studies on plants in arid and semi-arid regions in Western Australia. IV. Comparison of the field physiology of the host *Acacia grasbyi* and its hemiparasite, *Amyema nestor* under optimal and stress conditions. *J. Ecol.* 59: 351–363.

Hew, C. S. 1984. *Drymoglossum* under water stress. *Am. Fern J.* 74: 37–39.

Hoffmann, A. J., Fuentes, E. R., Cortés, I., Liberona, F., and Costa, V. 1986. *Tristerix tetrandrus* (Loranthaceae) and its host-plants in the Chilean matorral: patterns and mechanisms. *Oecologia* 69: 202–206.

Holthe, P. A., Sternberg, L. daS. L., and Ting, I. P. 1987. Developmental control of CAM in *Peperomia scandens. Plant Physiol.* 84: 743–747.

Hosokawa, T. 1943. Studies on the life forms of vascular epiphytes and the epiphyte flora of Ponape, Micronesia. *Trans. Nat. Hist. Soc. of Taiwan* 33: 35–55, 71–89, 113–141.

1955. On the vascular epiphyte communities in tropical rainforests of Micronesia. *8th Int. Bot. Congress,* Paris, July, 1954. Paris: Int. Cong. Bot.

Hull, R. J., and Leonard, O. A. 1964. Physiological aspects of parasitism in mistletoes (*Arceuthobium* and *Phoradendron*) I. The carbohydrate nutrition of mistletoe. *Plant Physiol.* 39: 996–1007.

Huxley, C. R. 1978. The ant-plants *Myrmecodia* and *Hydnophytum* (Rubiaceae), and the relationships between their morphology, ant occupants, physiology, and ecology. *New Phytol.* 80: 231–268.

1980. Symbiosis between ants and epiphytes. *Biol. Rev.* 55: 321–340.

1986. Evolution of benevolent ant–plant relationships. Pp. 275–282 in *Insects and the plant surface,* eds. B. Juniper and R. Southwood. London: Edward Arnold.

Istock, C. A., Tanner, K., and Zimmer, H. 1983. Habitat selection by the pitcher-plant mosquito, *Wyeomyia smithii:* behavioral and genetic aspects. Pp. 191–204 in *Phytotelmata: terrestrial plants as hosts for aquatic insect communities,* eds. J. H. Frank and L. P. Lounibos. Medford, NJ: Plexus.

Jane, G. T., and Green, T. G. A. 1985. Patterns of stomatal conductance in six evergreen tree species from a New Zealand cloud forest. *Bot. Gaz.* 146: 275–287.

Janzen, D. H. 1974. Epiphytic myrmecophytes in Sarawak: mutualism through the feeding of plants by ants. *Biotropica* 6: 237–259.

1977. Promising directions of study in tropical animal–plant interactions. *Ann. Mo. Bot. Gdns.* 64: 706–736.

Jebb, M. H. P. 1985. Taxonomy and tuber morphology of the rubiaceous ant-plants. Ph.D. thesis, Oxford Unviersity.

Johansson, D. R. 1974. Ecology of vascular epiphytes in West African rain forest. *Acta Phytogeogr. Suecica* 59: 1–136.

1975. Ecology of epiphytic orchids in West African rain forests. *Am. Orchid Soc. Bull.* 44: 125–136.

1977. Epiphytic orchids as parasites of their host trees. *Am. Orchid Soc. Bull.* 46: 703–707.

1978. A method to register the distribution of epiphytes on the host tree. *Am. Orchid Soc. Bull.* 47: 901–904.

Johnson, A., and Awan, B. 1972. The distribution of epiphytes on *Fragraea fragrans* and *Swietenia macrophylla. Malayan Forester* 35: 5–12.

Jordan, C. F. 1985. Pp. 24–27 in *Nutrient cycling in tropical forest ecosystems: principles and their application in management and conservation.* New York: Wiley.

Jordan, C. F., and Golley, F. 1980. Nutrient scavenging of rainfall by the canopy of an Amazonian rain forest. *Biotropica* 12: 61–66.

Jordan, P. W., and Nobel, P. S. 1984. Thermal and water relations of roots of desert succulents. *Ann. Bot.* 54: 705–717.

Junk, W. J., and Furch, K. 1985. The physical and chemical properties of Amazonian waters and their relationships with the biota. P. 7 in *Amazonia,* eds. G. T. Prance and T. E. Lovejoy. Oxford: Pergamon Press.

Kaul, R. B. 1977. The role of the multiple epidermis in foliar succulence of *Peperomia* (Piperaceae). *Bot. Gaz.* 138: 213–218.

Keeley, J. E. 1981. *Isoetes howellii:* a submerged aquatic CAM plant? *Am. J. Bot.* 68: 420–424.

Keigwin, L. D. 1978. Pliocene closing of the Isthmus of Panama, based on biostratigraphic evidence from nearby Pacific Ocean and Caribbean Sea cores. *Geology* 6: 630–634.

Kellman, M., Hudson, J, and Sanmugadas, K. 1982. Temporal variability in atmospheric nutrient influx to a tropical ecosystem. *Biotropica* 14: 1–9.

Kelly, D. L. 1985. Epiphytes and climbers of a Jamaican rain forest: vertical distribution, life forms and life histories. *J. Biogeogr.* 12: 223–241.

Kelly, D. L., Tanner, E. V. J., Kapos, V., Dickinson, T. A., Goodfriend, G. A., and Fairbairn, P. 1988. Jamaican limestone forests: floristics, structure and environment of three examples along a rainfall gradient. *J. Trop. Ecol.* 4: 121–156.

Kerr, L. R. 1925. A note on the symbiosis of *Loranthus* and *Eucalyptus. Proc. R. Soc. Victoria* 37: 248–251.

Kleinfeldt, S. E. 1978. Ant-gardens: the interaction of *Codonanthe crassifolia* (Gesneriaceae) and *Crematogaster longispina* (Formicidae). *Ecology* 59: 449–456.

Kluge, M., Friemert, V., Ong, B. L., Brulfert, J., and Goh, C. J. 1989. *In Situ* studies of crassulacean acid metabolism in *Drymoglossum piloselloides,* an epiphytic fern of the humid tropics. *J. Exp. Bot.* 40: 441–452.

Kluge, M., and Ting, I. P. 1978. *Crassulacean acid metabolism.* New York: Springer-Verlag.

Knab, F. 1912. New species of Anisopidae (Rhyphidae) from tropical America (Diptera, Nematocera). *Proc. Biol. Soc. Washington* 25: 111–114.

Knauft, R. L., and Arditti, J. 1969. Partial identification of dark $^{14}CO_2$ fixation products in leaves of *Cattleya* (Orchidaceae). *New Phytol.* 68: 657–661.

Knutson, D. M. 1983. Physiology of mistletoe parasitism and disease responses in the host. Pp. 295–316 in *The biology of mistletoes,* eds. D. M. Calder and P. Bernhardt. New York: Academic Press.

Koptur, S., Smith, R., and Baker, I. 1982. Nectaries in some neotropical species of *Polypodium* (Polypodiaceae): preliminary observations and analyses. *Biotropica* 14: 108–113.

Kress, W. J. 1986. A symposium: the biology of tropical epiphytes. Selbyana 9: 1–22.

Kuijt, J. 1969. Pp. 167, 198–199 in *The biology of parasitic flowering plants.* Berkeley: University of California Press.

Kuijt, J., and Toth, R. 1976. Ultrastructure of angiosperm haustoria: a review. *Ann. Bot. Lond.* 40: 1121–1130.

Laessle, A. M. 1961. A micro-limnological study of Jamaican bromeliads. *Ecology* 42: 499–517.

Lamont, B. 1983a. Germination of mistletoes. Pp. 129–143 in *The biology of mistletoes,* eds. D. M. Calder and P. Bernhardt. New York: Academic Press.

1983b. Mineral nutrition of mistletoes. Pp. 185–204 in *The biology of mistletoes,* eds. D. M. Calder and P. Bernhardt. New York: Academic Press.

Lamont, B., and Perry M. 1977. The effects of light, osmotic potential and atmospheric gases on germination of the mistletoe *Amyema preissii. Ann. Bot.* 41: 203–209.

Lamont, B. B., and Southall, K. J. 1982. Distribution of mineral nutrients between the mistletoe *Amyema preissii* and its host *Acacia acuminata. Ann. Bot.* 49: 721–725.

Lange, O. L., and Medina, E. 1979. Stomata of the CAM plant *Tillandsia recurvata* respond directly to humidity. *Oecologia* 40: 357–363.

Larcher, W. 1980. Pp. 31–35 in *Physiological plant ecology,* 2nd ed. New York: Springer-Verlag.

Lavelle, P., and Kohlmann, B. 1984. Étude quantitative de la microfaune du sol dans une forêt tropicale humide du Mexique (Bonampak, Chiapas). *Pedobiologia* 27: 377–393.

Lee, D. W., and Lowry, J. B. 1975. Physical basis and ecological significance of irridescence in blue plants. *Nature* 254: 50–51.

Lee, D. W., Lowry, J. B., and Stone, B. C. 1979. Abaxial anthocyanin layer in leaves of tropical rain forest plants: enhancer of light capture in deep shade. *Biotropica* 11: 70–77.

Liddy, J. 1983. Dispersal of Australian mistletoes: the Cowiebank study. Pp. 101–116 in *The biology of mistletoes,* eds. D. M. Calder and P. Bernhardt. New York: Academic Press.

Longino, J. T. 1986. Ants provide substrate for epiphytes. *Selbyana* 9: 100–103.

Lounibos, L. P., Frank, J. H., Machado-Allison, C. E., Ocanto, P., and Navarro, J. C. 1987. Survival, development and predatory effects of mosquito larvae in Venezuelan phytotelmata. *J. Trop. Ecol.* 3: 221–242.

Loveless, M. D., and Hamrick, J. L. 1984. Ecological determinants of genetic structure in plant populations. *Annu. Rev. Ecol. Syst.* 15: 65–96.

Lumer, C., and Schoer, R. D. 1986. Pollination of *Blakea austin-smithii* and *B. penduliflora* (Melastomataceae) by small rodents in Costa Rica. *Biotropica* 18: 363–364.

Lüttge, U. 1987. Carbon dioxide and water demand: crassulacean acid metabolism (CAM), a versatile ecological adaptation exemplifying the need for integration in ecophysiological work: *New Phytol.* 106: 593–629.

Lüttge, U. 1988. Day–night changes of citric-acid levels in crassulacean acid metabolism: phenomenon and ecophysiological significance. *Plant Cell Environ.* 11: 445–451.

Lüttge, U., and Ball, E. 1987. Dark respiration of CAM plants. *Plant Physiol. Biochem.* 25: 3–10.

Lüttge, U., Ball, E., Kluge, M., and Ong, B. L. 1986a. Photosynthetic light requirements of various tropical vascular epiphytes. *Physiol. Vég.* 24: 315–331.

Lüttge, U., Stimmel, K. -H., Smith, J. A. C., and Griffiths, H. 1986b. Comparative ecophysiology of CAM and C_3 bromeliads. II. Field measurements of gas exchange of CAM bromeliads in the humid tropics. *Plant Cell Environ.* 9: 377–383.

Madison, M. 1977. Vascular epiphytes: their systematic occurrence and salient features. *Selbyana* 2: 1–13.

1979a. Additional observations on ant-gardens in Amazonas. *Selbyana* 5: 107–115.

1979b. Distribution of epiphytes in a rubber plantation in Sarawak. *Selbyana* 5: 207–213.

Maguire, B. 1971. Phytotelmata: biota and community structure determination in plant-held waters. *Annu. Rev. Ecol. Syst.* 2: 439–464.

Martin, C. E., and Adams, W. W. 1987. Crassulacean acid metabolism, CO_2-recycling and tissue desiccation in the Mexican epiphyte *Tillandsia schiedeana* Steud. (Bromeliaceae). *Photosynth. Res.* 11: 237–244.

Martin, C. E., Christensen, N. L., and Strain, B. R. 1981. Seasonal pattern of

growth, tissue acid fluctuations and $^{14}CO_2$ uptake in the crassulacean acid metabolism epiphyte *Tillandsia usneoides* L. (Spanish moss). *Oecologia* 49: 322–328.
Martin, C. E., Eades, C. A., and Pitner, R. A. 1986. Effects of irradiance on crassulacean acid metabolism in the epiphyte *Tillandsia usneoides* (Bromeliaceae). *Plant Physiol.* 80: 23–26.
Martin, C. E., McLeod, K. W., Eades, C. A., and Pitzer, A. F. 1985. Morphological and physiological responses to irradiance in the CAM epiphyte *Tillandsia usneoides* L. (Bromeliaceae). *Bot. Gaz.* 146: 489–494.
Martin, C. E., and Peters, E. A. 1984. Functional stomata of the atmospheric epiphyte *Tillandsia usneoides* L. *Bot. Gaz.* 145: 502–507.
Martin, C. E., and Schmitt, A. K. 1989. Unusual water relations in the CAM atmospheric epiphyte *Tillandsia usneoides. Bot. Gaz.* 150: 1–8.
Martin, C. E., and Siedow, J. N. 1981. Crassulacean acid metabolism in the epiphyte *Tillandsia usneoides* L. (Spanish moss). *Plant Physiol.* 68: 335–339.
Maschwitz, U., and Hölldobler, B. 1970. Der Nestkartonbau bei *Lasius fuliginosus. Z. vergl. Physiol.* 66: 176–189.
McKey, D. 1975. The ecology of coevolved seed dispersal systems. Pp. 159–191 in *Coevolution of animals and plants,* eds. L. E. Gilbert and P. H. Raven. Austin: University of Texas Press.
Medina, E. 1974. Dark CO_2 fixation, habitat preference and evolution within the Bromeliaceae. *Evolution* 28: 677–686.
Medina, E., Delgado, M., Troughton, J. H., and Medina, J. D. 1977. Physiological ecology of CO_2 fixation in Bromeliaceae. *Flora* 166: 137–152.
Medina, E., Olivares, E., and Diaz, M. 1986. Water stress and light intensity effects on growth and nocturnal acid accumulation in a terrestrial CAM bromeliad (*Bromelia humilis* Jacq.) under natural conditons. *Oecologia* 70: 441–446. 446.
Mesler, M. R. 1975. The gametophytes of *Ophioglossum palmatum* L. *Am. J. Bot.* 62: 982–992.
Mez, C. 1904. Physiologische Bromeliaceen–Studien I. Die Wasser-Ökonomie der extrem atmosphärischen Tillandsien. *Jb. wiss. Bot.* 40: 157–229.
Michaloud, G., and Michaloud-Pelletier, S. 1987. *Ficus* hemi-epiphytes (Moraceae) et arbres supports. *Biotropica* 19: 125–136.
Miller, J. R., and Tocher, R. D. 1975. Photosynthesis and respiration of *Arceuthobium tsuqense* (Loranthaceae). *Am. J. Bot.* 62: 765–769.
Morat, Ph., Veillon, J. -M., and Mackee, H. S. 1984. Floristic relationships of New Caledonian rain forest phanerograms. Pp. 71–128 in *Biogeography of the tropical Pacific,* eds. R. Radovsky, P. H. Raven, and S. H. Sohmer. Honolulu: Association of Systematics Collections and Bernice P. Bishop Museum Special Publications No. 72.
Müller, L., Starnecker, G., and Winkler, S. 1981. Zur Ökologie epiphytischer Farne in Südbrasilien. I. Saugschuppen. *Flora* 171: 55–63.
Murray, K. G., Feinsinger, P., Busby, W. H., Linhart, Y. B., Beach, J. H., and Kinsman, S. 1987. Evaluation of character displacement among plants in two tropical pollination guilds. *Ecology* 68: 1283–1293.
Nadkarni, N. M. 1981. Canopy roots: convergent evolution in rainforest nutrient cycles. *Science* 214: 1023–1024.
1984. Epiphyte biomass and nutrient capital of a neotropical elfin forest. *Biotropica* 16: 249–256.

1986. The nutritional effects of epiphytes on host trees with special reference to alteration of precipitation chemistry. *Selbyana* 9: 44–51.

Nadkarni, N. M., and Matelson, T. J. in press. Bird use of epiphyte resources in neotropical montane forest and pasture tree crowns. *Condor.*

Newman, E. I., and Reddell, P. 1987. The distribution of mycorrhizas among familes of vascular plants. *New Phytol.* 106: 745–751.

Nishio, J. N., and Ting, I. P. 1987. Carbon flow and metabolic specialization in the tissue layers of the crassulacean acid metabolism plant *Peperomia camptotricha. Plant Physiol.* 84: 600–604.

Nyman, L. P., Davis, J. P., O'Dell, S. J., Arditti, J., Stephens, G. S., and Benzing, D. H. 1987. Active uptake of amino acids by leaves of an epiphytic vascular plant, *Tillandsia paucifolia* (Bromeliaceae). *Plant Physiol.* 83: 681–684.

Okahara, K. 1932. On the role of microorganisms in the digestion of insect bodies in insectivorous plants. *Bot. Mag. (Tokyo)* 46: 353–357.

Oliver, W. R. B. 1930. New Zealand epiphytes. *J. Ecol.* 18: 1–50.

Olmsted, I. C., and Dejean, A. 1987. Tree–epiphyte–ant relationships of the low inundated forest in Sian Ka'an Biosphere Reserve, Quintana Roo, Mexico. *Assoc. Trop. Biol. Abstr.* Columbus: Ohio State University.

Orús, M. I., Estévez, M. P., and Vicente, C. 1981. Manganese depletion in chloroplasts of *Quercus rotundifolia* during chemical simulation of lichen epiphytic states. *Physiol. Plant.* 52: 263–266.

Osmond, C. B. 1978. Crassulacean acid metabolism: a curiosity in context. *Annu. Rev. Plant Physiol.* 29: 379–414.

1987. Photosynthesis and carbon economy of plants. *New Phytol.* 106: 161–175.

Palaçios-Vargas, J. G. 1981. Collembola asociados a *Tillandsia* (Bromeliaceae) en el derrame lavico del Chichinautzin, Morelos, Mexico. *Southwest. Entomologist* 6: 87–98.

1982. Microartrópodos asociados a Bromeliáceas. Pp. 535–545 in *Zoología Neotropical. Actas del VIII Cong. Latin. de Zoología* (1980), ed. P. J. Salinas. Merida, Venezuela: Producciones Alfa.

Patel, A., and Ting, I. P. 1987. Relationship between respiration and CAM-cycling in *Peperomia camptotricha. Plant Physiol.* 84: 640–642.

Paterson, S. 1982. Observations on ant associations with rainforest ferns in Borneo. *Fern Gazette* 12: 243–245.

Pessin, L. J. 1925. An ecological study of the polypody fern *Polypodium polypodioides* as an epiphyte in Mississippi. *Ecology* 6: 17–38.

Peterson, C. A. 1988. Exodermal Casparian bonds: their significance for ion uptake by roots. *Physiol. Plant.* 72: 204–208.

Picado, C. 1913. Les Bromeliacées epiphytes considérée comme milieu biologique. *Bull. Sci. France et Belgique* 47: 215–360.

Pike, L. H. 1978. The importance of epiphytic lichens in mineral cycling. *Bryologist* 81: 247–257.

Pittendrigh, C. S. 1948. The bromeliad-*Anopheles*-malaria complex in Trinidad. I. The bromeliad flora. *Evolution* 2: 58–89.

Plummer, G. L., and Kethley, J. B. 1964. Foliar absorption of amino acids, peptides and other nutrients by the pitcher plant *Sarracenia flava. Bot. Gaz.* 125: 245–259.

Pócs, T. 1982. Tropical forest bryophytes. Pp. 59–104 in *Bryophyte ecology,* ed. A. J. E. Smith. London: Chapman & Hall.

Poole, H. A., and Sheehan, T. J. 1982. Mineral nutrition of orchids. Pp. 197–212 in *Orchid biology–reviews and perspectives II,* ed. J. Arditti. Ithaca: Cornell University Press.

Popp, M., Kramer, D., Lee, H., Diaz, M., Ziegler, H., and Lüttge, U. 1987. Crassulacean acid metabolism in tropical dicotyledonous trees. *Trees* 1: 238–247.

Press, M. C., Graves, J. D., and Stewart, G. R. 1988. Transpiration and carbon acquisition in root hemiparasitic angiosperms. *J. Exp. Bot.* 39: 1009–1014.

Pridgeon, A. M. 1981. Absorbing trichomes in the Pleurothallidinae (Orchidaceae). *Am. J. Bot.* 68: 64–71.

Pridgeon, A. M., Stern, W. L., and Benzing, D. H. 1983. Tilosomes in roots of Orchidaceae: morphology and systematic occurrence. *Am. J. Bot.* 70: 1365–1377.

Proctor, J., Lee, Y. F., Langley, A. M., Munro, W. R. C., and Nelson, T. 1988. Ecological studies on Gunang Silam, a small ultrabasic mountain in Sabah, Malaysia. I. Environment, forest structure and floristics. *J. Ecol.* 76: 320–340.

Putz, F. E. 1979. How trees avoid and shed lianas. *Biotropica* 16: 19–23.

Putz, F. E., and Holbrook, N. M. 1986. Notes on the natural history of hemiepiphytes. *Selbyana* 9: 61–69.

1987. Strangler fig rooting habits and nutrient relations in the Venezuelan llanos. *Assoc. Trop. Biol. Abstr.* Columbus: Ohio State University.

Ramírez, W. B. 1977. Evolution of the strangling habit in *Ficus* L., subgenus *Urostigma* (Moraceae). *Brenesia* 12/13: 11–19.

Raven, J. A. 1985. Regulation of pH and generation of osmolarity in vascular plants: a cost–benefit analysis in relation to efficiency of use of energy, nitrogen and water. *New Phytol* 101: 25–77.

1988. Acquisition of nitrogen by the shoots of land plants: its occurrence and implications for acid–base regulation. *New Phytol.* 109: 1–20.

Raven, P. H. 1976. Ethics and attitudes. Pp. 155–179 in *Conservation of threatened plants,* eds. J. Simmons et al. New York: Plenum Press.

Renner, S. S. 1986. The neotropical epiphytic Melastomataceae: phytogeographic patterns, fruit types, and floral biology. *Selbyana* 9: 105–111.

Richards, P. W. 1952. Chap. 5 in *The tropical rain forest: an ecological study.* Cambridge: Cambridge University Press.

1973. Africa, the "odd man out." Pp. 21–26 in *Tropical forest ecosystems in Africa and South America: a comparative review.* Washington, DC: Smithsonian.

Richardson, K., Griffiths, H., Reed, M. L., Raven, J. A., and Griffiths, N. M. 1984. Inorganic carbon assimilation in the Isoetids, *Isoetes lacustris* L. and *Lobelia dortmanna* L. *Oecologia* 61: 115–121.

Rickson, F. R. 1979. Absorption of animal tissue breakdown products into a plant stem–the feeding of a plant by ants. *Am. J. Bot.* 66: 87–90.

Rico-Gray, V., and Thien, L. B. 1989. Ant-mealybug interaction decreases reproductive fitness of *Schomburgkia tibicinis* (Orchidaceae) in Mexico. *J. Trop. Ecol.* 5: 109–112.

Rivero, J. A. 1984. Bromeliad frogs of Puerto Rico. *J. Bromeliad Soc.* 34: 64–66.

Rockwood, L. L. 1985. Seed weight as a function of life form, elevation and life zone in neotropical forests. *Biotropica* 17: 32–39.

Rubenstein, R., Hunter, D., McGowan, R. E., and Withner, C. L. 1976. Carbon dioxide metabolism in various orchid leaves. P. 10, abstr. 65 in *Northeastern*

Meetings, Plant Physiol. (*N. American Society of Plant Physiologists.* (Cited in *Ann. Bot. 54,* 586.)
Ruinen, J. 1953. Epiphytosis. A second view on epiphytism. *Ann. Bogor.* 1: 101–157.
St. John, B. J., Smith, S. E., Nicholas, D. J. D., and Smith, F. A. 1985. Enzymes of ammonium assimilation in the mycorrhizal fungus *Pezizella ericae* Read. *New Phytol.* 100: 579–584.
Sallé, G. 1983. Germination and establishment of *Viscum album* L. Pp. 145–160 in *The biology of mistletoes,* eds. D. M. Calder and P. Bernhardt. New York: Academic Press.
Sanford, R. L. 1987. Apogeotropic roots in an Amazon rain forest. *Science* 235: 1062–1064.
Sanford, W. W. 1969. The distribution of epiphytic orchids in Nigeria in relation to each other and to geographic location and climate, type of vegetation and tree species. *Biol. J. Linn. Soc.* 1: 247–285.
1974. The ecology of orchids. Pp. 1–100 in *The orchids: scientific studies,* ed. C. L. Withner. New York: Wiley.
Sanford, W. W., and Adanlawo, I. 1973. Velamen and exodermis characters of West African epiphytic orchids in relation to taxonomic grouping and habitat tolerance. *Bot. J. Linn. Soc.* 66: 307–321.
Schäfer, C., and Lüttge, U. 1986. Effects of water stress on gas exchange and water relations of a succulent epiphyte, *Kalanchoe uniflora. Oecologia* 71: 127–132.
1988. Effects of high irradiances on photosynthesis, growth and crassulacean acid metabolism in the epiphyte *Kalanchoe uniflora. Oecologia* 75: 567–574.
Schaffer, W. M., and Gadgil, M. D. 1975. Selection for optimal life histories in plants. Pp. 142–157 in *Ecology and evolution of communities,* eds. M. L. Cody and J. M. Diamond. Cambridge, MA: Harvard University Press.
Schimper, A. F. W. 1884. Ueber Bau and Lebenweise der Epiphyten Westindiens. *Bot. Zbl.* 17: 192–195 et seq.
1888. *Die epiphytische Vegetation Amerikas. Bot. Mitt. Tropen* II. Jena: Fischer.
1903. *Plant geography on a physiological basis.* Oxford: Clarendon Press.
Schmid, R., and Schmid, M. J. 1977. Pp. 25–46 in *Orchid biology–reviews and perspectives* I, ed. J. Arditti. Ithaca: Cornell University Press.
Schmidt, J. E., and Kaiser, W. M. 1987. Response of the succulent leaves of *Peperomia magnoliaefolia. Plant. Physiol.* 83: 190–194.
Schnell, R. 1952. Végétation et flore des monts Nimba. *Vegetatio* 3: 350–406.
Schrimpff, E. 1984. Air pollution patterns in two cities of Colombia, S. A. according to trace substances content of an epiphyte (*Tillandsia recurvata* L.). *Water, Air, and Soil Pollution* 21: 279–315.
Schulze, E. D., and Ehleringer, J. R. 1984. The effect of nitrogen supply on growth and water-use efficiency of xylem-tapping mistletoes. *Planta* 162: 268–275.
Seidel, J. L. 1988. The monoterpenes of *Gutierrezia sarothrae:* chemical interactions between ants and plants in neotropical ant-gardens. Ph.D. dissertation, University of Utah.
Sengupta, B., Nandi, A. S., Samanta, R. K., Pal, D., Sengupta, D. N., and Sen, S. P. 1981. Nitrogen fixation in the phyllosphere of tropical plants: occurrence of phyllosphere nitrogen-fixing microorganisms in eastern India and their utility for the growth and nitrogen nutrition of host plants. *Ann. Bot.* 48: 705–716.
Shacklette, H. T., and Connor, J. J. 1973. Airborne chemical elements in Spanish

moss. *Geol. Survey Professional Paper 574-E.* Washington, DC: U.S. Govt. Print. Office.

Simpson, B. B., and Haffer, J. 1978. Speciation patterns in the Amazonian forest biota. *Annu. Rev. Ecol. Syst.* 9: 497–518.

Sinclair, R. 1983a. Water relations of tropical epiphytes. I. Relationships between stomatal resistance, relative water content and the components of water potential. *J. Exp. Bot.* 34: 1652–1663.

1983b. Water relations of tropical epiphytes. II. Performance during droughting. *J. Exp. Bot.* 34: 1664–1675.

1984. Water relations of tropical epiphytes. III. Evidence for crassulacean acid metabolism. *J. Exp. Bot.* 35: 1–7.

Singh, J. S., and Chaturvedi, O. P. 1982. Photosynthetic pigments on plant bearing surfaces in the Himalaya. *Photosynthetica* 16: 101–114.

Sipes, D. L., and Ting, I. P. 1985. Crassulacean acid metabolism and crassulacean acid metabolism modifications in *Peperomia camptotricha. Plant Physiol.* 77: 59–63.

Smith, J. A. C., Griffiths, H., Bassett, M., and Griffiths, N. M. 1985. Day–night changes in the leaf water relations of epiphytic bromeliads in the rain forests of Trinidad. *Oecologia* 67: 474–485.

Smith, J. A. C., Griffiths, H., and Lüttge, U. 1986a. Comparative ecophysiology of CAM and C_3 bromeliads. I. The ecology of the Bromeliaceae in Trinidad. *Plant Cell Environ.* 9: 359–376.

Smith, J. A. C., Griffiths, H., Lüttge, U., Crook, C. E., Griffiths, N. M., and Stimmel, K. -H. 1986b. Comparative ecophysiology of CAM and C_3 bromeliads. IV. Plant water relations. *Plant Cell Environ.* 9: 395–410.

Smith, L. B. 1934. Geographical evidence on the lines of evolution in the Bromeliaceae. *Bot. Jahr.* 66: 446–468.

Smith, L. B., and Downs, R. J. 1977. Tillandsioideae (Bromeliaceae). *Flora Neotropica Monograph 14, Part 2.* New York: Hafner Press.

Smith, S. E. 1966. Physiology and ecology of orchid mycorrhizal fungi with reference to seedling nutrition. *New Phytol.* 65: 488–499.

1967. Carbohydrate translocation in orchid mycorrhizas. *New Phytol.* 66: 371–378.

Snow, B. K., and Snow, D. W. 1971. The feeding ecology of tanagers and honeycreepers in Trinidad. *Auk* 88: 291–322.

Soltis, D. E., Gilmartin, A. J., Rieseberg, L., and Gardner, S. 1987. Genetic variation in the epiphytes *Tillandsia ionantha* and *T. recurvata* (Bromeliaceae). *Am. J. Bot.* 74: 531–537.

Spanner, L. 1939. Untersuchungen über den Wärme und Wasserhaushalt von *Myrmecodia* und *Hydnophytum. Jb. Wiss. Bot.* 88: 243–283.

Stearns, S. C. 1976. Life-history tactics: a review of the ideas. *Q. Rev. Biol.* 51: 3–47.

Sternberg, L. da S. L., Ting, I. P., Price, D., and Hann, J. 1987. Photosynthesis in epiphytic and rooted *Clusia rosea* Jacq. *Oecologia* 72: 457–460.

Sternberg, L. O., DeNiro, M. J., and Ting, I. P. 1984. Carbon, hydrogen, and oxygen isotope ratios of cellulose from plants having intermediary photosynthetic modes. *Plant Physiol.* 74: 104–107.

Stewart, G. R., and Orebamjo, T. O. 1980. Nitrogen status and nitrate reductase activity of the parasitic angiosperm *Tapinanthus bangwensis* (Engl. & Krause) Danser growing on different hosts. *Ann. Bot.* 45: 587–589.

Stiles, F. G. 1978. Temporal organization of flowering among the hummingbird foodplants of a tropical wet forest. *Biotropica* 10: 194–210.
Stribley, D. P., and Read, D. J. 1975. Some nutritional aspects of the biology of ericaceous mycorrhizas. Pp. 195–207 in *Endomycorrhizas,* eds. F. E. Sanders, B. Mosse, and P. B. Tinker. New York: Academic Press.
Strong, D. R. 1977. Epiphyte loads, tree falls, and perennial forest disruption: a mechanism for maintaining higher tree species richness in the tropics without animals. *J. Biogeogr.* 4: 215–218.
Strong, D. R., and Ray, T. S. 1975. Host tree location behavior of a tropical vine *(Monstera gigantea)* by skototropism. *Science* 190: 804–806.
Stuart, T. S. 1969. The revival of respiration and photosynthesis in dried leaves of *Polypodium polypodioides. Planta* 83: 185–206.
Sudd, J. H., and Franks, N. R. 1987. Pp. 55–64 in *The behavioural ecology of ants.* London: Chapman & Hall/Methuen & Co.
Sugden, A. M. 1981. Aspects of the ecology of vascular epiphytes in two Colombian cloud forests. II. Habitat preferences of Bromeliaceae in the Serranía de Macuira. *Selbyana* 5: 264–273.
Sugden, A. M., and Robins, R. J. 1979. Aspects of the ecology of vascular epiphytes in Colombian cloud forests, I. The distribution of the epiphytic flora. *Biotropica* 11: 173–188.
Tanner, E. V. J. 1977. Four montane rain forests of Jamaica: a quantitative characterization of the floristics, the soils and the foliar mineral levels, and a discussion of the interrelations. *J. Ecol.* 65: 883–918.
1980. Studies on the biomass and productivity in a series of montane rain forests in Jamaica. *J. Ecol.* 68: 573–588.
Templeton, A. R. 1981. Mechanisms of speciation–a population genetic approach. *Annu. Rev. Ecol. Syst.* 12: 23–48.
Tewari, M., Upreti, N., Pandey, P., and Singh, S. P. 1985. Epiphytic succession on tree trunks in a mixed oak–cedar forest in the Kumaun Himalayas. *Vegetatio* 63: 105–112.
Thomas, D. W. 1988. The influence of aggressive ants on fruit removal in the tropical tree, *Ficus capensis* (Moraceae). *Biotropica* 20: 49–53.
Thompson, J. N. 1981. Reversed animal–plant interactions: the evolution of insectivorous and ant-fed plants. *Biol. J. Linn. Soc.* 16: 147–155.
Tiffney, B. H. 1984. Seed size, dispersal syndromes and the rise of angiosperms: evidence and hypothesis. *Ann. Mo. Bot. Gdns.* 71: 551–576.
Ting, I. P., Bates, L., Sternberg, L. O., and DeNiro, M. J. 1985a. Physiological and isotopic aspects of photosynthesis in *Peperomia. Plant Physiol.* 78: 246–249.
Ting, I. P., Hann, J., Holbrook, N. M., Putz, F. E., Sternberg, L. da S. L., Price, D., and Goldstein, G. 1987. Photosynthesis in hemiepiphytic species of *Clusia* and *Ficus. Oecologia* 74: 339–346.
Ting, I. P., Lord, E. M., Sternberg, L. da S. L., and DeNiro, M. J. 1985b. Crassulacean acid metabolism in the strangler *Clusia rosea* Jacq. *Science* 229: 969–971.
Todzia, C. 1986. Growth habits, host tree species, and density of hemiepiphytes on Barro Colorado Island, Panama. *Biotropica* 18: 22–27.
Tomlinson, P. B. 1969. *Anatomy of the monocotyledons: III. Commelinales–Zingiberales.* Oxford: Oxford University Press.
1986. Pp. 32–33 in *The botany of mangroves.* Cambridge: Cambridge University Press.

Tryon, R. 1985. Spores of myrmecophytic ferns. *Proc. R. Soc. Edinburgh* 86B: 105–110.
Tukey, H. B. 1970. The leaching of substances from plants. *Annu. Rev. Plant Physiol.* 21: 305–324.
Turner, R. M., Alcorn, S. M., and Olin, G. 1969. Mortality of transplanted saguaro seedlings. *Ecology* 50: 835–844.
Ule, E. 1906. Ameisenpflanzen. *Bot. Jahrb. Syst.* 37: 335–352.
Ullmann, I., Lange, O. L., Ziegler, H., Ehleringer, J., Schulze, E. -D., and Cowan, I. R. 1985. Diurnal courses of leaf conductance and transpiration of mistletoes and their hosts in Central Australia. *Oecologia* 67: 577–587.
Valdivia, P. E. 1977. Estudio botánico y ecológico de la región del Río Uxpanapa, Veracruz. IV. Las epifitas. *Biotica* 2: 55–81.
Van Oye, P. 1924. Sur l'écologie des épiphytes de la surface des troncs d'arbres à Java. *Revue Générale Botanique* 36: 12–30, 68–83.
Wagner, W. H., and Gómez, L. D. 1983. Pteridophytes (Helechos, Ferns). Pp. 311–318 in *Costa Rican natural history,* ed. D. H. Janzen. Chicago: University of Chicago Press.
Wake, D. B. 1987. Adaptive radiation of salamanders in Middle American cloud forests. *Ann. Mo. Bot. Gdns.* 74: 242–264.
Walbot, V., and Cullis, C. A. 1985. Rapid genomic change in higher plants. *Annu. Rev. Plant Physiol.* 36: 367–396.
Walker, T. G. 1985. Spore filaments in the ant-fern *Lecanopteris mirabilis*-an alternative viewpoint. *Proc. R. Soc. Edinburgh* 86B: 111–114.
Wallace, B. J. 1981. *The Australian vascular epiphytes: flora and ecology.* Ph.D. thesis, University of New England, New South Wales, Australia.
Walter, H. 1971. Pp. 134–138 in *Ecology of tropical and subtropical vegetation.* Edinburgh: Oliver Boyd.
Watson, M. A., and Casper, B. B. 1984. Morphogenetic constraints on patterns of carbon distribution in plants. *Annu. Rev. Ecol. Syst.* 15: 233–258.
Weir, J. S., and Kiew, R. 1986. A reassessment of the relations in Malaysia between ants *(Crematogaster)* on trees (*Leptospermum* and *Dacrydium*) and epiphytes of the genus *Dischidia* (Asclepiadaceae) including "ant-plants." *Biol. J. Linn. Soc.* 27: 113–132.
Welker, J. M., Rykiel, E. J. Jr., Briske, D. D., and Goeschl, J. D. 1985. Carbon import among vegetative tillers within two bunchgrasses: assessment with carbon-11 labelling. *Oecologia* 67: 209–212.
Went, F. W. 1940. Soziologie der Epiphyten eines tropischen Urwaldes. *Ann. Jard. Bot. Buitenz.* 50: 1–98.
Wheeler, W. M. 1921. A new case of parabiosis and the 'ant gardens' of British Guiana. *Ecology* 2: 89–103.
Whittington, J., and Sinclair, R. 1988. Water relations of the mistletoe, *Amyema miquelii,* and its host *Eucalyptus fasciculosa. Aust. J. Bot.* 36: 239–255.
Wiehler, H. 1978. The genera *Episcia, Alsobia, Nautilocalyx,* and *Paradrymonia* (Gesneriaceae). *Selbyana* 5: 11–60.
Wilder, G. J. 1986. Anatomy of first-order roots in the Cyclanthaceae (Monocotyledoneae). I. Epidermis, cortex and pericycle. *Can. J. Bot.* 64: 2622–2644.
Williams, N. H., and Whitten. W. M. 1983. Orchid floral fragrances and male euglossine bees: methods and advances in the last sesquidecade. *Biol. Bull.* 164: 355–395.

Wilson, E. O. 1987. The arboreal ant fauna of Peruvian Amazon forests: a first assessment. *Biotropica* 19: 245–282.
Winter, K. 1985. Crassulacean acid metabolism. Chap. 8 in *Photosynthetic mechanisms and the environment,* eds. J. Barber and N. R. Baker. New York: Elsevier Scientific.
Winter, K., Medina, E., Garcia, V., Mayoral, M. A., and Muñiz, R. 1985. Crassulacean acid metabolism in roots of a leafless orchid *Campylocentrum tyrridion* Garay & Dunsterv. *J. Plant Physiol.* 118: 73–78.
Winter, K., Osmond, C. B., and Hubick, K. T. 1986. Crassulacean acid metabolism in the shade. Studies on an epiphytic fern, *Pyrrosia longifolia,* and other rainforest species from Australia. *Oecologia* 68: 224–230.
Winter, K., Wallace, B. J., Stocker, G. C., and Roksandic, Z. 1983. Crassulacean acid metabolism in Australian vascular epiphytes and some related species. *Oecologia* 57: 129–141.
Wong, S. C., and Hew, C. S. 1976. Diffusive resistance, titratable acidity, and CO_2 fixation in two tropical epiphytic ferns. *Am. Fern J.* 66: 121–124.
Yeaton, R. I., and Gladstone, D. E. 1982. The pattern of colonization of epiphytes on calabash trees (*Crescentia alata* HBK.) in Guanacaste Province, Costa Rica. *Biotropica* 14: 137–140.
Zeil, W. 1979. *The Andes: a geological review.* Berlin: Gebrüder Borntraeger.
Zimmerman, M. H. 1983. Pp. 81–82 in *Xylem structure and the ascent of sap.* New York: Springer-Verlag.

Author index

All authors whose works are cited in the text, whether principal authors or coauthors, appear in this index. Parentheses indicate the main author of a coauthored work.

Subject index

Taxonomic index